THEORY AND DECISION LIBRARY

General Editors: W. Leinfellner (*Vienna*) and G. Eberlein (*Munich*)

Series A: Philosophy and Methodology of the Social Sciences

Series B: Mathematical and Statistical Methods

Series C: Game Theory, Mathematical Programming and Operations Research

Series D: System Theory, Knowledge Engineering and Problem Solving

SERIES C: GAME THEORY, MATHEMATICAL PROGRAMMING AND OPERATIONS RESEARCH

VOLUME 34

Scope: Particular attention is paid in this series to game theory and operations research, their formal aspects and their applications to economic, political and social sciences as well as to sociobiology. It will encourage high standards in the application of game-theoretical methods to individual and social decision making.

The titles published in this series are listed at the end of this volume.

Bezalel Peleg · Peter Sudhölter

Introduction to the Theory of Cooperative Games

Second Edition

Springer

Professor Bezalel Peleg
The Hebrew University of Jerusalem
Institute of Mathematics and
Center for the Study of Rationality
Givat-Ram, Feldman Building
91904 Jerusalem
Israel
pelegba@math.huji.ac.il

Professor Peter Sudhölter
University of Southern Denmark
Department of Business and Economics
Campusvej 55
5230 Odense M
Denmark
psu@sam.sdu.dk

Library of Congress Control Number: 2007931451

ISSN 0924-6126
ISBN 978-3-540-72944-0 Springer Berlin Heidelberg New York
ISBN 978-1-4020-7410-3 1st Edition Springer Berlin Heidelberg New York

Springer is a part of Springer Science+Business Media

springer.com

Production: LE-TeX Jelonek, Schmidt & Vöckler GbR, Leipzig, Germany
Cover design: WMX Design GmbH, Heidelberg

Spin 12073665 Printed on acid-free paper 43/3180/YL - 5 4 3 2 1 0

Preface to the Second Edition

The main purpose of the second edition is to enhance and expand the treatment of games with nontransferable utility. The main changes are:

(1) Chapter 13 is devoted entirely to the Shapley value and the Harsanyi solution. Section 13.4 is new and contains an axiomatization of the Harsanyi solution.

(2) Chapter 14 deals exclusively with the consistent Shapley value. Sections 14.2 and 14.3 are new and present an existence proof for the consistent value and an axiomatization of the consistent value respectively. Section 14.1, which was part of the old Chapter 13, deals with the consistent value of polyhedral games.

(3) Chapter 15 is almost entirely new. It is mainly devoted to an investigation of the Mas-Colell bargaining set of majority voting games. The existence of the Mas-Colell set is investigated and various limit theorems are proved for majority voting games. As a corollary of our results we show the existence of a four-person super-additive and non-levelled (NTU) game whose Mas-Colell bargaining set is empty.

(4) The treatment of the ordinal bargaining set was moved to the final chapter 16.

We also have used this opportunity to remove typos and inaccuracies from Chapters 2 – 12 which otherwise remained intact.

We are indebted to all our readers who pointed out some typo. In particular we thank Michael Maschler for his comments and Martina Bihn who personally supported this edition.

June 2007 Bezalel Peleg and Peter Sudhölter

Preface to the First Edition

In this book we study systematically the main solutions of cooperative games: the core, bargaining set, kernel, nucleolus, and the Shapley value of TU games, and the core, the Shapley value, and the ordinal bargaining set of NTU games. To each solution we devote a separate chapter wherein we study its properties in full detail. Moreover, important variants are defined or even intensively analyzed. We also investigate in separate chapters continuity, dynamics, and geometric properties of solutions of TU games. Our study culminates in uniform and coherent axiomatizations of all the foregoing solutions (excluding the bargaining set).

It is our pleasure to acknowledge the help of the following persons and institutions. We express our gratitude to Michael Maschler for his detailed comments on an early version, due to the first author, of Chapters 2 – 8. We thank Michael Borns for the linguistic edition of the manuscript of this book. We are indebted to Claus-Jochen Haake, Sven Klauke, and Christian Weiß for reading large parts of the manuscript and suggesting many improvements. Peter Sudhölter is grateful to the Center for Rationality and Interactive Decision Theory of the Hebrew University of Jerusalem and to the Edmund Landau Center for Research in Mathematical Analysis and Related Areas, the Institute of Mathematics of the Hebrew University of Jerusalem, for their hospitality during the academic year 2000-01 and during the summer of 2002. These institutions made the typing of the manuscript possible. He is also grateful to the Institute of Mathematical Economics, University of Bielefeld, for its support during several visits in the years 2001 and 2002.

December 2002 Bezalel Peleg and Peter Sudhölter

Contents

List of Figures

List of Tables

Notation and Symbols

We shall now list some of our notation.

The field of real numbers is denoted by $\mathbb{R}$ and $\mathbb{R}_+$ is the set of nonnegative reals. For a finite set S, the Euclidean vector space of real functions with the domain S is denoted by $\mathbb{R}^S$. An element x of $\mathbb{R}^S$ is represented by the vector $(x^i)_{i \in S}$. Also, $\mathbb{R}^S_+ = \{x \in \mathbb{R}^S \mid x^i \geq 0 \text{ for all } i \in S\}$ and $\mathbb{R}^S_{++} = \{x \in \mathbb{R}^S \mid x^i > 0 \text{ for all } i \in S\}$. If $x, y \in \mathbb{R}^N$, $S, T \subseteq N$, and $S \cap T = \emptyset$, then $x^S = (x^i)_{i \in S}$ and $z = (x^S, y^T) \in \mathbb{R}^{S \cup T}$ is given by $z^i = x^i$ for all $i \in S$ and $z^j = y^j$ for all $j \in T$.

The symbols in the following list are ordered according to the page numbers, the numbers in the first column, of their definitions or first occurrences.

2	$X \setminus Y$	set difference ($\{x \in X \mid x \notin Y\}$)
9	$\mathcal{U}$	the universe of players
9	(N, v)	coalitional TU game
10	$\Rightarrow$	implies, implication
11	$x(S)$	aggregate amount of S
11	$\alpha v + \beta$	strategically equivalent coalition function to v
12	$\mathcal{SYM}$	group of symmetries
12	$\|A\|$	cardinality of A
19	$X^*(N, v)$	feasible payoff vectors
19	Γ	set of TU games
19	σ	solution
19	$\mathcal{C}(N, v)$	the core
19	$\pi(x)$	image of x
19	πv	isomorphic coalition function
20	$X(N, v)$	set of preimputations
22	$v_{S,x}$	reduced coalition function
23	$\Gamma_{\mathcal{U}}, \Gamma^{\mathcal{C}}_{\mathcal{U}}$	set of all games, with nonempty cores

24	$\mathcal{P}(N)$	set of all pairs of players
27	2^N	set of all subsets
28	χ_S	characteristic vector
33	$x \cdot y$	scalar product
38	a_+	positive part of a
42	$\Gamma_{\mathcal{U}}^{tb}$	set of totally balanced games
45	$(N, v, \mathcal{R})$	TU game with coalition structure
46	Δ	set of TU games with coalition structures
47	$\Delta_{\mathcal{U}}, \Delta_{\mathcal{U}}^{\mathcal{C}}$	set of all games, with nonempty cores
47	$\mathcal{P}(\mathcal{R})$	set of partners in $\mathcal{R}$
52	$\mathcal{T}_{k\ell}(N)$	coalitions containing k and not ℓ
53	$\mathcal{PM}(N, v, \mathcal{R})$	unconstrained bargaining set
55	$\mathcal{M}(N, v, \mathcal{R})$	bargaining set
57	$\lVert \cdot \rVert$	Euclidean norm
58	$e(S, x, v)$	excess of S at x
58	$s_{k\ell}(x, v)$	maximum surplus
65	$\mathcal{M}_r, \mathcal{PM}_r$	reactive (pre-)bargaining set
66	$\mathcal{M}_{sr}, \mathcal{PM}_{sr}$	semi-reactive (pre-)bargaining set
67	$\geq, >, \gg$	weak and strict inequalities (between vectors)
69	$\mathcal{MB}, (\mathcal{PMB})$	Mas-Colell (pre-)bargaining set
82	$\mathcal{PK}(N, v, \mathcal{R})$	prekernel
84	$\mathcal{N}(N, v, \mathcal{R})$	nucleolus of a game with coalition structure
84	$\mathcal{PN}(\cdots), \nu(\cdots)$	prenucleolus, point
87	$\mathcal{D}(\alpha, x, v)$	coalitions whose excess is at least α
89	$k \succsim_v \ell$	desirability relation
95	$\mathcal{K}(N, v, \mathcal{R})$	kernel
110	$\binom{t}{k}$	binomial coefficient "t choose k"
113	$\prod$	(Cartesian) product
124	$\Psi(N, v, \mathcal{R})$	modiclus
133	$\mathcal{C}_\varepsilon(G)$	ε-core
134	$\mathcal{LC}(G)$	least-core
143	$[a, b]$	line segment between a and b
153	$\phi(v)$	Shapley value
159	$v_{S,\sigma}$	σ-reduced coalition function
171	$\phi_*(N, v, \mathcal{R})$	Aumann-Drèze value
173	$\overline{\phi}(N, v, \mathcal{R})$	Owen value
177	$\frac{\partial \overline{v}(\cdots)}{\partial x^j}$	partial derivative
181	$\varphi : X \rightrightarrows Y$	set-valued function
183	$\forall$	universal quantification, "for all"
206	$V_\alpha(\cdot, \cdot)$	NTU coalition function of α-effectiveness
207	$V_\beta(\cdot, \cdot)$	NTU coalition function of β-effectiveness
210	(N, V)	NTU coalitional game
210	(N, V_v)	NTU game corresponding to TU game
217	∂Z	boundary of Z

224	$(S, V_{S,x})$	reduced NTU game
226	$\widehat{\Gamma}$	set of (non-levelled) NTU games
233	$\exists$	existential quantification, "there exists"
235	$\Delta_{++}(N)$	the interior of the unit simplex
235	Δ_{++}^{V}	set of viable vectors
239	$\Phi(N, V)$	set of Shapley NTU values
244	$\boldsymbol{x} = (x_S)_{S \in 2^N \setminus \{\emptyset\}}$	payoff configuration
245	$\Delta_{++}^{V(N)}$	set of viable vectors for $V(N)$
244	(S, V^S)	NTU subgame
247	$\boldsymbol{\sigma}$	payoff configuration solution
247	$\boldsymbol{\Phi}^{\mathcal{H}}$	Harsanyi solution
254	$\phi(N, V)$	consistent Shapley value of a hyperplane game
258	$\boldsymbol{\Phi}^{\mathcal{MO}}(N, V)$	set of consistent Shapley solutions
282	$\mathcal{P}MB^*(N, V)$	extended Mas-Colell bargaining set
295	$\mathcal{PM}^o, \mathcal{M}^o$	ordinal (pre-)bargaining set
306	$\mathcal{BCPK}(N, V, \mathcal{R})$	bilateral consistent prekernel

1

Introduction

This chapter is divided into three sections. In the first section the different kinds of cooperative games are discussed. A verbal description of the contents of this book is given in the second section and, finally, Section 1.3 describes one of the main goals of this book and comments on some related aspects.

1.1 Cooperative Games

This book is devoted to a study of the basic properties of solutions of cooperative games in coalitional form. Only Chapter 11 is an exception: In Sections 11.1 and 11.2 we study cooperative games in strategic form. The reason for this exception will be explained below. A coalitional or a strategic game is cooperative if the players can make binding agreements about the distribution of payoffs or the choice of strategies, even if these agreements are **not** specified or implied by the rules of the game (see Harsanyi and Selten (1988)). Binding agreements are prevalent in economics. Indeed, almost every one-stage seller-buyer transaction is binding. Moreover, most multi-stage seller-buyer transactions are supported by binding contracts. Usually, an agreement or a contract is binding if its violation entails high monetary penalties which deter the players from breaking it. However, agreements enforceable by a court may be more versatile.

Cooperative coalitional games are divided into two categories: games with transferable utilities and games with nontransferable utilities. We shall now consider these two classes of coalitional games in turn.

Let N be a set of players. A coalitional game with transferable utilities (a *TU game*) on N is a function that associates with each subset S of N (a *coalition*, if nonempty), a real number $v(S)$, the worth of S. Additionally, it is required that v assign zero to the empty set. If a coalition S forms, then it can divide its worth, $v(S)$, in any possible way among its members. That is, S can achieve every payoff vector $x \in \mathbb{R}^S$ which is feasible, that is, which satisfies

$$\sum_{i \in S} x^i \leq v(S).$$

This is possible if money is available and desirable as a medium of exchange, and if the utilities of the players are linear in money (see Aumann (1960)).

Von Neumann and Morgenstern (1953) derive the TU coalition function from the strategic form of games with transferable utilities (i.e., utilities which are linear in money). The worth of a coalition S in a TU strategic game is its maximin value in the two-person zero-sum game, where S is opposed by its complement, $N \setminus S$, and correlated strategies of both S and $N \setminus S$ are used.

We consider the TU coalition function as a primitive concept, because in many applications of TU games coalition functions appear without any reference to a (TU) strategic game. This is, indeed, the case for many cost allocation problems. Furthermore, in a cooperative strategic game, any combination of strategies can be supported by a binding agreement. Hence the players focus on the choice of "stable" payoff vectors and not on the choice of a "stable" profile of strategies as in a noncooperative game. Clearly, the coalitional form is the suitable form for the analysis of the choice of a stable payoff distribution among the set of all feasible payoff distributions.

Coalitional games with nontransferable utilities (*NTU games*) were introduced in Aumann and Peleg (1960). They are suitable for the analysis of many cooperative and competitive phenomena in economics (see, e.g., Scarf (1967) and Debreu and Scarf (1963)). The axiomatic approach to NTU coalition functions, due to Aumann and Peleg (1960), has been motivated by a direct derivation of the NTU coalition function from the strategic form of the game. This approach is presented in Section 11.2.

1.2 Outline of the Book

We shall review the two parts consecutively.

1.2.1 TU Games

In Chapter 2 we first define coalitional TU games and some of their basic properties. Then we discuss market games, cost allocation games, and simple games. Games in the foregoing families frequently occur in applications. Finally, we systematically list the properties of the core. These properties, suitably modified, serve later, in different combinations, as axioms for the core itself, the prekernel, the prenucleolus, and the Shapley value.

Chapter 3 is devoted to the core. The main results are:

(1) A characterization of the set of all games with a nonempty core (the balanced games);

(2) a characterization of market games as totally balanced games; and

(3) an axiomatization of the core on the class of balanced games.

Various bargaining sets are studied in Chapter 4. We provide an existence theorem for bargaining sets which can be generalized to NTU games. Furthermore, it is proved that the Aumann-Davis-Maschler bargaining set of any convex game and of any assignment game coincides with its core.

Chapter 5 introduces the prekernel and the prenucleolus. We prove existence and uniqueness for the prenucleolus and, thereby, prove nonemptiness of the prekernel and reconfirm the nonemptiness of the aforementioned bargaining sets. The prekernel is axiomatized in Section 5.4. Moreover, we investigate individual rationality for the prekernel and, in addition, prove that it is reasonable. Finally, we prove that the kernel of a convex game coincides with its nucleolus.

Chapter 6 mainly focuses on:

(1) Sobolev's axiomatization of the prenucleolus;

(2) an investigation of the nucleolus of strong weighted majority games which shows, in particular, that the nucleolus of a strong weighted majority game is a representation of the game; and

(3) definition and verification of the basic properties of the modiclus; in particular, we show that the modiclus of any weighted majority game is a representation of the game.

In Chapter 7, ε-cores and the least-core are introduced, and their intuitive properties are studied. The main results are:

(1) A geometric characterization of the intersection of the prekernel of a game with an ε-core; and

(2) an algorithm for computing the prenucleolus.

Chapter 8 is entirely devoted to the Shapley value. Four axiomatizations of the Shapley value are presented:

(1) Shapley's axiomatization using additivity;

(2) Young's axiomatization using strong monotonicity;

(3) an axiomatization based on consistency by Hart and Mas-Colell; and

(4) Sobolev's axiomatization based on a special reduced game.

Moreover, Dubey's axiomatization of the Shapley value on the set of monotonic simple games is presented. We conclude with Owen's value of games with a priori unions and his formula relating the Shapley value of a game to the multilinear extension of the game.

Chapter 9 is devoted to continuity properties of solutions. All our solutions are upper hemicontinuous and closed-valued. The core and the nucleolus are actually continuous. The continuity of the Shapley value is obvious.

In Chapter 10 dynamic systems for the prekernel and various bargaining sets are introduced. Some results on stability and local asymptotic stability are obtained.

1.2.2 NTU Games

In Chapter 11 we define cooperative games in strategic form and derive their coalitional games. This serves as a basis for the axiomatic definition of coalitional NTU games.

Chapter 12 is entirely devoted to the core of NTU games. First we prove that suitably balanced NTU games have a nonempty core. Then we show that convex NTU games have a nonempty core. We conclude with various axiomatizations of the core.

In Chapter 13 we provide existence proofs and characterizations for the Shapley NTU value and the Harsanyi solution. We also give an axiomatic characterization of each solution.

Chapter 14 is devoted to the consistent Shapley value. First we investigate hyperplane games following Maschler and Owen (1989). Then we prove existence of the consistent value for p-smooth games. We conclude with an axiomatic analysis of the consistent value.

Chapter 15 investigates the classical and Mas-Colell bargaining sets for NTU games. We deal mainly with (NTU) majority voting games. We show that if there are at most five alternatives, then the Mas-Colell bargaining is nonempty. For majority games with six or more alternatives the Mas-Colell set may be empty. Using more elaborated examples we show that the Mas-Colell bargaining set of a non-levelled superadditive game may be empty. We conclude with some limit theorems for bargaining sets of majority games.

In Chapter 16 we conclude with an existence proof for the ordinal bargaining set of NTU games and with a discussion of related solutions.

1.2.3 A Guide for the Reader

We should like to make the following remarks.

Remark 1.2.1. The investigations of the various solutions are almost independent of each other. For example, you may study the core by reading Chapters 3 and 12 and browsing Sections 2.3 and 11.3. If you are interested only in the Shapley value, you should read Chapter 8 and Sections 13.1 and 13.2. Similar possibilities exist for the bargaining set, kernel, and nucleolus (see the Table of Contents).

Remark 1.2.2. If you plan an introductory course on game theory, then you may use Chapters 2, 3, and 8 for introducing cooperative games at the end of your course.

Remark 1.2.3. Chapters 2 - 12 may be used for a one-semester course on cooperative games. Part II may be used in a graduate course on cooperative games without side-payments.

Remark 1.2.4. Each section concludes with some exercises. The reader is advised to solve at least those exercises that are used in the text to complete the proofs of various results.

1.3 Special Remarks

The analysis of solutions of cooperative games emphasizes the axiomatic approaches which do not rely on interpersonal comparisons of utility. Moreover, we comment on the Nash program.

1.3.1 Axiomatizations

One of our main goals is to supply uniform and coherent axiomatizations for the main solutions of cooperative games. Indeed, this book is the first to include axiomatizations of the core, the prekernel, and the prenucleolus. Every axiom which we use is satisfied, sometimes after a suitable modification, by the core of TU games; the only exception is consistency (in the sense of Hart and Mas-Colell), which is satisfied only by the Shapley value. Table 8.11.1 shows our success for TU games.

1.3.2 Interpersonal Comparisons of Utility

For a definition of interpersonal comparisons of utility the reader is referred to Harsanyi (1992). In our view a solution is free of interpersonal comparisons of utility, if it has an axiomatization which does not use interpersonal comparisons of utility. As none of our axioms implies interpersonal comparisons of utility, all the solutions which we discuss do not rely on interpersonal comparisons of utility. (Covariance for TU games implies cardinal unit comparability. However, it is **not** used for actual comparisons of utilities (see Luce and Raiffa (1957), pp. 168 - 169).) The bargaining set, which is left unaxiomatized, does not involve interpersonal comparisons of utility by its definition.

1.3.3 Nash's Program

According to Harsanyi and Selten (1988), Section 1.11, "... analysis of any cooperative game G should be based on a formal bargaining model $B(G)$, involving bargaining moves and countermoves by the various players and resulting in an agreement about the outcome of the game. Formally, this bargaining model $B(G)$ would always be a noncooperative game in extensive form (or possibly in normal form), and the solution of the cooperative game G would be defined in terms of the equilibrium points of this noncooperative game $B(G)$." This claim is known as Nash's program. Peleg (1996) and (1997) shows that Nash's program cannot be implemented. Hence, we shall not further discuss it.

Part I

TU Games

2

Coalitional TU Games and Solutions

This chapter is divided into three sections. In the first section we define coalitional games and discuss some of their basic properties. In particular, we consider superadditivity and convexity of games. Also, constant-sum, monotonic, and symmetric games are defined.

Some families of games that occur frequently in applications are considered in Section 2.2. The first class of games that is discussed is that of market games. They model an exchange economy with money. Then we proceed to describe cost allocation games. We give in detail three examples: a water supply problem, airport games, and minimum cost spanning tree games. Finally, we examine the basic properties of simple games. These games describe parliaments, town councils, ad hoc committees, and so forth. They occur in many applications of game theory to political science.

The last section is devoted to a detailed discussion of properties of solutions of coalitional games. We systematically list all the main axioms for solutions, consider their plausibility, and show that they are satisfied by the core, which is an important solution for cooperative games.

2.1 Coalitional Games

Let $\mathcal{U}$ be a nonempty set of *players*. The set $\mathcal{U}$ may be finite or infinite. A *coalition* is a nonempty and **finite** subset of $\mathcal{U}$.

Definition 2.1.1. *A* **coalitional game with transferable utility** *(a TU game) is a pair (N, v) where N is a coalition and v is a function that associates a real number $v(S)$ with each subset S of N. We always assume that $v(\emptyset) = 0$.*

Remark 2.1.2. Let $G = (N, v)$ be a coalitional game. The set N is called the set of *players* of G and v the *coalition function.* Let S be a subcoalition of N. If S forms in G, then its members get the amount $v(S)$ of money (however, see Assumption 2.1.4). The number $v(S)$ is called the *worth* of S.

Remark 2.1.3. In most applications of coalitional games the players are persons or groups of persons, for example, labor unions, towns, nations, etc. However, in some interesting game-theoretic models of economic problems the players may not be persons. They may be objectives of an economic project, factors of production, or some other economic variables of the situation under consideration.

Assumption 2.1.4. At this stage we assume that the von Neumann-Morgenstern utility functions of the players are linear and increasing in money. (In Section 11.4 we show how this assumption can be somewhat relaxed.) Therefore, we may further assume that they all have the same positive slope. Now, if a coalition S forms, it may divide $v(S)$ among its members in any feasible way, that is, side payments are unrestricted. In view of the foregoing assumptions, there is a simple transformation from monetary side payments to the corresponding utility payoff vectors. Thus, technically, we may express all possible distributions of $v(S)$ (and lotteries on payoff distributions) as distributions of utility payoffs. In this sense coalitional games are transferable utility games. Henceforth, we shall be working with coalitional games where the payoffs are in utility units.

Definition 2.1.5. *A game* (N, v) *is* **superadditive** *if*

$$\Big(S, T \subseteq N \text{ and } S \cap T = \emptyset\Big) \Rightarrow v(S \cup T) \geq v(S) + v(T). \tag{2.1.1}$$

Condition 2.1.1 is satisfied in most of the applications of TU games. Indeed, it may be argued that if $S \cup T$ forms, its members can decide to act as if S and T had formed separately. Doing so they will receive $v(S) + v(T)$, which implies (2.1.1). Nevertheless, quite often superadditivity is violated. Anti-trust laws may exist, which reduce the profits of $S \cup T$, if it forms. Also, large coalitions may be inefficient, because it is more difficult for them to reach agreements on the distribution of their proceeds.

The following weak version of superadditivity is very useful.

Definition 2.1.6. *A game is* **weakly superadditive** *if*

$$v(S \cup \{i\}) \geq v(S) + v(\{i\}) \text{ for all } S \subseteq N \text{ and } i \notin S.$$

Definition 2.1.7. *A game* (N, v) *is* **convex** *if*

$$v(S) + v(T) \leq v(S \cup T) + v(S \cap T) \text{ for all } S, T \subseteq N.$$

Clearly, a convex game is superadditive. The following equivalent characterization of convex games is left to the reader (see Exercise 2.1.1): A game is convex if and only if, for all $i \in N$,

$$v(S \cup \{i\}) - v(S) \leq v(T \cup \{i\}) - v(T) \text{ for all } S \subseteq T \subseteq N \setminus \{i\}. \quad (2.1.2)$$

Thus, the game (N, v) is convex if and only if the marginal contribution of a player to a coalition is monotone nondecreasing with respect to set-theoretic inclusion. This explains the term convex. Convex games appear in some important applications of game theory.

Definition 2.1.8. *A game (N, v) is* **constant-sum** *if*

$$v(S) + v(N \setminus S) = v(N) \textit{ for all } S \subseteq N.$$

Constant-sum games have been extensively investigated in the early work in game theory (see von Neumann and Morgenstern (1953)). Also, very often political games are constant-sum.

Definition 2.1.9. *A game (N, v) is* **inessential** *if it is additive, that is, if $v(S) = \sum_{i \in S} v(\{i\})$ for every $S \subseteq N$.*

Clearly, an inessential game is trivial from a game-theoretic point of view. That is, if every player $i \in N$ demands at least $v(\{i\})$, then the distribution of $v(N)$ is uniquely determined.

Notation 2.1.10. Let N be a coalition and let $\mathbb{R}$ denote the real numbers. We denote by $\mathbb{R}^N$ the set of all functions from N to $\mathbb{R}$. If $x \in \mathbb{R}^N$ and $S \subseteq N$, then we write $x(S) = \sum_{i \in S} x^i$. Clearly, $x(\emptyset) = 0$.

Remark 2.1.11. Let N be a coalition and $x \in \mathbb{R}^N$. Applying the foregoing notation enables us to consider x as a coalition function as well. Thus, (N, x) is the coalitional game given by $x(S) = \sum_{i \in S} x^i$ for all $S \subseteq N$.

Definition 2.1.12. *Two games (N, v) and (N, w) are* **strategically equivalent** *if there exist $\alpha > 0$ and $\beta \in \mathbb{R}^N$ such that*

$$w(S) = \alpha v(S) + \beta(S) \textit{ for all } S \subseteq N. \quad (2.1.3)$$

Clearly, Definition 2.1.12 is compatible with the restriction on the utilities of the players of a coalitional game. Indeed, these are determined up to positive affine transformations, one for each player, and all with the same slope. In view of Remark 2.1.11, Eq. (2.1.3) can be expressed as $w = \alpha v + \beta$.

Definition 2.1.13. *A game (N, v) is* **zero-normalized** *(0-normalized) if $v(\{i\}) = 0$ for all $i \in N$.*

Clearly, every game is strategically equivalent to a 0-normalized game.

The following definition is useful.

Definition 2.1.14. *A game* (N, v) *is* **monotonic** *if*

$$S \subseteq T \subseteq N \Rightarrow v(S) \leq v(T).$$

We conclude this section with the following definition and notation.

Definition 2.1.15. *Let* $G = (N, v)$ *be a game and let* π *be a permutation of* N. *Then* π *is a* **symmetry** *of* G *if* $v(\pi(S)) = v(S)$ *for all* $S \subseteq N$. *The group of all symmetries is denoted by* $\mathcal{SYM}(G)$. *The game* G *is* **symmetric** *if* $\mathcal{SYM}(G)$ *is the group* $\mathcal{SYM}_N$ *of all permutations of* N.

Notation 2.1.16. If A is a finite set, then we denote by $|A|$ the number of members of A.

Exercises

Exercise 2.1.1. Prove that a game (N, v) is convex, if and only if (2.1.2) is satisfied.

Exercise 2.1.2. Prove that *strategic equivalence* is an equivalence relation, that is, it is reflexive, symmetric, and transitive.

Exercise 2.1.3. Let the games (N, v) and (N, w) be strategically equivalent. Prove that if (N, v) is superadditive (respectively weakly superadditive, convex, constant-sum, or inessential), then (N, w) is superadditive (respectively weakly superadditive, convex, constant-sum, or inessential).

Exercise 2.1.4. Prove that every game is strategically equivalent to a monotonic game.

Exercise 2.1.5. Prove that a game is weakly superadditive, if and only if it is strategically equivalent to a 0-normalized monotonic game. (Note that the terms *zero-monotonicity* (0-monotonicity) and *weak superadditivity* are synonymous.)

Exercise 2.1.6. Prove that a game (N, v) is symmetric, if and only if

$$|S| = |T| \Rightarrow v(S) = v(T) \quad \text{for all } S, T \subseteq N.$$

Exercise 2.1.7. Let (N, v) be a game and let $\pi \in \mathcal{SYM}_N$. Prove that $\pi \in \mathcal{SYM}(N, v)$ if for each $S \subseteq N$ there exists $\pi^* \in \mathcal{SYM}(N, v)$ such that $\pi^*(S) = \pi(S)$.

Exercise 2.1.8. Prove the following converse of Exercise 2.1.7. Let N be a coalition and let $\mathcal{S}$ be a subgroup of $\mathcal{SYM}_N$ which has the following property: If $\pi \in \mathcal{SYM}_N$ and for each $S \subseteq N$ there exists $\pi^* \in \mathcal{S}$ such that $\pi^*(S) = \pi(S)$, then $\pi \in \mathcal{S}$. Show that there exists a superadditive game (N, v) such that $\mathcal{SYM}(N, v) = \mathcal{S}$.

2.2 Some Families of Games

In this section we introduce some important classes of coalitional games.

2.2.1 Market Games

Let $\mathcal{U}$ be the set of players. A *market* is a quadruple $(N, \mathbb{R}^m_+, A, W)$. Here N is a coalition (the set of *traders*); $\mathbb{R}^m_+$ is the nonnegative orthant of the m-dimensional Euclidean space (the *commodity space*); $A = (a^i)_{i \in N}$ is an indexed collection of points in $\mathbb{R}^m_+$ (the *initial endowments*); and $W = (w^i)_{i \in N}$ is an indexed collection of continuous concave functions on $\mathbb{R}^m_+$ (the *utility functions*).

We make the assumption that our markets have transferable utility, that is, there exists an additional commodity, money, and each trader measures his utility for goods in terms of this money. Formally, the utility of trader $i \in N$ for $x \in \mathbb{R}^m_+$ and the amount $\xi \in \mathbb{R}$ of money is $W^i(x, \xi) = w^i(x) + \xi$. The amount ξ of money may be negative in the foregoing equality. Also, it is no loss of generality to assume that, initially, each trader has no money. Indeed, if W^i is a utility function for trader i, then so is $W^i + b$, where $b \in \mathbb{R}$. (See also Shapley and Shubik (1966) for a discussion of these assumptions.)

Let $(N, \mathbb{R}^m_+, A, W)$ be a market and let $\emptyset \neq S \subseteq N$. A trade among the members of S results in an indexed collection $(x^i, \xi^i)_{i \in S}$ such that $x^i \in \mathbb{R}^m_+$ for all $i \in S$, $\sum_{i \in S} x^i = \sum_{i \in S} a^i$, and $\sum_{i \in S} \xi^i = 0$. The total utility to the coalition S as a result of the foregoing transaction is

$$\sum_{i \in S} W^i(x^i, \xi^i) = \sum_{i \in S} w^i(x^i) + \sum_{i \in S} \xi^i = \sum_{i \in S} w^i(x^i).$$

Thus, we are led to the following definitions. A *feasible S-allocation* is an indexed collection $x_S = (x^i)_{i \in S}$ such that $x^i \in \mathbb{R}^m_+$ for all $i \in S$ and $\sum_{i \in S} x^i = \sum_{i \in S} a^i$. We denote by X^S the set of all feasible S-allocations.

Definition 2.2.1. *A game (N, v) is a* **market game**, *if there exists a market $(N, \mathbb{R}^m_+, A, W)$ such that*

$$v(S) = \max \left\{ \sum_{i \in S} w^i(x^i) \,\middle|\, x_S \in X^S \right\}$$

for every $S \subseteq N$.

Definition 2.2.1 is due to Shapley and Shubik (1969a).

Example 2.2.2. Let $N = N_1 \cup N_2$, where $N_1 \cap N_2 = \emptyset$ and $|N_j| \geq 1$ for $j = 1, 2$, and let $m = 2$. For $i \in N_1$ let $a^i = (1, 0)$ and for $i \in N_2$ let $a^i = (0, 1)$.

Finally, let $w^i(x_1, x_2) = \min\{x_1, x_2\}$ for all $i \in N$. Then $(N, \mathbb{R}^2_+, A, W)$ is a market. The coalition function v of the corresponding market game is given by

$$v(S) = \min\{|S \cap N_1|, |S \cap N_2|\} \text{ for all } S \subseteq N.$$

This game was introduced in Shapley (1959). See also Shapley and Shubik (1969b).

2.2.2 Cost Allocation Games

Let $\mathcal{U}$ be a set of players. A *cost allocation problem* is a game (N, c) where N is a coalition and c, the coalition function, is the *cost function* of the problem. Intuitively, N represents a set of potential customers of a public service or public facility. Each customer will either be served at some preassigned level or not served at all. Let $S \subseteq N$. Then $c(S)$ represents the least cost of serving the members of S by the most efficient means. The game (N, c) is called a *cost game.*

Although a cost game (N, c) is, formally, a game, it is not so from the point of view of applications, because the cost function is not interpreted as an ordinary coalition function. It is possible to associate with a cost game (N, c) an ordinary game (N, v), called the *savings game*, which is given by $v(S) = \sum_{i \in S} c(\{i\}) - c(S)$ for all $S \subseteq N$.

Let (N, c) be a cost game and (N, v) be the corresponding savings game. Then (N, c) is *subadditive*, that is,

$$\Big(S, T \subseteq N \text{ and } S \cap T = \emptyset\Big) \Rightarrow c(S) + c(T) \geq c(S \cup T),$$

iff (N, v) is superadditive, and (N, c) is *concave*, that is,

$$c(S) + c(T) \geq c(S \cup T) + c(S \cap T) \text{ for all } S, T \subseteq N,$$

iff (N, v) is convex. In applications cost games are usually subadditive.

See Lucas (1981), Young (1985a), and Tijs and Driessen (1986) for surveys concerning cost allocation games.

Example 2.2.3 (A municipal cost-sharing problem).
A group N of towns considers the possibility of building a common water treatment facility. Each municipality requires a minimum supply of water that it can either provide from its own distribution system or from a system shared with some or all of the other municipalities. The *alternative* or *stand-alone* cost $c(S)$ of a coalition $S \subseteq N$ is the minimum cost of supplying the members of S by the most efficient means available. In view of the fact that a set $S \subseteq N$ can be served by several separate subsystems, we obtain a subadditive cost game. Such games have been investigated by Suzuki and Nakayama (1976), Young, Okada, and Hashimoto (1982), and others.

Example 2.2.4 (Airport games).
Consider an airport with one runway. Suppose that there are m different types of aircrafts and that c_k, $1 \leq k \leq m$, is the cost of building a runway to accommodate an aircraft of type k. Let N_k be the set of aircraft landings of type k in a given time period, and let $N = \bigcup_{k=1}^{m} N_k$. Thus, the "players" (the members of N) are landings of aircrafts. The cost function of the corresponding cost game, which is an *airport* game, is given by

$$c(S) = \max\{c_k \mid S \cap N_k \neq \emptyset\} \text{ and } c(\emptyset) = 0.$$

We remark that an airport game is concave. The foregoing model has been investigated by Littlechild (1974), Littlechild and Owen (1973), and others.

Example 2.2.5 (Minimum cost spanning tree games).
A group N of customers who are geographically separated has to be connected to a certain supplier 0. For example, the customers may be cities and the supplier an electricity plant. A user can be linked directly to the supplier or via other users. Let $N_* = N \cup \{0\}$. We consider the complete (undirected) graph whose node set is N_*. The cost of connecting $i, j \in N_*$, $i \neq j$, by an edge $e_{\{i,j\}}$ is $c_{\{i,j\}}$. We frequently write e_{ij} for $e_{\{i,j\}}$ and c_{ij} for $c_{\{i,j\}}$. Now the minimum cost spanning tree game is defined as follows. Let $S \subseteq N$. A minimum cost spanning tree $\Gamma_S = (S \cup \{0\}, E_S)$ is a tree with node set $S \cup \{0\}$ and a set of edges E_S, that connects the members of S to the common supplier 0, such that the total cost of all connections is minimal. The cost function c of the cost game (N, c) is now defined by

$$c(S) = \sum_{e_{ij} \in E_S} c_{ij} \text{ for all } S \subseteq N \quad (c(\emptyset) = 0).$$

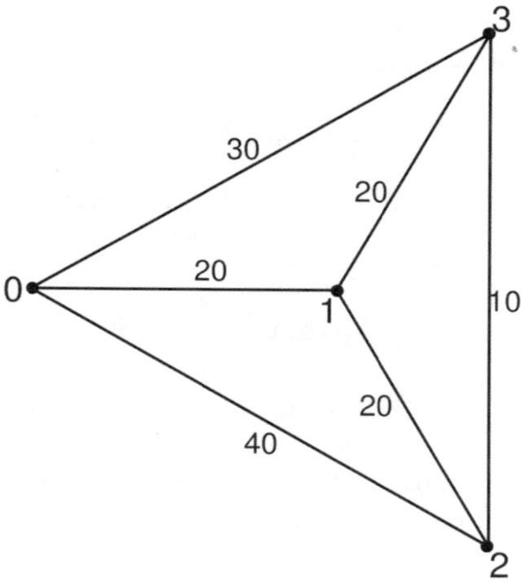

Fig. 2.2.1. Connection Cost

Now we consider the following particular example. Let $N = \{1, 2, 3\}$ and let the cost of the various links be as shown in Figure 2.2.1.

The cost function is given by the following formula:

$$c(S) = \begin{cases} 0 \ , \text{if } S = \emptyset \\ 20 \ , \text{if } S = \{1\} \\ 30 \ , \text{if } S = \{3\} \\ 50 \ , \text{if } S = N \\ 40 \ , \text{otherwise.} \end{cases}$$

Minimum cost spanning tree games have been investigated by Granot and Huberman (1981), Granot and Huberman (1984), Megiddo (1978), and others.

2.2.3 Simple Games

Let $\mathcal{U}$ be a set of players.

Definition 2.2.6. *A* **simple game** *is a pair* $(N, \mathcal{W})$ *where* N *is a coalition and* $\mathcal{W}$ *is a set of subsets of* N *satisfying:*

$$N \in \mathcal{W} \tag{2.2.1}$$

$$\emptyset \notin \mathcal{W} \tag{2.2.2}$$

$$\Big(S \subseteq T \subseteq N \text{ and } S \in \mathcal{W}\Big) \Rightarrow T \in \mathcal{W}. \tag{2.2.3}$$

The collection $\mathcal{W}$ *of coalitions is the set of* **winning coalitions**.

Property 2.2.3 is the *monotonicity property* of simple games. Intuitively, a simple game $g = (N, \mathcal{W})$ represents a committee: The coalition N is the set of members of the committee and $\mathcal{W}$ is the set of coalitions that fully control the decision of g. We observe that every parliament is a committee; every town council is a committee; the UN Security Council is a committee, and so forth.

We shall be interested in properties of simple games.

Definition 2.2.7. *Let* $g = (N, \mathcal{W})$ *be a simple game.*

$$\text{The simple game } g \text{ is } \left\{ \begin{array}{c} \textbf{proper} \\ \textbf{strong} \\ \textbf{weak} \end{array} \right\} \text{ if } \left\{ \begin{array}{c} S \in \mathcal{W} \Rightarrow N \setminus S \notin \mathcal{W} \\ S \notin \mathcal{W} \Leftrightarrow N \setminus S \in \mathcal{W} \\ V = \bigcap_{S \in \mathcal{W}} S \neq \emptyset \end{array} \right\}.$$

The members of V *are called* **veto players** *or* **vetoers**. *The simple game* g *is* **dictatorial** *if there exists* $j \in N$ *("the" dictator) such that*

$$S \in \mathcal{W} \Leftrightarrow j \in S.$$

Remark 2.2.8. Let $g = (N, \mathcal{W})$ be a simple game. In many applications it is convenient to associate with g the coalitional game $G = (N, v)$ where $v(S) = 1$ if $S \in \mathcal{W}$ and $v(S) = 0$ otherwise. For example, this is the case if the committee g has to allocate a fixed amount of money among its members. This fact leads to the following definition.

Definition 2.2.9. *Let $g = (N, \mathcal{W})$ be a simple game. The* **associated coalitional game (with a simple game)** (N, v) *is given by:*

$$v(S) = \begin{cases} 1, & \text{if } S \in \mathcal{W} \\ 0, & \text{otherwise.} \end{cases}$$

Let $g = (N, \mathcal{W})$ be a simple game and let $G = (N, v)$ be the associated coalitional game. Then G is monotonic. Also, G is superadditive if and only if g is proper, and G is constant-sum if and only if g is strong.

Note that any monotonic coalitional game (N, v) which satisfies $v(S) \in \{0, 1\}$ for all $S \subseteq N$ and $v(N) = 1$ is the associated game of some simple game.

Definition 2.2.10. *A simple game is* **symmetric** *if the associated game is symmetric.*

Thus, a simple game $g = (N, \mathcal{W})$ is symmetric if

$$\Big(S \in \mathcal{W}, T \subseteq N, \text{ and } |T| = |S|\Big) \Rightarrow T \in \mathcal{W}$$

(see Exercise 2.1.6).

Definition 2.2.11. *A simple game $(N, \mathcal{W})$ is a* **weighted majority game** *if there exist a* **quota** $q > 0$ *and* **weights** $w^i \geq 0$ *for all $i \in N$ such that for all $S \subseteq N$*

$$S \in \mathcal{W} \Leftrightarrow w(S) \geq q \quad \textit{(see Notation 2.1.10).}$$

Let $g = (N, \mathcal{W})$ be a weighted majority game with quota $q > 0$ and weights $w^i \geq 0$ for all $i \in N$. The $(|N|+1)$-tuple $\big(q; (w^i)_{i \in N}\big)$ is called a *representation* of g, and we write $g \widehat{=} \big(q; (w^i)_{i \in N}\big)$.

Notation 2.2.12. If $g = (N, \mathcal{W})$ is a simple game we denote by

$$\mathcal{W}^m = \{S \in \mathcal{W} \mid T \subsetneq S \Rightarrow T \notin \mathcal{W}\}$$

the set of *minimal* winning coalitions.

Definition 2.2.13. *Let $g = (N, \mathcal{W})$ be a weighted majority game. The representation $\big(q; (w^i)_{i \in N}\big)$ of g is a* **homogeneous** *representation of g if*

$$S \in \mathcal{W}^m \Rightarrow w(S) = q.$$

A weighted majority game is **homogeneous** *if it has a homogeneous representation.*

Remark 2.2.14. A symmetric simple game $g = (N, \mathcal{W})$ has the homogeneous representation $(k; 1, \ldots, 1)$, where k denotes the common size of every minimal winning coalition. Such a game is also denoted by (n, k) where $n = |N|$.

Example 2.2.15. The UN Security Council is given by the game

$$g \mathrel{\widehat{=}} (39; 7, 7, 7, 7, 7, \underbrace{1, \ldots, 1}_{10 \text{ times}}).$$

This game is weak (the vetoers are the Big Five) and homogeneous.

For a comprehensive study of simple games, the reader is referred to Shapley (1962a).

Exercises

Exercise 2.2.1. Prove that every market game is superadditive and give an example of a market game which is not convex.

Exercise 2.2.2. Let (V, E) be the complete graph on a nonempty finite set V of vertices and let $c : E \to \mathbb{R}$ be a cost function. Prove that a spanning tree (V, E^*) is a minimum cost spanning tree (m.c.s.t.) iff for every path $(v_j, v_{j+1})_{j=1,\ldots,k-1}$ in E^* the following inequalities are true:

$$c(v_1, v_k) \geq c(v_j, v_{j+1}),\ j = 1, \ldots, k-1.$$

Exercise 2.2.3. Using Exercise 2.2.2 verify that the following algorithm yields an m.c.s.t. of (V, E) after $|V| - 1$ steps.
Step 1: Choose a cheapest edge.
Step k: Let $E_{k-1} = \{e_1, \ldots, e_{k-1}\}$ be the set of edges chosen in Steps $1, \ldots, k-1$. Choose a cheapest edge e_k in $E \setminus E_{k-1}$ such that

$$\left(V, E_{k-1} \cup \{e_k\}\right)$$

is acyclic. (See Kruskal (1956).)

Exercise 2.2.4. Prove the following assertions:
(1) A weak simple game is proper.
(2) A simple game is dictatorial if and only if it is both weak and strong.

Exercise 2.2.5. Find all strong weighted majority games with five players. (If we do not distinguish between games that are obtained from one another by renaming the players, then there exist seven games.)

Exercise 2.2.6. Find a strong weighted majority game that is **not** homogeneous. (Six players are sufficient.)

2.3 Properties of Solutions

Let $\mathcal{U}$ be a set of players and let (N, v) be a game. We denote

$$X^*(N, v) = \left\{x \in \mathbb{R}^N \mid x(N) \leq v(N)\right\}.$$

The set $X^*(N, v)$ is the set of *feasible payoff vectors* for the game (N, v).

Definition 2.3.1. *Let Γ be a set of games. A* **solution** *on Γ is a function σ which associates with each game $(N, v) \in \Gamma$ a subset $\sigma(N, v)$ of $X^*(N, v)$.*

Intuitively, a solution is determined by a system of "reasonable" restrictions on the correspondence $X^*(\cdot, \cdot)$. For example, we may impose certain inequalities that guarantee the "stability" of the members of $\sigma(N, v)$ in some sense. Alternatively, σ may be characterized by a set of axioms. We remark that *each* member of $\sigma(N, v)$ is considered a possible final payoff distribution for (N, v).

In this section we shall deal only with the following solution.

Definition 2.3.2. *The* **core** *of a game (N, v), denoted by $\mathcal{C}(N, v)$, is defined by*

$$\mathcal{C}(N, v) = \{x \in X^*(N, v) \mid x(S) \geq v(S) \text{ for all } S \subseteq N\}.$$

Let $x \in X^*(N, v)$. Then $x \in \mathcal{C}(N, v)$ if and only if no coalition can improve upon x. Thus, each member of the core is a highly stable payoff distribution.

We shall now define some properties of solutions that are satisfied by the core. They will enable us to investigate other solutions in subsequent chapters.

Definition 2.3.3. *Let σ be a solution on a set Γ of games. We say that σ is* **covariant under strategic equivalence** *(COV) if the following condition is satisfied: If $(N, v), (N, w) \in \Gamma, \alpha > 0, \beta \in \mathbb{R}^N$, and $w = \alpha v + \beta$ (see Remark 2.1.11), then*

$$\sigma(N, w) = \alpha\sigma(N, v) + \beta.$$

COV expresses the following simple condition. If the two games (N, v) and (N, w) are strategically equivalent, that is, there exist $\alpha > 0$ and $\beta \in \mathbb{R}^N$ such that $w = \alpha v + \beta$, then their solution sets are related by the same transformation on the utilities of the players, that is, $\sigma(N, w) = \alpha\sigma(N, v) + \beta$. Thus, COV is a basic property of solutions which we may consider a necessary condition. As the reader may easily verify, the core satisfies COV.

Let (N, v) be a game and let $\pi : N \to \mathcal{U}$ be an injection. The game $(\pi(N), \pi v)$ is defined by $\pi v(\pi(S)) = v(S)$ for all $S \subseteq N$. Also, if $x \in \mathbb{R}^N$, then $y = \pi(x) \in \mathbb{R}^{\pi(N)}$ is given by $y^{\pi(i)} = x^i$ for all $i \in N$. A game (N', w) is *equivalent* or *isomorphic* to (N, v) if there exists an injection $\pi : N \to \mathcal{U}$ such that $\pi(N) = N'$ and $\pi v = w$.

Definition 2.3.4. *Let σ be a solution on a set Γ of games. We say that σ is* **anonymous** *(AN) if the following condition is satisfied: If $(N, v) \in \Gamma$, $\pi : N \to \mathcal{U}$ is an injection, and if $(\pi(N), \pi v) \in \Gamma$, then $\sigma(\pi(N), \pi v) = \pi(\sigma(N, v))$.*

AN simply says that σ is independent of the names of the players. Thus, AN also is a necessary condition for solutions. As the reader can easily verify, the core satisfies AN.

Remark 2.3.5. A solution σ on a set Γ of games is *symmetric* (SYM) if $\sigma(N, v) = \pi(\sigma(N, v))$ for all games $(N, v) \in \Gamma$ and all symmetries π of (N, v) (see Definition 2.1.15). Clearly, SYM follows from AN.

The following notation is needed in the sequel. Let (N, v) be a game. We denote

$$X(N, v) = \left\{x \in \mathbb{R}^N \mid x(N) = v(N)\right\}.$$

The set $X(N, v)$ is the set of *Pareto optimal* feasible payoffs or the set of *preimputations.*

Definition 2.3.6. *A solution σ on a set Γ of games is* **Pareto optimal** *(PO) if $\sigma(N, v) \subseteq X(N, v)$ for every game $(N, v) \in \Gamma$.*

PO is equivalent to the following condition: If $x, y \in X^*(N, v)$ and $x^i > y^i$ for all $i \in N$, then $y \notin \sigma(N, v)$. This formulation seems quite plausible, and similar versions to it are used in social choice (see Arrow (1951)) and bargaining theory (see Nash (1950)). Nevertheless, PO is actually quite a strong condition in the context of cooperative game theory. Indeed, players may fail to agree on a choice of a Pareto optimal point, because different players may have different preferences over the Pareto optimal set.

Clearly, the core satisfies PO.

Definition 2.3.7. *A solution σ on a set Γ of games is* **individually rational** *(IR) if it satisfies the following condition: If $(N, v) \in \Gamma$ and $x \in \sigma(N, v)$, then $x^i \geq v(\{i\})$ for all $i \in N$.*

IR says that every player i gets, at every point of the solution set, at least his solo worth $v(\{i\})$. If, indeed, all the singleton coalitions $\{i\}$, $i \in N$, may be formed, then IR follows from the usual assumption of utility maximization. We remark that the core satisfies IR. The set of *imputations* of (N, v), $I(N, v)$, is defined by

$$I(N, v) = \{x \in X(N, v) \mid x^i \geq v(\{i\}) \text{ for all } i \in N\}.$$

The following notation is needed for the next definition. If N is a coalition and $A, B \subseteq \mathbb{R}^N$, then

$$A + B = \{a + b \mid a \in A \text{ and } b \in B\}.$$

Definition 2.3.8. *A solution σ on a set Γ of games is* **superadditive** *(*SUPA*) if*

$$\sigma(N, v_1) + \sigma(N, v_2) \subseteq \sigma(N, v_1 + v_2)$$

when $(N, v_1), (N, v_2)$, and $(N, v_1 + v_2)$ are in Γ.

Clearly, SUPA is closely related to additivity. Indeed, for one-point solutions SUPA is equivalent to additivity. Plausibility arguments for additivity can be based on games that consist of two games played separately by the same players (e.g., at different times, or simultaneously using agents). However, these arguments are not always valid. If σ satisfies COV, which is usually assumed, then SUPA may be justified by considering the action of σ on probability combinations of games.

Intuitively, SUPA is somewhat weaker than additivity. As the reader can easily verify the core satisfies SUPA.

Let (N, v) be a game and $i \in N$. We denote

$$\begin{aligned} b^i_{\max}(N, v) &= \max_{S \subseteq N \setminus \{i\}} \Big(v(S \cup \{i\}) - v(S)\Big), \\ b^i_{\min}(N, v) &= \min_{S \subseteq N \setminus \{i\}} \Big(v(S \cup \{i\}) - v(S)\Big). \end{aligned}$$

Thus, $b^i_{\max}(N, v)$ (or $b^i_{\min}(N, v)$, respectively) is i's maximum (or minimum, respectively) incremental contribution to a coalition with respect to (N, v).

Definition 2.3.9. *A solution σ on a set Γ of games is*

(1) **reasonable from above** *(*REAB*) if*

$$\Big((N, v) \in \Gamma \text{ and } x \in \sigma(N, v)\Big) \Rightarrow x^i \leq b^i_{\max}(N, v) \text{ for all } i \in N;$$

(2) **reasonable from below** *(*REBE*) if*

$$\Big((N, v) \in \Gamma \text{ and } x \in \sigma(N, v)\Big) \Rightarrow x^i \geq b^i_{\min}(N, v) \text{ for all } i \in N;$$

(3) **reasonable** *from both sides (*RE*) if it satisfies* REAB *and* REBE.

REAB is due to Milnor (1952). Arguments supporting REAB and REBE are very simple: It seems unreasonable to pay any player more than his maximal incremental contribution to any coalition, because that seems to be the strongest threat that he can employ against a particular coalition. Conversely, he may refuse to join any coalition that offers him less than his minimal incremental contribution. Moreover, player i can demand $b^i_{\min}(N, v)$ and nevertheless join any coalition without hurting its members by this demand. Note that IR implies REBE, which is discussed in Sudhölter (1997) (see also Kikuta (1976)).

We prove that the core is reasonable from both sides.

Lemma 2.3.10. *The core satisfies* RE.

Proof: In view of the fact that the core satisfies IR, we only have to show REAB. Let (N, v) be a game, let $x \in \mathcal{C}(N, v)$, and let $i \in N$. Then $x(N) = v(N)$ and $x(N \setminus \{i\}) \geq v(N \setminus \{i\})$. Hence

$$x^i = v(N) - x(N \setminus \{i\}) \leq v(N) - v(N \setminus \{i\}) \leq b^i_{\max}(N, v).$$

q.e.d.

Definition 2.3.11. *Let (N, v) be a game, $S \subseteq N$, $S \neq \emptyset$, and let $x \in X^*(N, v)$. The* **reduced game** *with respect to S and x is the game $(S, v_{S,x})$ defined by*

$$v_{S,x}(T) = \begin{cases} 0 & , \text{ if } T = \emptyset \\ v(N) - x(N \setminus S) & , \text{ if } T = S \\ \max_{Q \subseteq N \setminus S} \Big(v(T \cup Q) - x(Q) \Big) & , \text{ otherwise.} \end{cases}$$

Definition 2.3.11 is due to Davis and Maschler (1965).

Let M be a coalition and let $x \in \mathbb{R}^M$. If $T \subseteq M$, then we denote by x^T the restriction of x to T.

Remark 2.3.12. The reduced game $(S, v_{S,x})$ describes the following situation. Assume that **all** members of N agree that the members of $N \setminus S$ will get $x^{N \setminus S}$. Then, the members of S may get $v(N) - x(N \setminus S)$. Furthermore, suppose that the members of $N \setminus S$ continue to cooperate with the members of S (subject to the foregoing agreement). Then, for every $T \subsetneqq S$ which is nonempty, the amount $v_{S,x}(T)$ is the (maximal) total payoff that the coalition T expects to get. However, we notice that the expectations of different disjoint subcoalitions may not be compatible with each other, because they may require cooperation of the **same** subset of $N \setminus S$ (see Example 2.3.13). Thus, $(S, v_{S,x})$ is not a game in the ordinary sense; it serves only to determine the distribution of $v_{S,x}(S)$ to the members of S.

Example 2.3.13. Let (N, v) be the game associated with the simple majority three-person game represented by $(2; 1, 1, 1)$. Moreover, let $x = (1/2, 1/2, 0)$ and let $S = \{1, 2\}$. In order to obtain $v_{S,x}(\{i\}) = 1$, player i, $i = 1, 2$, needs the cooperation of player 3.

Definition 2.3.14. *A solution σ on a set Γ of games has the* **reduced game property** *(*RGP*) if it satisfies the following condition: If $(N, v) \in \Gamma$, $S \subseteq N$, $S \neq \emptyset$, and $x \in \sigma(N, v)$, then $(S, v_{S,x}) \in \Gamma$ and $x^S \in \sigma(S, v_{S,x})$.*

For one-point solutions Definition 2.3.14 is due to Sobolev (1975). The present definition is a set-valued extension due to Peleg (1986).

Remark 2.3.15. RGP is a condition of **self-consistency**: If (N, v) is a game and $x \in \sigma(N, v)$, that is, x is a solution to (N, v), then for every $S \subseteq N$, $S \neq$

$\emptyset$, the proposal x^S solves $(S, v_{S,x})$ and, therefore, it is consistent with the expectations of the members of S as reflected by the reduced game $(S, v_{S,x})$.

We denote $\Gamma_{\mathcal{U}}^{\mathcal{C}} = \{(N,v) \in \Gamma_{\mathcal{U}} \mid \mathcal{C}(N,v) \neq \emptyset\}$ where $\Gamma_{\mathcal{U}}$ denotes the set of all games.

Lemma 2.3.16. *The core has* RGP *on* $\Gamma_{\mathcal{U}}^{\mathcal{C}}$.

Proof: Let $(N,v) \in \Gamma_{\mathcal{U}}^{\mathcal{C}}$, $x \in \mathcal{C}(N,v)$, and let $\emptyset \neq S \subseteq N$. Further, let $T \subseteq S$ satisfy $T \neq \emptyset$. If $T = S$, then $v_{S,x}(T) - x(T) = x(S) - x(S) = 0$, because $x(N) = v(N)$. If $T \neq S$, then

$$\begin{aligned} v_{S,x}(T) - x(T) &= \max_{Q \subseteq N \setminus S} \Big(v(T \cup Q) - x(Q) \Big) \; - x(T) \\ &= \max_{Q \subseteq N \setminus S} \Big(v(T \cup Q) - x(T \cup Q) \Big) \qquad \leq 0. \end{aligned}$$

Thus, $x^S \in \mathcal{C}\,(S, v_{S,x})$ and the proof is complete. **q.e.d.**

The following weaker version of RGP is very useful.

Definition 2.3.17. *A solution σ on a set Γ of games has the* **weak** *reduced game property (*WRGP*) if it satisfies the following condition: If $(N,v) \in \Gamma$, $S \subseteq N$, $1 \leq |S| \leq 2$, and $x \in \sigma(N,v)$, then $(S, v_{S,x}) \in \Gamma$ and $x^S \in \sigma\,(S, v_{S,x})$.*

Clearly, RGP implies WRGP. The converse is not true in general.

A further kind of "reduced game property" is of interest.

Definition 2.3.18. *A solution σ on a set Γ of games satisfies the* **reconfirmation property** *(*RCP*), if the following condition is satisfied for every $(N,v) \in \Gamma$, every $x \in \sigma(N,v)$ and every $\emptyset \neq S \subseteq N$: If $(S, v_{S,x}) \in \Gamma$ and $y^S \in \sigma\,(S, v_{S,x})$, then $\left(y^S, x^{N \setminus S}\right) \in \sigma(N,v)$.*

RCP occurs in Balinski and Young (1982) as one condition inside one property, Shimomura (1992) uses the term "flexibility" for a similar property, and Hwang and Sudhölter (2000) use the present definition.

RCP is a stability property: Any member of the solution of the reduced game when combined with $x^{N \setminus S}$, the payoff vector of the "passive" players, yields a member of $\sigma(N,v)$, that is, it reconfirms that σ will be used for (N,v). Thus, σ is stable for behavior in reduced games which is specified by σ itself.

In some sense RGP is a "reduced game property from above". Indeed, if a solution satisfies RGP, then the restriction of any member of the solution of a game belongs to the solution of the corresponding reduced game. RCP reflects, in some sense, the opposite direction. Every member of the solution of a reduced game yields an element of the solution of the game, whenever it is combined with the corresponding restriction of the initial element of the solution. More precisely, on $\Gamma_{\mathcal{U}}$ the reduced game properties can be described as follows. A solution σ satisfies RGP or RCP respectively, if for every game $(N, v) \in \Gamma_{\mathcal{U}}$, every $x \in \sigma(N, v)$, and every coalition $S \subseteq N$,

$$\left\{y^S \in \mathbb{R}^S \mid (y^S, x^{N\setminus S}) \in \sigma(N, v)\right\} \subseteq \sigma\left(S, v_{S,x}\right)$$

or

$$\left\{y^S \in \mathbb{R}^S \mid (y^S, x^{N\setminus S}) \in \sigma(N, v)\right\} \supseteq \sigma\left(S, v_{S,x}\right)$$

holds true respectively.

Remark 2.3.19. The properties RGP and RCP are equivalent for one-point solutions on $\Gamma_{\mathcal{U}}$.

Lemma 2.3.20. *The core satisfies* RCP *on every set* Γ *of games.*

Proof: Let $(N, v) \in \Gamma$, $x \in \mathcal{C}(N, v)$, $\emptyset \neq S \subseteq N$, $u = v_{S,x}$, and $y^S \in \mathcal{C}(S, u)$. With $z = \left(y^S, x^{N\setminus S}\right)$ it remains to show that $z \in \mathcal{C}(N, v)$. Let $T \subseteq N$ and distinguish the following cases. If $T \cap S = \emptyset$ or if $T \cap S = S$, then $v(T) - z(T) = v(T) - x(T)$ by Pareto optimality of z. Thus $v(T) - z(T) \leq 0$, because $x \in \mathcal{C}(N, v)$. In the remaining case, that is, if $\emptyset \neq S \cap T \neq S$, we obtain

$$v(T) - z(T) = v(T) - x(T \setminus S) - y(T \cap S) \leq v_{S,x}(T \cap S) - y(T \cap S) \leq 0.$$

q.e.d.

From a practical (or, at least, computational) point of view the following problem may be interesting. Let σ be a solution, let (N, v) be a game, and let $x \in \sigma(N, v)$. Further, let $\mathcal{P}$ be a set of nonempty subsets of N. Then we ask whether or not σ satisfies

$$\left(x^S \in \sigma\left(S, v_{S,x}\right) \text{ for all } S \in \mathcal{P}\right) \Rightarrow x \in \sigma(N, v).$$

The foregoing question motivates the following definition due to Peleg (1986). If N is a coalition, then we denote

$$\mathcal{P} = \mathcal{P}(N) = \{S \subseteq N \mid |S| = 2\}. \tag{2.3.1}$$

Definition 2.3.21. *A solution σ on a set Γ of games has the* **converse reduced game property** *(*CRGP*) if the following condition is satisfied: If $(N, v) \in \Gamma$, $|N| \geq 2$, $x \in X(N, v)$, $(S, v_{S,x}) \in \Gamma$, and $x^S \in \sigma\left(S, v_{S,x}\right)$ for every $S \in \mathcal{P}(N)$, then $x \in \sigma(N, v)$.*

CRGP has the following simple interpretation. Let x be a Pareto optimal payoff vector (that is, $x \in X(N, v)$). Then x is an "equilibrium" payoff distribution if every pair of players is in "equilibrium".

Lemma 2.3.22. *The core satisfies* CRGP *on every set* Γ *of games.*

Proof: Assume that $(N,v) \in \Gamma$, $x \in X(N,v)$, and assume that, for every $S \in \mathcal{P}(N)$, $(S, v_{S,x}) \in \Gamma$ and $x^S \in \mathcal{C}(S, v_{S,x})$. Let $T \subseteq N$ satisfy $\emptyset \neq T \neq N$. Choose $i \in T$ and $j \in N \setminus T$, and let S={i,j}. The fact that $x^S \in \mathcal{C}(S, v_{S,x})$ implies that

$$0 \geq v_{S,x}(\{i\}) - x^i \geq v(T) - x(T).$$

Hence $x \in \mathcal{C}(N,v)$. **q.e.d.**

The monotonicity of solutions will be extensively investigated in Chapter 8. In the present section we shall only note the following modest monotonicity property which is satisfied by the core.

Definition 2.3.23. *A solution σ on a set Γ of games satisfies* **strong aggregate monotonicity** *(*SAM*) if the following condition holds: If $(N,u),(N,v) \in \Gamma$, $u(N) > v(N)$, $u(S) = v(S)$ for all $S \subsetneqq N$, and $x \in \sigma(N,v)$, then there exists $y \in \sigma(N,u)$ such that $y^i > x^i$ for all $i \in N$.*

The foregoing definition has the following intuitive and simple interpretation: If the members of N increase, by unanimous efforts, the size of the cake available for consumption, and if the bargaining power of any proper subcoalition of N remains unchanged, then everybody should be able to benefit from the increase of $v(N)$.

Clearly, the core satisfies SAM.

Definition 2.3.24. *A solution σ on a set Γ of games satisfies* **nonemptiness** *(*NE*) if $\sigma(N,v) \neq \emptyset$ for every $(N,v) \in \Gamma$.*

Exercises

Assume that $|\mathcal{U}| \geq 4$.

Exercise 2.3.1. Let Γ be a set of games. Prove that the core
(1) satisfies WRGP on Γ whenever $\{(N,v) \in \Gamma_{\mathcal{U}}^{\mathcal{C}} \mid |N| \leq 2\} \subseteq \Gamma$; and
(2) does not satisfy RGP on $\{(N,v) \in \Gamma_{\mathcal{U}}^{\mathcal{C}} \mid (N,v) \text{ is superadditive}\}$.

Exercise 2.3.2. Let N be a coalition. A subset $\mathcal{Q} \subseteq \{S \subseteq N \mid |S| = 2\} = \mathcal{P}(N)$ is *sufficient* if the graph $(N, \mathcal{Q})$ (see Example 2.2.5 for this notation) is connected. Let (N,v) be a game and let $\mathcal{Q} \subseteq \mathcal{P}(N)$ be sufficient. Prove the following statement: If $x \in X(N,v)$ and $x^S \in \mathcal{C}(S, v_{S,x})$ for all $S \in \mathcal{Q}$, then $x \in \mathcal{C}(N,v)$ (compare with Definition 2.3.21 and Lemma 2.3.22).

Exercise 2.3.3. Let (N,v) be a game, $\emptyset \neq S \subseteq N$, and let $x \in X^*(N,v)$. The *Moulin reduced game* with respect to S and x is the game $(S, v^M_{S,x})$ defined by

$$v^M_{S,x}(T) = \begin{cases} 0 & \text{, if } T = \emptyset \\ v\Big(T \cup (N \setminus S)\Big) - x(N \setminus S) & \text{, if } \emptyset \neq T \subseteq S. \end{cases}$$

The Moulin RGP and the Moulin RCP, denoted by RGP^M and RCP^M respectively, are obtained by replacing $v_{S,x}$ by $v^M_{S,x}$ in Definition 2.3.14 and Definition 2.3.18 respectively. Prove that the core
(1) satisfies RGP^M on $\Gamma^{\mathcal{C}}_{\mathcal{U}}$; and
(2) does not satisfy RCP^M on $\Gamma^{\mathcal{C}}_{\mathcal{U}}$.

2.4 Notes and Comments

Coalitional games with transferable utility were introduced and extensively studied in von Neumann and Morgenstern (1953). The coalitional function of a game was called the "characteristic function". They tried to clarify the assumption of "transferable utility" (see von Neumann and Morgenstern (1953), p. 8). Indeed, they made a fundamental contribution to utility theory (see von Neumann and Morgenstern (1953), pp. 15 - 29). However, the precise (technical) meaning of "transferable utility" was left undetermined (see Luce and Raiffa (1957), Section 8.1). It was Aumann (1960) who proved that transferability of utility in games with side payments implies linearity of the utility functions (in money).

Most of the concepts and the examples in Sections 2.1 and 2.2 are due to von Neumann and Morgenstern (1953). However, Definition 2.1.7 is due to Shapley (1971). Also, Definition 2.1.6 and Subsection 2.2.2 are of a later vintage. The core, which was investigated in Section 2.3, was defined in Gillies (1959). Finally, additivity of solutions of coalitional games was first considered in Shapley (1953).

3

The Core

This chapter contains some basic results on the core of coalitional TU games. First the Bondareva-Shapley theorem which gives necessary and sufficient conditions for the nonemptiness of the core is proved. As an application of the foregoing theorem, we show that the core of a market game is nonempty. If the core of a game is nonempty, then the game is called balanced. A game is totally balanced if all of its subgames are balanced. The player set and the coalition function of a subgame are a subcoalition and the corresponding restriction of the coalition function of the game. In Section 3.3 we show that a coalitional game is a market game if and only if it is totally balanced. We prove in Section 3.4 that minimum cost spanning tree games and permutation games are totally balanced.

The "largeness" of the core of convex games is briefly discussed in Section 3.5. The next section contains an axiomatic characterization of the core on the set of balanced games. The following four independent axioms characterize the core: nonemptiness, individual rationality, the weak reduced game property, and superadditivity. In Section 3.7 it is shown that the core on the class of totally balanced games is axiomatized by the converse reduced game property and the foregoing axioms. The last section is devoted to an investigation of the core of games with coalition structures. Throughout this chapter we assume that $\mathcal{U}$ is a nonempty set of players.

3.1 The Bondareva-Shapley Theorem

Let N be a coalition and $\mathcal{V}(N) = \mathcal{V}$ be the set of all coalition functions on $2^N = \{S \mid S \subseteq N\}$. (Throughout this section we frequently identify a game (N, v) with its coalition function v.) We denote

$$\mathcal{V}_\mathcal{C} = \{v \in \mathcal{V} \mid \mathcal{C}(N, v) \neq \emptyset\}.$$

In this section we shall present a minimal finite set of linear inequalities that determine $\mathcal{V}_C$. Furthermore, we shall provide a combinatorial interpretation of the foregoing inequalities.

Let $v \in \mathcal{V}$. We consider the following linear programming problem:

$$\begin{cases} \min x(N) \\ \text{subject to } x(S) \geq v(S) \text{ for all } S \subseteq N,\ S \neq \emptyset. \end{cases} \tag{3.1.1}$$

Clearly, $v \in \mathcal{V}_C$ if and only if the value of (3.1.1) is $v(N)$. The following notation helps to formulate the dual of (3.1.1).

Notation 3.1.1. Let $S \subseteq N$. The *characteristic vector* χ_S of S is the member of $\mathbb{R}^N$ which is given by

$$\chi_S^i = \begin{cases} 1, \text{ if } i \in S \\ 0, \text{ if } i \in N \setminus S. \end{cases}$$

Using the foregoing notion the dual program is:

$$\begin{cases} \max \sum_{S \subseteq N} \delta_S v(S) \\ \text{subject to} \begin{cases} \sum_{S \subseteq N} \delta_S \chi_S = \chi_N \text{ and} \\ \delta_S \geq 0 \text{ for all } S \subseteq N,\ S \neq \emptyset. \end{cases} \end{cases} \tag{3.1.2}$$

As both programs (3.1.1) and (3.1.2) are feasible, it follows from the duality theorem (see, e.g., Franklin (1980), p. 62) that (3.1.1) and (3.1.2) have the same value. Hence, (N, v) has a nonempty core if and only if

$$v(N) \geq \sum_{S \subseteq N} \delta_S v(S) \text{ for all feasible vectors } (\delta_S)_{S \subseteq N} \text{ of (3.1.2).} \tag{3.1.3}$$

Assertion (3.1.3) leads to the weak form of the Bondareva-Shapley theorem 3.1.4 as proved independently in Bondareva (1963) and Shapley (1967).

Definition 3.1.2. *A collection $\mathcal{B} \subseteq 2^N$, $\emptyset \notin \mathcal{B}$, is called* **balanced** *(over N) if positive numbers δ_S, $S \in \mathcal{B}$, exist such that*

$$\sum_{S \in \mathcal{B}} \delta_S \chi_S = \chi_N.$$

The collection $(\delta_S)_{S \in \mathcal{B}}$ is called a system of **balancing weights**.

Remark 3.1.3. Every partition of N is a balanced collection. Hence, balanced collections may be considered as generalized partitions. Indeed, if $\mathcal{B}$ is a balanced collection with balancing weights δ_S, $S \in \mathcal{B}$, then it may be interpreted as a "fractional partition" in the following way. Each $i \in N$ devotes the fraction δ_S of his time to each coalition $S \in \mathcal{B}$ that contains him. As $\mathcal{B}$ is balanced

$$\sum_{S \in \mathcal{B}: i \in S} \delta_S = 1 \text{ for every } i \in N.$$

Thus the balanced collection $\mathcal{B}$ is a generalized partition in the foregoing sense.

Theorem 3.1.4 (The Bondareva-Shapley Theorem, weak form). *A necessary and sufficient condition that the core of a game (N, v) is not empty is that for each balanced collection $\mathcal{B}$ and each system $(\delta_S)_{S \in \mathcal{B}}$ of balancing weights*

$$v(N) \geq \sum_{S \in \mathcal{B}} \delta_S v(S). \tag{3.1.4}$$

Proof: Definition 3.1.2 and (3.1.3) prove the theorem. **q.e.d.**

Motivated by Theorem 3.1.4 a game with a nonempty core is called *balanced.*

Remark 3.1.5. We say that $v \in \mathcal{V}$ is *superadditive at* N if, for every partition $\mathcal{P}$ of N, $v(N) \geq \sum_{S \in \mathcal{P}} v(S)$. (If v is superadditive, then v is superadditive at N.) Thus, in view of Remark 3.1.3, inequality (3.1.4) may be interpreted as *strong* superadditivity of v at N.

Now, let F be the set of feasible vectors of (3.1.2), that is,

$$F = \left\{ (\delta_S)_{S \in 2^N \setminus \{\emptyset\}} \,\middle|\, \sum_{S \in 2^N \setminus \{\emptyset\}} \delta_S \chi_S = \chi_N \text{ and } \delta_S \geq 0 \text{ for } S \in 2^N \setminus \{\emptyset\} \right\}.$$

Clearly F is a (compact nonempty) convex polytope and, thus, it is the convex hull of its extreme points. Let $\mathrm{EXT}(F)$ denote the set of extreme points of F. Hence, for every $v \in \mathcal{V}$, $v \in \mathcal{V}_{\mathcal{C}}$ if and only if

$$v(N) \geq \sum_{S \in 2^N \setminus \{\emptyset\}} \delta_S v(S) \text{ for every } (\delta_S)_{S \in 2^N \setminus \{\emptyset\}} \in \mathrm{EXT}(F). \tag{3.1.5}$$

We shall now provide a combinatorial characterization of the extreme points of F.

Definition 3.1.6. *A balanced collection is called* **minimal** *balanced if it does not contain a proper balanced subcollection.*

Lemma 3.1.7. *A balanced collection is minimal balanced if and only if it has a unique system of balancing weights.*

Proof: Let $\mathcal{B}$ be a balanced collection and $(\delta_S)_{S \in \mathcal{B}}$ be a system of balancing weights.

(1) Sufficiency: If $\mathcal{B}^* \subsetneqq \mathcal{B}$ is a balanced collection with a system $(\delta^*_S)_{S \in \mathcal{B}^*}$ of balancing weights, then, as the reader can easily verify, $\mathcal{B}$ has infinitely many systems of balancing weights $(\delta^\gamma_S)_{S \in \mathcal{B}}$ defined by

$$\delta_S^\gamma = \begin{cases} \gamma\delta_S + (1-\gamma)\delta_S^* & \text{, if } S \in \mathcal{B}^* \\ \gamma\delta_S & \text{, if } S \in \mathcal{B} \setminus \mathcal{B}^* \end{cases}$$

where $0 < \gamma \leq 1$.

(2) Necessity: Assume that $(\delta'_S)_{S\in\mathcal{B}}$ is a system of balancing weights distinct from $(\delta_S)_{S\in\mathcal{B}}$. Then there exists $S \in \mathcal{B}$ such that $\delta'_S > \delta_S$, so $\tau = \min\{\frac{\delta_S}{\delta'_S - \delta_S} \mid \delta'_S > \delta_S\}$ is well defined. Let $(\widetilde{\delta}_S)_{S\in\mathcal{B}}$ be defined by $\widetilde{\delta}_S = (1+\tau)\delta_S - \tau\delta'_S$ for all $S \in \mathcal{B}$. Then $\mathcal{B}^* = \{S \in \mathcal{B} \mid \widetilde{\delta}_S > 0\}$ is a proper balanced subcollection of $\mathcal{B}$. **q.e.d.**

We shall need the following characterization of extreme points of convex polyhedral sets.

Lemma 3.1.8. *Let P be the convex polyhedral set in $\mathbb{R}^k$ given by*

$$P = \left\{ x \in \mathbb{R}^k \,\middle|\, \sum_{j=1}^{k} x^j a_{tj} \geq b_t,\ t = 1, \ldots, m \right\}.$$

For $x \in P$ let $S(x) = \left\{ t \in \{1, \ldots, m\} \,\middle|\, \sum_{j=1}^{k} x^j a_{tj} = b_t \right\}$. The point $x \in P$ is an extreme point of P if and only if the system of linear equations

$$\sum_{j=1}^{k} y^j a_{tj} = b_t \text{ for all } t \in S(x) \tag{3.1.6}$$

has x as its unique solution.

Proof: If (3.1.6) has more than one solution, then the solution set contains a straight line through x. Therefore, x is a midpoint of a straight line segment contained in P.

Conversely, if $x + z$ and $x - z$ are in P for some $z \neq 0$, then $S(x) = S(x+z) \cap S(x-z)$ and (3.1.6) has more than one solution ($x + z$ and $x - z$ solve (3.1.6)). **q.e.d.**

Corollary 3.1.9. *Let $(\delta_S)_{S\in 2^N\setminus\{\emptyset\}} \in F$ and $\mathcal{B} = \{S \subseteq N \mid \delta_S > 0\}$. Then $(\delta_S)_{S\in 2^N\setminus\{\emptyset\}}$ is an extreme point of F if and only if $\mathcal{B}$ is minimal balanced.*

Proof: Lemmata 3.1.7 and 3.1.8. **q.e.d.**

In view of Corollary 3.1.9 extreme points of F and minimal balanced collections can be identified. Bondareva (1963) and Shapley (1967) prove the following theorem.

Theorem 3.1.10 (The Bondareva-Shapley Theorem, sharp form). *A necessary and sufficient condition that the core of a game (N, v) is not empty is that for each minimal balanced collection $\mathcal{B}$*

$$v(N) \geq \sum_{S \in \mathcal{B}} \delta_S v(S), \tag{3.1.7}$$

where $(\delta_S)_{S \in \mathcal{B}}$ is the system of balancing weights for $\mathcal{B}$. None of the conditions stated in (3.1.7) is redundant, except for the collection $\{N\}$.

Proof: The first part of the theorem follows from (3.1.5) and Corollary 3.1.9. Thus, it remains to show that if $\mathcal{B}^* \neq \{N\}$ is a minimal balanced collection with balancing weights $(\delta^*_S)_{S \in \mathcal{B}^*}$, then there exists $v \in \mathcal{V}$ such that $v(N) < \sum_{S \in \mathcal{B}^*} \delta^*_S v(S)$ and $v(N) \geq \sum_{S \in \mathcal{B}} \delta_S v(S)$ for every minimal balanced collection $\mathcal{B} \neq \mathcal{B}^*$ with balancing weights $(\delta_S)_{S \in \mathcal{B}}$. Let Q be the convex hull of the extreme points of F that correspond to the minimal balanced collections $\mathcal{B} \neq \mathcal{B}^*$ and let $(\delta^*_S)_{S \in 2^N \setminus \{\emptyset\}}$ correspond to $\mathcal{B}^*$. Then $(\delta^*_S)_{S \in 2^N \setminus \{\emptyset\}} \notin Q$. Hence there exists a hyperplane that separates Q from $(\delta^*_S)_{S \in 2^N \setminus \{\emptyset\}}$. That is, there exist $(\gamma(S))_{S \in 2^N \setminus \{\emptyset\}}$ and $r \in \mathbb{R}$ such that

$$\textstyle\sum_{S \in 2^N \setminus \{\emptyset\}} \gamma(S) \delta^*_S > r; \text{ and}$$
$$\textstyle\sum_{S \in 2^N \setminus \{\emptyset\}} \gamma(S) \delta_S \leq r \text{ for all } (\delta_S)_{S \in 2^N \setminus \{\emptyset\}} \in Q.$$

Define now $v \in \mathcal{V}$ by $v(S) = \gamma(S)$ for all $\emptyset \neq S \subsetneqq N$, $v(N) = r$, and $v(\emptyset) = 0$. Then for each minimal balanced collection $\mathcal{B} \neq \mathcal{B}^*$ with balancing weights $(\delta_S)_{S \in \mathcal{B}}$

$$\sum_{S \in \mathcal{B}} \delta_S v(S) \leq r = v(N)$$

and

$$\sum_{S \in \mathcal{B}^*} \delta^*_S v(S) > r = v(N).$$

q.e.d.

Remark 3.1.11. Peleg (1965) contains an inductive method for finding all the minimal balanced collections over N.

Exercises

Exercise 3.1.1. A minimal balanced collection is called *proper* if no two of its elements are disjoint. Prove the following theorem: *A superadditive game (N, v) has a nonempty core if and only if for every proper minimal balanced collection $\mathcal{B}$ over N with balancing weights $(\delta_S)_{S \in \mathcal{B}}$ condition (3.1.7) holds* (see Shapley (1967)).

Exercise 3.1.2. Let $\mathcal{B}^* \neq \{N\}$ be a proper minimal balanced collection over N. Prove that there exists a superadditive game (N, v) with an **empty** core such that every proper minimal balanced collection $\mathcal{B} \neq \mathcal{B}^*$ of N satisfies (3.1.7) (see Charnes and Kortanek (1967)).

Exercise 3.1.3. Prove Theorem 3.1.4 by means of separating between two disjoint convex sets. (Hint: Let $v \in \mathcal{V}$ and define

$$B_1 = \left\{ \left(\sum_{S \in 2^N \setminus \{\emptyset\}} \lambda_S \chi_S, \sum_{S \in 2^N \setminus \{\emptyset\}} \lambda_S v(S) \right) \middle| \lambda_S \geq 0 \text{ for all } S \in 2^N \setminus \{\emptyset\} \right\}$$

and

$$B_2 = \{(\chi_N, v(N) + \varepsilon) \mid \varepsilon > 0\}.$$

If (3.1.4) is satisfied, then $B_1 \cap B_2 = \emptyset$. Manipulate the normal of a separating hyperplane between B_1 and B_2 to obtain a member of $\mathcal{C}(N, v)$. The necessity part of Theorem 3.1.4 is straightforward.)

3.2 An Application to Market Games

In this section we prove that every market game is balanced. The proof is an application of the Bondareva-Shapley theorem (Theorem 3.1.4).

Theorem 3.2.1 (Shapley and Shubik (1969a)). *Every market game is balanced.*

Proof: Let $(N, \mathbb{R}^m_+, A, W)$ be a market (see Subsection 2.2.1) and let (N, v) be the corresponding market game (see Definition 2.2.1). We shall prove that v satisfies (3.1.4). Thus, let $\mathcal{B}$ be a balanced collection (over N) and let $(\delta_S)_{S \in \mathcal{B}}$ be a system of balancing weights for $\mathcal{B}$. For each $S \in \mathcal{B}$, there exists a feasible S-allocation $x_S \in X^S$ such that

$$v(S) = \sum_{i \in S} w^i(x^i_S). \tag{3.2.1}$$

For $i \in N$ let $\mathcal{B}_i = \{S \in \mathcal{B} \mid i \in S\}$. Define

$$y^i = \sum_{S \in \mathcal{B}_i} \delta_S x^i_S. \tag{3.2.2}$$

The allocation $y_N = (y^i)_{i \in N}$ is a feasible N-allocation. Indeed,

$$\begin{aligned} \textstyle\sum_{i \in N} y^i &= \textstyle\sum_{i \in N} \sum_{S \in \mathcal{B}_i} \delta_S x^i_S = \sum_{S \in \mathcal{B}} \delta_S \sum_{i \in S} x^i_S \\ &= \textstyle\sum_{S \in \mathcal{B}} \delta_S \sum_{i \in S} a^i \ = \sum_{i \in N} a^i \sum_{S \in \mathcal{B}_i} \delta_S = \sum_{i \in N} a^i. \end{aligned}$$

(Notice that $\sum_{S \in \mathcal{B}_i} \delta_S = 1$ for every $i \in N$, because $\mathcal{B}$ is balanced.) By the definition of v

$$v(N) \geq \sum_{i \in N} w^i(y^i). \tag{3.2.3}$$

By the concavity of the functions w^i, it follows from (3.2.2) that

$$w^i(y^i) \geq \sum_{S \in \mathcal{B}^i} \delta_S w^i(x^i_S). \tag{3.2.4}$$

By (3.2.1), (3.2.3), and (3.2.4)

$$v(N) \geq \sum_{i \in N} w^i(y^i) \geq \sum_{i \in N} \sum_{S \in \mathcal{B}_i} \delta_S w^i(x^i_S) = \sum_{S \in \mathcal{B}} \delta_S \sum_{i \in S} w^i(x^i_S) = \sum_{S \in \mathcal{B}} \delta_S v(S).$$

Thus by the Bondareva-Shapley theorem $\mathcal{C}(N,v) \neq \emptyset$. **q.e.d.**

Definition 3.2.2. *Let (N,v) be a game. A* **subgame** *of (N,v) is a game (T,v^T) where $\emptyset \neq T \subseteq N$ and $v^T(S) = v(S)$ for all $S \subseteq T$. The subgame (T,v^T) will also be denoted by (T,v).*

Definition 3.2.3. *A game (N,v) is* **totally balanced** *if every subgame of (N,v) is balanced.*

Corollary 3.2.4. *Every market game is totally balanced.*

Proof: Definition 2.2.1 and Theorem 3.2.1. **q.e.d.**

Exercises

Exercise 3.2.1. Prove that every superadditive two-person game is totally balanced. Find a balanced game which is not a market game.

Exercise 3.2.2. Let $(N, \mathbb{R}^m_+, A, W)$ be a market. For $p, q \in \mathbb{R}^m$ let $p \cdot q$ denote the *scalar product* of p and q. An $n+1$-tuple $\left((x^i)_{i \in N}, p\right)$ is a *competitive equilibrium* if $(x^i)_{i \in N}$ is an N-allocation, $p \in \mathbb{R}^m$, and for each $i \in N$

$$w^i(x^i) - p \cdot (x^i - a^i) \geq w^i(x) - p \cdot (x - a^i) \text{ for all } x \in \mathbb{R}^m_+.$$

Let $\left((x^i)_{i \in N}, p\right)$ be a competitive equilibrium, let (N,v) be the market game of $(N, \mathbb{R}^m_+, A, W)$, and define $t^i = w^i(x^i) - p \cdot (x^i - a^i)$ for every $i \in N$. Prove that $t \in \mathcal{C}(N,v)$ (see Shapley and Shubik (1969a)).

Exercise 3.2.3. Let $(N, \mathbb{R}^m_+, A, W)$ be a market. Prove that there exists a competitive equilibrium, if $\sum_{i \in N} a^i_j > 0$ for all $j = 1, \ldots, m$. (Hint: Let

$$Y_1 = \left\{ \left(\sum_{i \in N} (a^i - x^i), t \right) \middle| \; t \leq \sum_{i \in N} w^i(x^i) - v(N), \; x^i \in \mathbb{R}^m_+ \text{ for all } i \in N \right\}$$

and $Y_2 = \{y \in \mathbb{R}^{m+1} \mid y = (0, \varepsilon), \; \varepsilon > 0\}$. Now the convex sets Y_1 and Y_2 can be separated by a hyperplane, because $Y_1 \cap Y_2 = \emptyset$.)

3.3 Totally Balanced Games

In this section we shall prove the converse of Corollary 3.2.4. For the proof the following class of markets is useful.

Definition 3.3.1. *A* **direct** *market is a market* $(N, \mathbb{R}^N_+, I_N, (w)_{i\in N})$ *where* $I_N = (\chi_{\{i\}})_{i\in N}$ *(that is,* I_N *is the indexed collection of unit vectors of* $\mathbb{R}^N$*) and* w *is the common utility function of the traders which is homogeneous of degree* 1*, concave, and continuous.*

Thus, in a direct market, each trader starts with one unit of a personal commodity (e.g., his labor or time). Feasible allocations of these commodities are valued by the common utility function w. As w is concave and homogeneous, it is also *superadditive*, that is,

$$w(x+y) \geq w(x) + w(y) \text{ for all } x, y \in \mathbb{R}^N_+.$$

Hence, if (N, v) is the game corresponding to the market

$$(N, \mathbb{R}^N_+, I_N, (w)_{i\in N}),$$

then

$$v(S) = w(\chi_S) \text{ for all } S \subseteq N. \tag{3.3.1}$$

Now we associate with every game a direct market in the following way. Let (N, v) be a game. The corresponding market is $(N, \mathbb{R}^N_+, I_N, (w)_{i\in N})$ where $w(x)$, $x \in \mathbb{R}^N_+$, is given by

$$\begin{cases} w(x) = \max \sum_{S\subseteq N} \delta_S v(S) \\ \text{subject to } \sum_{S\subseteq N} \delta_S \chi_S = x \text{ and } \delta_S \geq 0 \text{ for all } S \subseteq N. \end{cases} \tag{3.3.2}$$

Indeed, w is homogeneous of degree 1, concave, and continuous (Exercise 3.3.1).

Remark 3.3.2. The foregoing market associated with the game (N, v) has the following interpretation. Each coalition $S \subseteq N$ has the activity χ_S which yields $v(S)$ dollars (that is, if S is formed, then it gets the amount $v(S)$). More generally, S earns $\delta_S v(S)$ dollars if each member of S devotes the fraction δ_S of his time to S. Thus, (3.3.2) is simply a linear program which yields an optimal assignment of activity levels to the various χ_S's, subject to the constraint that each player i distributes exactly the amount x^i of his time among his activities.

Definition 3.3.3. *Let* (N, v) *be a game and let* $(N, \mathbb{R}^N_+, I_N, (w)_{i\in N})$ *be the associated direct market. The game* $(N, \bar{v})$ *which is defined by the market* $(N, \mathbb{R}^N_+, I_N, (w)_{i\in N})$ *is called the* **totally balanced cover** *of* (N, v).

By (3.3.1) and (3.3.2), $\bar{v}(R)$, $R \subseteq N$, is given by

$$\begin{cases} \bar{v}(R) = \max \sum_{S\subseteq R} \delta_S v(S) \\ \text{subject to } \sum_{S\subseteq N} \delta_S \chi_S = \chi_R \text{ and } \delta_S \geq 0 \text{ for all } S \subseteq R. \end{cases} \tag{3.3.3}$$

Remark 3.3.4. The totally balanced cover of a game is a useful mathematical concept. The program (3.3.3) may serve as a direct definition of it.

It follows from (3.3.3) that

$$\bar{v}(R) \geq v(R) \text{ for all } R \subseteq N. \tag{3.3.4}$$

Lemma 3.3.5. *A game (N, v) is balanced if and only if $v(N) = \bar{v}(N)$.*

Proof: By Theorem 3.1.4 and (3.3.3), $v(N) \geq \bar{v}(N)$ if and only if (N, v) is balanced. The proof is completed by (3.3.4). **q.e.d.**

Corollary 3.3.6. *A game (N, v) is totally balanced if and only if $v = \bar{v}$.*

Proof: Let $\emptyset \neq R \subseteq N$. The totally balanced cover of the subgame (R, v) is $(R, \bar{v})$. Hence Lemma 3.3.5 completes the proof. **q.e.d.**

Theorem 3.3.7 (Shapley and Shubik (1969a)). *A game is a market game if and only if it is totally balanced.*

Proof: By Corollary 3.2.4 every market game is totally balanced. Also, by Corollary 3.3.6, a totally balanced game is equal to its totally balanced cover, which is a market game. **q.e.d.**

Exercises

Exercise 3.3.1. Prove that w, which is defined by (3.3.2), is homogeneous of degree 1, concave, and continuous.

Exercise 3.3.2. Characterize all totally balanced symmetric games with player set N.

Exercise 3.3.3 (Minimum Games). Let (N, v) be a game. Show that (N, v) is totally balanced, if and only if it is a minimum of finitely many inessential games, that is, there is a finite nonempty set $\mathcal{X} \subseteq \mathbb{R}^N$ such that $v(S) = \min_{x \in \mathcal{X}} x(S)$ for all $S \subseteq N$ (see Kalai and Zemel (1982)).

3.4 Some Families of Totally Balanced Games

3.4.1 Minimum Cost Spanning Tree Games

Let N be a nonempty finite set of customers, let $0 \notin N$ be a supplier, let $N_* = N \cup \{0\}$, and let $c_{ij} \in \mathbb{R}$ be the cost of connecting $i, j \in N_*$, $i \neq j$, by the edge e_{ij} (see Example 2.2.5). The tuple $\left(N_*, (c_{ij})_{\{i,j\}\subseteq N_*, i\neq j}\right)$ is called a *minimum cost spanning tree* (m.c.s.t.) *problem*.

Let $\left(N_*, (c_{ij})_{\{i,j\}\subseteq N_*, i\neq j}\right)$ be a m.c.s.t. problem and let (N, c) be the corresponding cost game. We claim that the core of the *cost game* (N, c) (that is, $-\mathcal{C}(N, v) + (c(\{i\}))_{i\in N}$ where (N, v) is the associated savings game) is nonempty. Indeed, let $\Gamma_N = (N_*, E_N)$ be a m.c.s.t. for $\left(N_*, (c_{ij})_{\{i,j\}\subseteq N_*, i\neq j}\right)$ and let $i \in N$. Then there exists a unique path $(0, i_1, \ldots, i_r)$ in Γ_N such that $i_r = i$. Define $x \in \mathbb{R}^N$ by $x^i = c_{i_{r-1}, i_r}$ for all $i \in N$ (with the convention $i_0 = 0$). Then $x(N) = c(N)$ by construction of x. In order to show that x belongs to the core of (N, c) it remains to verify that $x(S) \leq c(S)$ for all $S \subseteq N$. Let $\emptyset \neq S \subseteq N$ and let $\Gamma_S = (S \cup \{0\}, E_S)$ be a m.c.s.t. for $\left(S \cup \{0\}, (c_{ij})_{\{i,j\}\subseteq S\cup\{0\}, i\neq j}\right)$. Expand Γ_S to a graph $\widehat{\Gamma}_N = (N_*, \widehat{E}_N)$ by adding, for each $i \in N \setminus S$, the edge $(j(i), i) \in E_N$ (on the path in E_N from 0 to i). The graph $\widehat{\Gamma}$ has $|S| + |N \setminus S|$ edges and it is connected. Hence it is a tree. The observation that

$$c(S) + x(N \setminus S) = \sum_{e_{ij}\in \widehat{E}_N} c_{ij} \geq \sum_{e_{ij}\in E_N} c_{ij} = c(N) = x(N)$$

implies that $c(S) \geq x(S)$.

Moreover, we observe that a subgame of a m.c.s.t. game is a m.c.s.t. game. Hence, we have proved the following result.

Theorem 3.4.1. *Every* m.c.s.t. *game is totally balanced.*

Remark 3.4.2. (1) The foregoing proof of Theorem 3.4.1 is due to Bird (1976).

(2) The *monotonic* m.c.s.t. game $(N, \widetilde{c})$ of an m.c.s.t. problem arises from its m.c.s.t. game (N, c) by defining $\widetilde{c}(S) = \min_{N\supseteq T\supseteq S} c(T)$ for all $S \subseteq N$. Granot and Huberman (1981) show that the core of a monotonic m.c.s.t. game is nonempty.

3.4.2 Permutation Games

Let N be a coalition and let $p : N \times N \to \mathbb{R}$ be a "profit" function. Further, for $S \subseteq N$ let

$$\Pi_S = \{\pi \in \mathcal{SYM}_N \mid \pi(i) = i \text{ for all } i \in N \setminus S\}.$$

A game (N, v) is the *permutation game* with respect to (N, p), if $v(S)$, for every $\emptyset \neq S \subseteq N$, is given by

$$v(S) = \max_{\pi\in\Pi_S} \sum_{i\in S} p(i, \pi(i)). \tag{3.4.1}$$

Clearly, (3.4.1) is given by the following integer programming problem:

$$\begin{cases} \max \sum_{i \in N} \sum_{j \in N} p(i,j) x_{ij} \\ \text{subject to} \begin{cases} \sum_{j \in N} x_{ij} = \chi_S^i & , i \in N \\ \sum_{i \in N} x_{ij} = \chi_S^j & , j \in N \\ x_{ij} \in \{0,1\} & , i,j \in N. \end{cases} \end{cases} \tag{3.4.2}$$

By the Birkhoff-von Neumann theorem the permutation matrices (of S), which are feasible for (3.4.2), are the extreme points of the set of doubly stochastic matrices (on S) which are feasible solutions of the following programming problem:

$$\begin{cases} \max \sum_{i \in N} \sum_{j \in N} p(i,j) x_{ij} \\ \text{subject to} \begin{cases} \sum_{j \in N} x_{ij} = \chi_S^i , i \in N \\ \sum_{i \in N} x_{ij} = \chi_S^j , j \in N \\ x_{ij} \geq 0 \quad , i,j \in N. \end{cases} \end{cases} \tag{3.4.3}$$

The dual program of (3.4.3) is

$$\begin{cases} \min \sum_{i \in S} (y^i + z^i) \\ \text{subject to } y^i + z^j \geq p(i,j), i,j \in N. \end{cases} \tag{3.4.4}$$

Let $(\hat{y}^N, \hat{z}^N)$ be an optimal solution to (3.4.4) with $S = N$. Let $x = \hat{y}^N + \hat{z}^N \in \mathbb{R}^N$. Then $x(N) = v(N)$. As $(\hat{y}^S, \hat{z}^S)$ is a feasible solution to (3.4.4) for every $\emptyset \neq S \subseteq N$, we obtain

$$x(S) = \sum_{i \in S} (\hat{y}^i + \hat{z}^i) \geq v(S).$$

Thus, $x \in \mathcal{C}(N,v)$. Clearly, a subgame of a permutation game is a permutation game. Hence, we have proved the following result.

Theorem 3.4.3. *Every permutation game is totally balanced.*

Remark 3.4.4. Let (N,v) be a permutation game. In order to find a member x of $\mathcal{C}(N,v)$ we only have to solve the optimization problem (see (3.4.4)) for $S = N$ defining $v(N)$. Analogously, in order to find a member of the core of an m.c.s.t. game (N,c) we only have to construct an m.c.s.t. (see Exercise 2.2.2) to determine $c(N)$. (See also Exercise 3.2.3 which yields a similar result for market games.)

Exercises

Exercise 3.4.1 (Assignment Game). Shapley and Shubik (1972) discuss the following model. Let $N = S \cup B, S, B \neq \emptyset$, and $S \cap B = \emptyset$. Each $i \in S$

is a seller who has a house which for him is worth a_i (units of money). Each $j \in B$ is a potential buyer whose reservation price for i's house, $i \in S$, is b_{ij}. For all $i \in S$ and $j \in B$ define $c(\{i,j\}) = (b_{ij} - a_i)_+$, where $a_+ = \max\{a, 0\}$ for every $a \in \mathbb{R}$. Hence $c(\{i,j\})$ is the joint "net profit" of $\{i,j\}$. Let $T \subseteq N$. An *assignment* (*matching*) for T is a set $\mathcal{T} \subseteq 2^T$ satisfying

$$P \cap Q = \emptyset \text{ and } |P \cap S| = |P \cap B| = 1 \text{ for all } P, Q \in \mathcal{T} \text{ with } P \neq Q.$$

(Hence in an assignment $\mathcal{T}$ for T, every seller in $S \cap T \cap \bigcup_{P \in \mathcal{T}} P$ sells his house (is matched to) some buyer in $B \cap T \cap \bigcup_{P \in \mathcal{T}} P$.) The *assignment game* (N, v) (with respect to $(a_i)_{i \in S}$ and $(b_{ij})_{i \in S, j \in B}$) is defined by

$$v(T) = \max\left\{\sum_{P \in \mathcal{T}} c(P) \,\middle|\, \mathcal{T} \text{ is an assignment for } T\right\} \text{ for all } T \subseteq N.$$

(By convention the "empty sum" is 0.) Prove that the assignment game (N, v) is a permutation game.

Exercise 3.4.2 (Glove Game). Let $L, R \neq \emptyset, L \cap R = \emptyset$ and $N = L \cup R$ be a coalition. Each member of L is assumed to own one left-hand glove and each member of R is a right-hand glove owner. The market price of a pair of gloves is 1. The game (N, v) defined by

$$v(S) = \min\{|S \cap L|, |S \cap R|\} \text{ for all } S \subseteq N$$

is the *glove game* (with respect to L and R). Show that a glove game is an assignment game.

Exercise 3.4.3. Show that a game is a permutation game, if it is strategically equivalent to a permutation game.

Exercise 3.4.4. Find a permutation game which is not strategically equivalent to an assignment game.

Exercise 3.4.5. Find a permutation game which is not strategically equivalent to the savings game of an m.c.s.t. game. (Hint: Example 2.2.5 can be used to show that if (N, c) is an m.c.s.t. game, then there exists $i \in N$ such that

$$c(N) - c(N \setminus \{i\}) \geq \min\Big\{c(\{i\}), \min\Big\{c(\{i,j\}) - c(\{j\}) \mid j \in N \setminus \{i\}\Big\}\Big\}.$$

Deduce that the corresponding savings game (N, v) satisfies

$$v(N) - v(N \setminus \{i\}) \leq \max\Big\{v(\{i\}), \max\Big\{v(\{i,j\}) - v(\{j\}) \mid j \in N \setminus \{i\}\Big\}\Big\}$$

and construct a (symmetric, convex, three-person) permutation game which does not satisfy this property.)

3.5 A Characterization of Convex Games

Let N be a coalition and $\mathcal{V}$ be the set of all coalition functions on 2^N. Throughout this section we shall assume that $N = \{1, \ldots, n\}$. If $\pi \in \mathcal{SYM}_N$ (see Definition 2.1.15) and $i \in N$, then we denote

$$P_\pi^i = \{j \in N \mid \pi(j) < \pi(i)\}.$$

P_π^i is the set of members of N which precede i with respect to the order π. Let $v \in \mathcal{V}$ and $\pi \in \mathcal{SYM}_N$. We define $a_\pi(v) \in \mathbb{R}^N$ by

$$a_\pi^i(v) = v(P_\pi^i \cup \{i\}) - v(P_\pi^i) \text{ for every } i \in N.$$

The following theorem is due to Shapley (1971) and Ichiishi (1981).

Theorem 3.5.1. *A game $v \in \mathcal{V}$ is convex if and only if $a_\pi(v) \in \mathcal{C}(N, v)$ for every $\pi \in \mathcal{SYM}_N$.*

Proof: Necessity. Assume that $v \in \mathcal{V}$ is convex. Let $\emptyset \neq S \subseteq N$ and $\pi \in \mathcal{SYM}_N$. We have to show that

$$\sum_{i \in S} a_\pi^i(v) \geq v(S). \tag{3.5.1}$$

Let $i_1, \ldots, i_s \in S$, where $s = |S|$, be chosen such that $S = \{i_1, \ldots, i_s\}$ and $\pi(i_1) < \pi(i_2) < \cdots < \pi(i_s)$. Hence $\{i_1, \ldots, i_{j-1}\} \subseteq P_\pi^{i_j}$ for every $j = 1, \ldots, s$. Thus, by Exercise 2.1.1,

$$v(P_\pi^{i_j} \cup \{i_j\}) - v(P_\pi^{i_j}) \geq v(\{i_1, \ldots, i_j\}) - v(\{i_1, \ldots, i_{j-1}\}) \tag{3.5.2}$$

for $j = 1, \ldots, s$. Summing up the inequalities (3.5.2), we obtain (3.5.1).

Sufficiency. Let $v \in \mathcal{V}$ and assume that $a_\pi \in \mathcal{C}(N, v)$ for every $\pi \in \mathcal{SYM}_N$. Let $\emptyset \neq S, T \subseteq N$. We denote $S \cap T = \{i_1, \ldots, i_r\}$, $T \setminus S = \{i_{r+1}, \ldots, i_t\}$, $S \setminus T = \{i_{t+1}, \ldots, i_q\}$, and $N \setminus (S \cup T) = \{i_{q+1}, \ldots, i_n\}$, where $r = |S \cap T|$ and $q = |S \cup T|$. Define $\pi \in \mathcal{SYM}_N$ by $\pi(i_j) = j$, $j = 1, \ldots, n$. Hence

$$\begin{aligned} v(S) &\leq \textstyle\sum_{i \in S} a_\pi^i(v) = \sum_{i \in S} \left(v\left(P_\pi^i \cup \{i\}\right) - v\left(P_\pi^i\right)\right) \\ &= \textstyle\sum_{j=1}^r \Big(v(\{i_1, \ldots, i_j\}) - v(\{i_1, \ldots, i_{j-1}\})\Big) \\ &\quad + \textstyle\sum_{j=t+1}^q \Big(v(T \cup \{i_{t+1}, \ldots, i_j\}) - v(T \cup \{i_{t+1}, \ldots, i_{j-1}\})\Big) \\ &= v(S \cap T) + v(S \cup T) - v(T). \end{aligned}$$

Thus, $v(S) + v(T) \leq v(S \cup T) + v(S \cap T)$, and (N, v) is convex. **q.e.d.**

Corollary 3.5.2. *A convex game is totally balanced.*

Proof: A convex game is balanced by Theorem 3.5.1. A subgame of a convex game is convex, so a convex game is totally balanced. **q.e.d.**

Exercises

Exercise 3.5.1. Let (N, v) be a game and let $\mathcal{W}(N, v)$ be the convex hull of $\{a_\pi(v) \mid \pi \in \mathcal{SYM}_N\}$. Prove that $\mathcal{C}(N, v) \subseteq \mathcal{W}(N, v)$ (see Weber (1988) and, for a short proof based on the separation theorem, Derks (1992)).

Exercise 3.5.2. Show that the glove game with respect to L and R (see Exercise 3.4.2) is convex, if and only if $|L| = |R| = 1$.

Exercise 3.5.3. Find an m.c.s.t. game which is not concave. (A cost game is *concave*, if the associated savings game is convex.)

3.6 An Axiomatization of the Core

In this section we shall assume that the universe $\mathcal{U}$ of players contains at least three members. We recall that

$$\Gamma_{\mathcal{U}}^{\mathcal{C}} = \{(N, v) \mid N \subseteq \mathcal{U},\ \mathcal{C}(N, v) \neq \emptyset\}.$$

Theorem 3.6.1. *The core is the unique solution on* $\Gamma_{\mathcal{U}}^{\mathcal{C}}$ *that satisfies* NE, IR, WRGP, *and* SUPA.

We postpone the proof of Theorem 3.6.1 and shall now prove two useful lemmata.

Lemma 3.6.2. *Let* σ *be a solution on a set* Γ *of games. If* σ *satisfies* IR *and* WRGP, *then it also satisfies* PO.

Proof: Assume, on the contrary, that there exist $(N, v) \in \Gamma$ and $x \in \sigma(N, v)$ such that $x(N) < v(N)$. Let $i \in N$. By WRGP, $\left(\{i\}, v_{\{i\},x}\right) \in \Gamma$ and $x^i \in \sigma\left(\{i\}, v_{\{i\},x}\right)$. By IR $x^i \geq v_{\{i\},x}(\{i\})$. On the other hand (see Definition 2.3.11),

$$v_{\{i\},x}(\{i\}) = v(N) - x(N \setminus \{i\}) > x^i.$$

Thus, the desired contradiction has been obtained. **q.e.d.**

Lemma 3.6.3. *Let* σ *be a solution on a set* Γ *of games. If* σ *satisfies* IR *and* WRGP, *then* $\sigma(N, v) \subseteq \mathcal{C}(N, v)$ *for every* $(N, v) \in \Gamma$.

Proof: Let $(N, v) \in \Gamma$ and $n = |N|$. If $n = 1$, then $\sigma(N, v) \subseteq \mathcal{C}(N, v)$ by IR. By Lemma 3.6.2 σ satisfies PO. Hence, if $n = 2$, then

$$\sigma(N, v) \subseteq \{x \in X(N, v) \mid x^i \geq v(\{i\}) \text{ for all } i \in N\} = \mathcal{C}(N, v).$$

If $n \geq 3$ and $x \in \sigma(N, v)$, then WRGP implies that $x^S \in \sigma(S, v_{S,x})$ for all $S \in \mathcal{P}(N)$, so $x^S \in \mathcal{C}(S, v_{S,x})$ for every $S \in \mathcal{P}(N)$. (See (2.3.1).) By Lemma 2.3.22, $x \in \mathcal{C}(N, v)$. **q.e.d.**

Corollary 3.6.4. *Let σ be a solution on $\Gamma_{\mathcal{U}}^{\mathcal{C}}$ that satisfies* NE, IR, *and* WRGP. *If the core of a game (N, v) consists of a unique point, then $\sigma(N, v) = \mathcal{C}(N, v)$.*

Proof of Theorem 3.6.1: The core on $\Gamma_{\mathcal{U}}^{\mathcal{C}}$ satisfies NE, IR, SUPA (see Section 2.3), and WRGP (see Lemma 2.3.16). Thus, we only have to prove the uniqueness part of the theorem. Let σ be a solution on $\Gamma_{\mathcal{U}}^{\mathcal{C}}$ that satisfies NE, IR, WRGP, and SUPA and let $(N, v) \in \Gamma_{\mathcal{U}}^{\mathcal{C}}$ be an n-person game. By Lemma 3.6.3 $\sigma(N, v) \subseteq \mathcal{C}(N, v)$. Thus, we only have to show $\mathcal{C}(N, v) \subseteq \sigma(N, v)$. Let $x \in \mathcal{C}(N, v)$. Two possibilities may occur:

(1) $n \geq 3$. Define a coalition function w on 2^N by the following rule: $w(\{i\}) = v(\{i\})$ for all $i \in N$ and $w(S) = x(S)$ for all $S \subseteq N$ with $|S| \neq 1$. As $n \geq 3$, $\mathcal{C}(N, w) = \{x\}$. Hence, by Corollary 3.6.4, $\sigma(N, w) = \{x\}$. Let $u = v - w$. Then $u(\{i\}) = 0$ for all $i \in N$, $u(N) = 0$, and $u(S) \leq 0$ for all $S \subseteq N$. Therefore $\mathcal{C}(N, u) = \{0\}$ and, again by Corollary 3.6.4, $\sigma(N, u) = \{0\}$. Hence, by SUPA,

$$\{x\} = \sigma(N, u) + \sigma(N, w) \subseteq \sigma(N, v).$$

We conclude that $x \in \sigma(N, v)$, and thus $\mathcal{C}(N, v) \subseteq \sigma(N, v)$.

(2) $n \leq 2$. If $n = 1$, then $x \in \sigma(N, v)$ by NE and IR. Thus, assume that $n = 2$; let us say $N = \{i, j\}$. Let $k \in \mathcal{U} \setminus N$. (Indeed, k exists, because $\mathcal{U} \setminus N \neq \emptyset$ by the assumption that $|\mathcal{U}| \geq 3$.) We define a coalition function u on 2^M where $M = \{i, j, k\}$ by the following rules:

$$u(S) = \begin{cases} \sum_{h \in S \cap N} v(\{h\}) & , \text{ if } S \subsetneq M \\ v(N) & , \text{ if } S = M. \end{cases}$$

Then $y \in \mathbb{R}^M$, defined by $y^k = 0$ and $y^N = x$, is in $\mathcal{C}(M, u)$. As $|M| = 3$, $\mathcal{C}(M, u) \subseteq \sigma(M, u)$. Thus $y \in \sigma(M, u)$. Also, $u_{N,y} = v$. Hence, by WRGP, $x \in \sigma(N, v)$, and thus $\mathcal{C}(N, v) \subseteq \sigma(N, v)$. **q.e.d.**

Theorem 3.6.1 is due to Peleg (1986).

Now we comment on the logical independence of the axioms that characterize the core.

Example 3.6.5. Let $\sigma(N, v) = \emptyset$ for every $(N, v) \in \Gamma_{\mathcal{U}}^{\mathcal{C}}$. Then σ satisfies IR, RGP, and SUPA. Obviously, σ violates NE.

Example 3.6.6. Let the solution σ on $\Gamma_{\mathcal{U}}^{\mathcal{C}}$ be defined by $\sigma(N, v) = \mathcal{C}(N, v)$ if $|N| \geq 2$ and by $\sigma(\{i\}, v) = X^*(\{i\}, v)$ for every one-person game $(\{i\}, v)$. Then σ satisfies NE, RGP, and SUPA. On one-person games it violates IR.

Example 3.6.7. Define a solution σ on $\Gamma_{\mathcal{U}}^{\mathcal{C}}$ by

$$\sigma(N, v) = \{x \in X(N, v) \mid x^i \geq v(\{i\}) \text{ for all } i \in N\}.$$

Then σ satisfies NE, IR, and SUPA. By Lemma 3.6.3, σ violates WRGP.

Remark 3.6.8. In Chapter 5 (Example 5.2.9) we shall prove that SUPA is logically independent of NE, IR, and WRGP (on $\Gamma_{\mathcal{U}}^{\mathcal{C}}$).

Exercises

Exercise 3.6.1. Prove that Theorem 3.6.1 does not hold if $|\mathcal{U}| = 2$.

Exercise 3.6.2. Find an example of a game which explicitly shows that the solution σ defined in Example 3.6.7 violates RGP.

Exercise 3.6.3. Let a solution σ on $\Gamma_{\mathcal{U}}^{\mathcal{C}}$ be defined by

$$\sigma(N,v) = \left\{x \in X^*(N,v) \,\middle|\, (S, v_{S,x}) \in \Gamma_{\mathcal{U}}^{\mathcal{C}} \text{ for all } S \in \mathcal{P}(N)\right\}.$$

Prove that σ satisfies NE, SUPA, RGP, and CRGP and that it violates IR. (Hint: In order to show SUPA let $(N,v), (N,w)$ be games, let $x \in X^*(N,v)$, $y \in X^*(N,w)$, and let $\emptyset \neq S \subseteq N$. Prove that $v_{S,x}(T) + w_{S,x}(T) \geq (v+w)_{S,x+y}(T)$ for every $T \subseteq S$ and that the foregoing inequality is an equality for $T = S$.)

3.7 An Axiomatization of the Core on Market Games

Let

$$\Gamma_{\mathcal{U}}^{tb} = \{(N,v) \mid N \subseteq \mathcal{U},\ (N,v) \text{ is totally balanced}\}$$

and recall (see Theorem 3.3.7) that $\Gamma_{\mathcal{U}}^{tb}$ is the set of market games. In this section we shall assume that the universe $\mathcal{U}$ of players contains at least four members. For simplicity we assume that $M = \{1, \ldots, 4\} \subseteq \mathcal{U}$.

Theorem 3.7.1. *The core is the unique solution on* $\Gamma_{\mathcal{U}}^{tb}$ *that satisfies* NE, IR, WRGP, CRGP, *and* SUPA.

We postpone the proof of Theorem 3.7.1 and shall now discuss the four-person game (M,u) defined by the following formula:

$$u(S) = \begin{cases} 0 & \text{, if } S \in \{M, \{1,2\}, \{2,3\}, \{3,4\}, \{4,1\}, \emptyset\}\ , \\ -1 & \text{, if } |S| = 3\ , \\ -4 & \text{, otherwise.} \end{cases} \tag{3.7.1}$$

This game will be used in the proof of Theorem 3.7.1.

Remark 3.7.2. Note that the symmetry group $\mathcal{SYM}(M,u)$ is generated by the cyclic permutation, which maps 1 to 2, 2 to 3, 3 to 4, and 4 to 1; thus $\mathcal{SYM}(M,u)$ is transitive.

Lemma 3.7.3. *The game (M, u) defined by (3.7.1) is totally balanced and*

$$\mathcal{C}(M, u) = \{(\gamma, -\gamma, \gamma, -\gamma) \mid -1 \leq \gamma \leq 1\}.$$

Proof: For $\gamma \in \mathbb{R}$ define $x_\gamma = (\gamma, -\gamma, \gamma, -\gamma)$. Every x_γ, $-1 \leq \gamma \leq 1$, is in $\mathcal{C}(M, u)$. Every core element assigns zero to M and, thus, to the members of the partitions $\{\{1,2\}, \{3,4\}\}$ and $\{\{2,3\}, \{4,1\}\}$. Therefore the core is contained in the line $\{x_\gamma \mid \gamma \in \mathbb{R}\}$. The facts $x_\gamma(\{1,2,3\}) < -1$ for all $\gamma < -1$ and $x_\gamma(\{2,3,4\}) < -1$ for all $\gamma > 1$ show that the core has the claimed shape.

The restrictions of the vectors x_γ for $\gamma = 1$ and $\gamma = -1$, respectively, to the coalitions $\{1,2,4\}, \{2,3,4\}$ and $\{1,2,3\}, \{1,3,4\}$, respectively, show that the three-person subgames are balanced. All one- and two-person subgames are balanced as well. We conclude that (M, u) is totally balanced. **q.e.d.**

The easy proof of the following statement is left to the reader (see Exercise 3.7.1):

$$\Big(S \in \mathcal{P}(M) \text{ and } x \in \mathcal{C}(M, u)\Big) \Rightarrow (S, u_{S,x}) \text{ is inessential.} \tag{3.7.2}$$

Proof of Theorem 3.7.1: The core on $\Gamma^{tb}_{\mathcal{U}}$ satisfies NE, IR, SUPA (see Section 2.3), WRGP (see Exercise 2.3.1), and CRGP (by Lemma 2.3.22). Thus, we only have to prove the uniqueness part of the theorem. Let σ be a solution on $\Gamma^{tb}_{\mathcal{U}}$ that satisfies NE, IR, WRGP, CRGP, and SUPA and let $(N, v) \in \Gamma^{tb}_{\mathcal{U}}$ be an n-person game. By Lemma 3.6.3, we only have to show that $\mathcal{C}(N, v) \subseteq \sigma(N, v)$. Let $x \in \mathcal{C}(N, v)$.

(1) $n = 1$. Then $x \in \sigma(N, v)$ by NE and IR.

(2) $n = 2$. If (N, v) is inessential (see Definition 2.1.9), then the proof is finished by NE and IR. Hence we assume that (N, v) is not inessential, so $v(N) > \sum_{i \in N} v(\{i\})$. Also, we assume without loss of generality that $N = \{1, 2\}$. Define (N, w) by

$$w(S) = \begin{cases} -2 \ , & \text{if } |S| = 1 \ , \\ 0 \ , & \text{if } S = \emptyset, N \ , \end{cases}$$

and observe that $v = \alpha w + \beta$ where $\alpha = \frac{v(N) - v(\{1\}) - v(\{2\})}{4}$ and $\beta = \left(\frac{v(N)+v(\{1\})-v(\{2\})}{2}, \frac{v(N)+v(\{2\})-v(\{1\})}{2}\right) \in \mathbb{R}^N$. As $\alpha > 0$, (N, v) is strategically equivalent to (N, w). Put $y := \frac{x - \beta}{\alpha}$ and observe that $y \in \mathcal{C}(N, w)$ by COV of the core. We first prove that $y \in \sigma(N, w)$. With $\gamma := y^1/2$ the vector y can be expressed as $y = (2\gamma, -2\gamma)$. Also, $-1 \leq \gamma \leq 1$, because $y \in \mathcal{C}(N, w)$.

Choose two members of $\mathcal{U} \setminus N$, let us say $3, 4$, which is possible by the assumption that $|\mathcal{U}| \geq 4$, let (M, u) be defined by (3.7.1), let π be the permutation of M which exchanges 1 and 2, and let $(M, \pi u)$ be the "permuted" game (see Definition 2.3.4). By Lemma 3.7.3, $x_\gamma, x_{-\gamma} \in \mathcal{C}(M,u)$ and, similarly, $\pi x_{-\gamma} = (\gamma, -\gamma, -\gamma, \gamma) \in \mathcal{C}(M, \pi u)$. An application of CRGP shows that $x_\gamma \in \sigma(M,u)$ and $\pi x_{-\gamma} \in \sigma(M, \pi u)$ by (3.7.2) and the first part of this step of the proof. SUPA implies that $z := (y, 0, 0) = x_\gamma + \pi x_{-\gamma} \in \sigma(M, u + \pi u)$. WRGP yields $y \in \sigma(N, (u + \pi u)_{N,z})$. The reduced coalition function $\widehat{u} = (u+\pi u)_{N,z}$ coincides with w. Indeed, $\widehat{u}(N) = (u + \pi u)(M) - z(\{3,4\}) = 0 = w(N)$ by definition of the reduced game. Moreover, the unique two-person coalitions S with $u(S) = (\pi u)(S) = 0$ are $\{1,2\}$ and $\{3,4\}$, so $(u + \pi u)(\{i,j\}) = -4$ for $i \in N$ and $j \in M \setminus N$. Hence

$$\widehat{u}(\{i\}) = (u + \pi u)(\{i,3,4\}) - z(\{3,4\}) = -2 = w(\{i\}) \text{ for } i = 1, 2,$$

so $y = z^N \in \sigma(N, w)$.

In order to show $x \in \sigma(N, v)$ we can proceed similarly. Only the coalition functions u and πu are replaced by $\alpha u + \frac{1}{2}(\beta, 0, 0)$ and $\alpha(\pi u) + \frac{1}{2}(\beta, 0, 0)$. The proof is finished by COV of the core.

(3) $n \geq 3$. The core satisfies WRGP, so $x^S \in \mathcal{C}(S, v_{S,x})$ for all $S \in \mathcal{P}(N)$. By the preceding step, $x^S \in \sigma(S, v_{S,x})$ for all $S \in \mathcal{P}(N)$, so $x \in \sigma(N, v)$ by CRGP of σ. **q.e.d.**

The proof of Theorem 3.7.1 is due to Sudhölter and Peleg (2002). In Peleg (1993) it is shown that the core of market games can be characterized by the foregoing axioms *and* a weak variant of anonymity (see also Peleg (1989)).

Examples 3.6.5, 3.6.6, 3.6.7, and Exercise 3.7.2 show that the axioms NE, IR, WRGP, and CRGP are each logically independent of the remaining axioms. The independence of SUPA will be proved in Chapter 5 (Example 5.2.10).

Exercises

Exercise 3.7.1. Prove (3.7.2).

Exercise 3.7.2. Find a solution σ on $\Gamma_{\mathcal{U}}^{tb}$ which satisfies NE, IR, WRGP, SUPA, and which does **not** coincide with the core (see Peleg (1989), Example 5.5).

3.8 The Core for Games with Various Coalition Structures

Let $\mathcal{U}$ be a set of players and let N be a coalition.

Definition 3.8.1. *A* **coalition structure** *for N is a partition of N.*

If (N, v) is a game and $\mathcal{R}$ is a coalition structure for N, then the triple $(N, v, \mathcal{R})$ is called a *game with coalition structure.* Let $(N, v, \mathcal{R})$ be a game with coalition structure. Then

$$X^*(N, v, \mathcal{R}) = \{x \in \mathbb{R}^N \mid x(R) \leq v(R) \text{ for every } R \in \mathcal{R}\}$$

denotes the set of feasible payoff vectors for $(N, v, \mathcal{R})$.

Definition 3.8.2. *Let $(N, v, \mathcal{R})$ be a game with coalition structure. The* **core** *$\mathcal{C}(N, v, \mathcal{R})$ of $(N, v, \mathcal{R})$ is defined by*

$$\mathcal{C}(N, v, \mathcal{R}) = \{x \in X^*(N, v, \mathcal{R}) \mid x(S) \geq v(S) \text{ for all } S \subseteq N\}.$$

Clearly, $\mathcal{C}(N, v) = \mathcal{C}(N, v, \{N\})$. Hence, we shall write (N, v) instead of $(N, v, \{N\})$. In order to investigate the core of games with coalition structures we now define the superadditive cover of a game.

Definition 3.8.3. *Let (N, v) be a game. The* **superadditive cover** *of (N, v) is the game $(N, \hat{v})$ defined by*

$$\hat{v}(S) = \max\left\{\sum_{T \in \mathcal{T}} v(T) \,\middle|\, \mathcal{T} \text{ is a partition of } S\right\} \text{ for all } \emptyset \neq S \subseteq N$$

and $\hat{v}(\emptyset) = 0$.

Obviously, the superadditive cover of a game is superadditive.

Theorem 3.8.4. *Let $(N, v, \mathcal{R})$ be a game with coalition structure. Then*

(1) $\mathcal{C}(N, v, \mathcal{R}) \neq \emptyset$ if and only if $\mathcal{C}(N, \hat{v}) \neq \emptyset$ and $\hat{v}(N) = \sum_{R \in \mathcal{R}} v(R)$; and

(2) if $\mathcal{C}(N, v, \mathcal{R}) \neq \emptyset$, then $\mathcal{C}(N, v, \mathcal{R}) = \mathcal{C}(N, \hat{v})$.

Proof: (1) Let $x \in \mathcal{C}(N, v, \mathcal{R})$, $\emptyset \neq S \subseteq N$, and let $\mathcal{T}$ be a partition of S. Then $x(T) \geq v(T)$ for every $T \in \mathcal{T}$. Thus, $x(S) = \sum_{T \in \mathcal{T}} x(T) \geq \sum_{T \in \mathcal{T}} v(T)$. We conclude that $x(S) \geq \hat{v}(S)$. In particular, $x(N) = \sum_{R \in \mathcal{R}} v(R) \geq \hat{v}(N)$. By Definition 3.8.3, $\hat{v}(N) \geq \sum_{R \in \mathcal{R}} v(R)$. Hence $x(N) = \hat{v}(N) = \sum_{R \in \mathcal{R}} v(R)$ and $x \in \mathcal{C}(N, \hat{v})$.

Conversely, assume that $x \in \mathcal{C}(N, \hat{v})$ and $\hat{v}(N) = \sum_{R \in \mathcal{R}} v(R)$. As $x(R) \geq v(R)$ for every $R \in \mathcal{R}$ and $x(N) = \hat{v}(N) = \sum_{R \in \mathcal{R}} v(R)$, it follows that $x(R) = v(R)$ for every $R \in \mathcal{R}$. Hence, $x \in \mathcal{C}(N, v, \mathcal{R})$.

(2) Assume that $\mathcal{C}(N, v, \mathcal{R}) \neq \emptyset$. Then $\mathcal{C}(N, \hat{v}) \neq \emptyset$ and $\hat{v}(N) = \sum_{R \in \mathcal{R}} v(R)$. Thus, by the proof of (1), $\mathcal{C}(N, \hat{v}) = \mathcal{C}(N, v, \mathcal{R})$. **q.e.d.**

Another interesting property of the core of games with coalition structures is discovered with the help of the following definition.

Definition 3.8.5. *Let (N,v) be a game and let $i,j \in N$. Players i and j are* **substitutes** *if*

$$v(S \cup \{i\}) = v(S \cup \{j\}) \text{ for all } S \subseteq N \setminus \{i,j\}.$$

Clearly, two players i and j are substitutes of (N,v) if and only if the transposition (i,j) is a symmetry of (N,v).

Theorem 3.8.6. *Let $(N,v,\mathcal{R})$ be a game with coalition structure, let i and j be substitutes of (N,v), and let $x \in \mathcal{C}(N,v,\mathcal{R})$. If i and j belong to different members of $\mathcal{R}$, then $x^i = x^j$.*

Proof: Let $i \in R \in \mathcal{R}$. Then $j \notin R$. The observation that

$$0 \geq v((R\setminus\{i\})\cup\{j\}) - x((R\setminus\{i\})\cup\{j\}) = v(R) - (x(R) + x^j - x^i) = x^i - x^j$$

shows that $x^j \geq x^i$. Exchanging the roles of i and j yields $x^i \geq x^j$, so $x^i = x^j$.
q.e.d.

Theorems 3.8.4 and 3.8.6 are due to Aumann and Drèze (1974). The rest of this section is devoted to an axiomatic characterization of the core of games with coalition structures.

Definition 3.8.7. *Let Δ be a set of games with coalition structures. A* **solution** *on Δ is a function σ that associates with each game with coalition structure $(N,v,\mathcal{R})$ a subset $\sigma(N,v,\mathcal{R})$ of $X^*(N,v,\mathcal{R})$.*

Let N be a coalition, let $\mathcal{R}$ be a coalition structure for N, and let $\emptyset \neq S \subseteq N$. We use the following notation:

$$\mathcal{R}_{|S} = \{R \cap S \mid R \in \mathcal{R} \text{ and } R \cap S \neq \emptyset\}.$$

We proceed with the following definition.

Definition 3.8.8. *Let $(N,v,\mathcal{R})$ be a game with coalition structure, let $\emptyset \neq S \subseteq N$, and let $x \in X^*(N,v,\mathcal{R})$. The* **reduced game** *with respect to S and x is the game with coalition structure $\left(S, v^{\mathcal{R}}_{S,x}, \mathcal{R}_{|S}\right)$ defined by the following formula:*

$$v^{\mathcal{R}}_{S,x}(T) = \begin{cases} 0, \text{ if } T = \emptyset \\ v(R) - x(R \setminus T), \text{ if } \emptyset \neq T = S \cap R \text{ for some } R \in \mathcal{R} \\ \max_{Q \subseteq N\setminus S}\left(v(T \cup Q) - x(Q)\right), \text{ otherwise.} \end{cases}$$

Let Δ be a set of games with coalition structures.

Definition 3.8.9. *A solution σ on Δ has the* **reduced game property** *(RGP) if it satisfies the following condition: If $(N,v,\mathcal{R}) \in \Delta$, $\emptyset \neq S \subseteq N$, and $x \in \sigma(N,v,\mathcal{R})$, then $\left(S, v^{\mathcal{R}}_{S,x}, \mathcal{R}_{|S}\right) \in \Delta$ and $x^S \in \sigma\left(S, v^{\mathcal{R}}_{S,x}, \mathcal{R}_{|S}\right)$.*

Remark 3.8.10. Let $\Delta_{\mathcal{U}}^{\mathcal{C}} = \{(N, v, \mathcal{R}) \in \Delta_{\mathcal{U}} \mid \mathcal{C}(N, v, \mathcal{R}) \neq \emptyset\}$ where $\Delta_{\mathcal{U}}$ denotes the set of all games with coalition structures. The core satisfies RGP on $\Delta_{\mathcal{U}}^{\mathcal{C}}$. A slight modification of the proof of Lemma 2.3.16 yields this result.

Let N be a coalition and let $\mathcal{R}$ be a coalition structure for N. Two players $i, j \in N$ with $i \neq j$, are *partners* in $\mathcal{R}$ if there exists $R \in \mathcal{R}$ such that $i, j \in R$. We denote

$$\mathcal{P}(\mathcal{R}) = \{\{i, j\} \mid i \neq j \text{ and } i \text{ and } j \text{ are partners in } \mathcal{R}\}. \tag{3.8.1}$$

Also, we need the following notation. If $(N, v, \mathcal{R})$ is a game with coalition structure, then

$$X(N, v, \mathcal{R}) = \left\{x \in \mathbb{R}^N \middle| \, x(R) = v(R) \text{ for every } R \in \mathcal{R}\right\}.$$

Definition 3.8.11. *A solution σ on a set Δ of games with coalition structures has the* **converse reduced game property** *(*CRGP*) if the following condition is satisfied: If $(N, v, \mathcal{R}) \in \Delta$, $\mathcal{P}(\mathcal{R}) \neq \emptyset$, $x \in X(N, v, \mathcal{R})$, $(S, v_{S,x}) \in \Delta$, and $x^S \in \sigma(S, v_{S,x})$ for every $S \in \mathcal{P}(\mathcal{R})$, then $x \in \sigma(N, v, \mathcal{R})$.*

Lemma 3.8.12. *The core satisfies* CRGP *on $\Delta_{\mathcal{U}}^{\mathcal{C}}$.*

Proof: Let $(N, v, \mathcal{R}) \in \Delta_{\mathcal{U}}^{\mathcal{C}}$, let $\mathcal{F}$ be the field generated by $\mathcal{R}$, that is,

$$\mathcal{F} = \left\{ \bigcup_{R \in \widehat{\mathcal{R}}} R \;\middle|\; \widehat{\mathcal{R}} \subseteq \mathcal{R} \right\},$$

and let $T = \bigcup_{R \in \widehat{\mathcal{R}}} R \in \mathcal{F}$. If $x \in \mathcal{C}(N, v, \mathcal{R})$, then

$$v(T) \leq x(T) = \sum_{R \in \widehat{\mathcal{R}}} x(R) = \sum_{R \in \widehat{\mathcal{R}}} v(R). \tag{3.8.2}$$

Assume that $\mathcal{R}$ satisfies $\mathcal{P}(\mathcal{R}) \neq \emptyset$ and let $y \in X(N, v, \mathcal{R})$ satisfy $y^S \in \mathcal{C}(S, v_{S,y})$ for every $S \in \mathcal{P}(\mathcal{R})$. In view of (3.8.2), $y(T) \geq v(T)$ for every $T \in \mathcal{F}$. Now let $T \in 2^N \setminus \mathcal{F}$. Then there exists $R \in \mathcal{R}$ such that $\emptyset \neq T \cap R \neq R$. Choose $i \in T \cap R$ and $j \in R \setminus T$ and let $S = \{i, j\}$. Then $S \in \mathcal{P}(\mathcal{R})$. Thus, by our assumption, $y^S \in \mathcal{C}(S, v_{S,y})$. Hence, in particular,

$$0 \geq v_{S,y}(\{i\}) - y^i \geq v(T) - y(T).$$

Thus, $y(T) \geq v(T)$ and the proof is complete. **q.e.d.**

For the sake of completeness we now formulate the superadditivity property for solutions on classes of games with coalition structures.

Definition 3.8.13. *A solution σ on a set Δ of games with coalition structures is* **superadditive** *if*

$$\sigma(N, v, \mathcal{R}) + \sigma(N, w, \mathcal{R}) \subseteq \sigma(N, v + w, \mathcal{R})$$

when $(N, v, \mathcal{R})$, $(N, w, \mathcal{R})$ and $(N, v + w, \mathcal{R})$ are in Δ.

We remark that nonemptiness (NE), individual rationality (IR), and the weak reduced game property (WRGP) are generalized in a straightforward manner to solutions for games with coalition structures (see Definitions 2.3.7, 2.3.24, and 2.3.17). Now we may formulate a generalization of Theorem 3.6.1. Again, assume that $\mathcal{U}$ contains at least three members.

Theorem 3.8.14. *The core is the unique solution on* $\Delta_{\mathcal{U}}^{\mathcal{C}}$ *that satisfies* NE, IR, WRGP, *and* SUPA.

The proof of Theorem 3.8.14 is similar to that of Theorem 3.6.1 and left as Exercise 3.8.2.

Exercises

Exercise 3.8.1. Let (N, v) be an assignment game with respect to S and B (see Exercise 3.4.1). Let (N, v_0) be defined by $v_0(\{i,j\}) = v(\{i,j\})$ for all $i \in S$ and $j \in B$ and $v_0(T) = 0$ for all other $T \subseteq N$. Show that (N, v) is the superadditive cover of (N, v_0).

Exercise 3.8.2. Prove Theorem 3.8.14.

Exercise 3.8.3. Let (N, v) be a convex game and let $\mathcal{R}$ be a coalition structure for N. Show that $\mathcal{C}(N, v, \mathcal{R}) \neq \emptyset$ iff (N, v) is *decomposable with respect to* $\mathcal{R}$, that is, if $v(S) = \sum_{R \in \mathcal{R}} v(S \cap R)$ for all $S \subseteq N$ (see Shapley (1971)).

3.9 Notes and Comments

(1) Theorem 3.1.4 may be proved by means of the Minimax Theorem without reference to linear programming (see Aumann (1989)). A proof of the Krein-Milman theorem, which is used in the proof of Theorem 3.1.10, is contained in Klein (1973).

(2) The core is axiomatized by Theorems 3.6.1 and 3.7.1. Other axiomatizations can be found in the literature. We mention four of them. Tadenuma (1992) presents a characterization of the core which employs NE, IR, and RGP^M, the reduced game property with respect to Moulin reduced games. An approach which uses a different reduced game can be found in Voorneveld and van den Nouveland (1998). A different kind of CRGP is employed in Serrano and Volij (1998). Moreover, Hwang and Sudhölter (2000) show that the core is axiomatized by AN, COV, WRGP, RCP, CRGP, REBE, and the requirement of nonemptiness for inessential games, on many interesting sets of games that may contain non-balanced games.

(3) Theorem 3.8.4 is not sharp. Indeed, as the reader may verify, it is possible to replace the superadditive cover $(N, \hat{v})$ of (N, v) by the totally balanced

cover $(N, \bar{v})$ of (N, v) in the formulation of the theorem. Clearly, $\bar{v}(S) \geq \hat{v}(S)$ for every $S \subseteq N$.

(4) The definition of the core is simple and highly intuitive. Nevertheless, some examples of cores have been criticized by several authors. First, consider Example 2.2.2 and let $n_i = |N_i|$, $i = 1, 2$. If $n_1 > n_2$, then the only payoff vector in the core is χ_{N_2}. Similarly, if $n_1 < n_2$, then $\mathcal{C}(N, v) = \{\chi_{N_1}\}$. However, if $n_1 = n_2$, then

$$\mathcal{C}(N, v) = \left\{ t\chi_{N_1} + (1 - t)\chi_{N_2} \;\middle|\; 0 \leq t \leq 1 \right\}.$$

Thus, the core is unreasonable when $\min\{n_1, n_2\}$ is large, $|n_1 - n_2|$ is small, and $n_1 \neq n_2$ (see also Shapley and Shubik (1969b)).

Another counterintuitive example is due to S. Zamir.

Example 3.9.1. Let $M = \{1, 2, 3\}$ and let (M, u) be defined by $u(S) = 1$ if $|S| \geq 2$ and $u(S) = 0$ otherwise, where $S \subseteq M$. Then $\mathcal{C}(M, u) = \emptyset$. Now let $N = M \cup \{4\}$ and let (N, v) be defined by $v(S) = u(S)$ if $S \subseteq M$, $v(S) = 0$ otherwise, where $S \subsetneqq N$, and $v(N) = 3/2$. Then

$$\mathcal{C}(N, v) = \left\{ \left(\frac{1}{2}, \frac{1}{2}, \frac{1}{2}, 0 \right) \right\}.$$

Thus, player 4, whose addition to (M, u) rendered the core nonempty, is not paid (in the core) for his contribution.

Of course, one should not reject the core because of the foregoing examples. Indeed, there is no solution that yields intuitive results for *all* coalitional games. The core is very useful because it is acceptable for many classes of games. Furthermore, it has many nice properties and may be justified by means of its axiomatic characterizations.

4

Bargaining Sets

The "classical" bargaining set $\mathcal{M}$ and some relatives are introduced and studied in this chapter. In the first section we provide the basic definitions of objections and counterobjections. Also, we check whether or not $\mathcal{M}$ satisfies the usual properties of solutions.

Section 4.2 is devoted to an existence theorem for the bargaining set. First of all, justified objections, that is, objections that cannot be countered, are studied. Player k is stronger than player ℓ at the proposal x, if k has a justified objection against ℓ at x. Secondly, we prove that the binary relation "stronger than" is continuous and acyclic. Finally, the foregoing results enable us to prove the desired existence theorem. We have chosen a proof that later will be used for the investigation of bargaining sets of cooperative games without transferable utility.

Section 4.3 mainly deals with convex games and with assignment games. We prove that the bargaining set (for the grand coalition) of a game of these classes coincides with the core. However, the general result can be applied to many families of superadditive balanced games.

In Section 4.4 related bargaining sets are introduced and discussed. It is shown that "small" modifications in the definition of "justified objection" yield bargaining sets which are contained in $\mathcal{M}$ and which contain the core. The existence result can also be applied to the new bargaining sets which are (in contrast to $\mathcal{M}$) reasonable, when restricted to superadditive games. Also, a bargaining set based on the notion of "global" objections and counterobjections is briefly discussed.

In Section 4.5 we show that the bargaining sets based on "individual" objections and counterobjections do not satisfy aggregate monotonicity. Also, we prove that according to the classical bargaining set a dummy may have to decrease his demand, if he is going to contribute to the grand coalition.

Section 4.6 contains an example of a market game due to Postlewaite and Rosenthal (1974). If we adopt the core as a solution concept in this example, then syndication may not be profitable. Following Maschler (1976) we also look at the bargaining set of the foregoing example. As Maschler has shown, according to $\mathcal{M}$ syndication is advantageous. The aforementioned relatives of the bargaining set coincide with $\mathcal{M}$. Hence they can be used as well to emphasize that syndication may be advantageous.

The section is concluded with notes and comments.

4.1 The Bargaining Set $\mathcal{M}$

A disadvantage of the core is that many games, which appear in various applications, are not balanced. For example, every non-trivial constant-sum game has an empty core. Indeed, the following result is true.

Lemma 4.1.1. *A constant-sum game is balanced if and only if it is inessential.*

Proof: Let (N, v) be a constant-sum game. If $x \in \mathcal{C}(N, v)$, then

$$v(N) = x(N) = x(S) + x(N \setminus S) \geq v(S) + v(N \setminus S) = v(N) \text{ for all } S \subseteq N,$$

so $v(S) = x(S)$ for all $S \subseteq N$. Hence (N, v) is inessential in this case. If (N, v) is inessential, let us say $v(S) = x(S)$ for some $x \in \mathbb{R}^N$, then $x \in \mathcal{C}(N, v)$. **q.e.d.**

A simple game without vetoers is another example of a game with an empty core (see Exercise 4.1.1).

Players who face a game with an empty core will find out that if they desire the kind of stability that is implicit in the core concept, then they will be unable to reach any agreement. If they nevertheless want to profit from the game, then they have no choice but to relax their stability requirements. In this section we shall present a solution which, on the one hand, enables the players to reach agreements and, on the other hand, maintains some stability of the outcome (which is weaker than the stability implied by the core).

Let $\mathcal{U}$ be a set of players and let N be a coalition. If $k, \ell \in N,\ k \neq \ell$, then we denote

$$\mathcal{T}_{k\ell}(N) = \mathcal{T}_{k\ell} = \{S \subseteq N \setminus \{\ell\} \mid k \in S\}.$$

Hence, $\mathcal{T}_{k\ell}$ is the set of coalitions containing k and not containing ℓ.

Definition 4.1.2. *Let $(N, v, \mathcal{R})$ be a game with coalition structure, $x \in X(N, v, \mathcal{R})$, and let $k, \ell \in R \in \mathcal{R},\ k \neq \ell$. An* **objection** *of k against ℓ at x (with respect to $(N, v, \mathcal{R})$) is a pair (P, y) satisfying*

$$P \in \mathcal{T}_{k\ell} \text{ and } y \in \mathbb{R}^P; \tag{4.1.1}$$
$$y^i \geq x^i \text{ for all } i \in P \text{ and } y^k > x^k; \tag{4.1.2}$$
$$y(P) \leq v(P). \tag{4.1.3}$$

Thus, an objection (P, y) of k against ℓ is a potential threat by a coalition P, which contains k but not ℓ, to deviate from x. The threat is feasible by (4.1.1)–(4.1.3). The purpose of presenting an objection is not to disrupt $\mathcal{R}$, but to demand a transfer of money from ℓ to k, that is, to modify x within $X(N, v, \mathcal{R})$. It is assumed that the players (tentatively) agreed upon the formation of $\mathcal{R}$ and only the problem of choosing a point x out of $X(N, v, \mathcal{R})$ has been left open.

Definition 4.1.3. *Let (P, y) be an objection of k against ℓ at $x \in X(N, v, \mathcal{R})$ with respect to a game with coalition structure $(N, v, \mathcal{R})$. A* **counterobjection** *to (P, y) is a pair (Q, z) satisfying*

$$Q \in \mathcal{T}_{\ell k} \text{ and } z \in \mathbb{R}^Q; \tag{4.1.4}$$
$$z^i \geq x^i \text{ for all } i \in Q; \tag{4.1.5}$$
$$z^i \geq y^i \text{ for all } i \in P \cap Q; \tag{4.1.6}$$
$$z(Q) \leq v(Q). \tag{4.1.7}$$

In a counterobjection ℓ has to prove that he can protect his share x^ℓ in spite of the existing objection of k.

Definition 4.1.4. *Let $(N, v, \mathcal{R})$ be a game with coalition structure. A vector $x \in X(N, v, \mathcal{R})$ is* **stable** *if for each objection at x there is a counterobjection. The* **unconstrained bargaining set**, *$\mathcal{PM}(N, v, \mathcal{R})$, is the set of all stable members of $X(N, v, \mathcal{R})$.*

In the literature the *unconstrained bargaining* set is usually called the *prebargaining* set. Henceforth we shall use both terms.

Remark 4.1.5. Let (N, v) be a game and $\mathcal{R}$ be a coalition structure. We may ask the following question: Which payoff vectors in $X(N, v, \mathcal{R})$ should the players expect, if $\mathcal{R}$ is formed? The prebargaining set $\mathcal{PM}(N, v, \mathcal{R})$ is a possible answer which guarantees minimum stability.

Remark 4.1.6. Each player in a game may compute $\mathcal{PM}$ for each coalition structure. Thereby, the players might form preference orderings on coalition structures. Thus, $\mathcal{PM}$ might have implications on coalition formation.

Remark 4.1.7. If $(N, v, \mathcal{R})$ is a game with coalition structure, then

$$\mathcal{C}(N, v, \mathcal{R}) \subseteq \mathcal{PM}(N, v, \mathcal{R}).$$

Indeed, if $x \in \mathcal{C}(N, v, \mathcal{R})$ and $k \in N$, then k has no objection against any other player at x (see (4.1.2) and (4.1.3)).

In view of our convention to identify a game (N, v) and $(N, v, \{N\})$ we denote $\mathcal{PM}(N, v, \{N\}) = \mathcal{PM}(N, v)$.

Example 4.1.8. Consider the market given by

$$N = \{1, \ldots, 6\},\ m = 2,\ a^1 = a^2 = a^3 = (1, 0),\ a^4 = a^5 = a^6 = (0, 1),$$

and $w^i(x_1, x_2) = \min\{2x_1, x_2\}$, $i = 1, \ldots, 6$. Let v be the coalition function of the market. Then $v(N) = 3$ and $v(S) = 2$ if $|S \cap \{1, 2, 3\}| = 1$ and $|S \cap \{4, 5, 6\}| = 2$. Hence $x = (1, 1, 1, 0, 0, 0) \in X(N, v) \setminus \mathcal{C}(N, v)$. It is straightforward to show that $x \in \mathcal{PM}(N, v)$. Thus, $\mathcal{PM}(N, v)$ may be strictly larger than $\mathcal{C}(N, v)$ even when $\mathcal{C}(N, v) \neq \emptyset$.

We now enquire which properties (of solutions) are satisfied by $\mathcal{PM}$. First, we need the following generalizations of COV and AN.

Definition 4.1.9. *A solution σ on a set Δ of games with coalition structures is* **covariant under strategic equivalence** *(*COV*) if the following condition is satisfied: If $(N, v, \mathcal{R}), (N, w, \mathcal{R}) \in \Delta, \alpha > 0, \beta \in \mathbb{R}^N$, and $w = \alpha v + \beta$, then*

$$\sigma(N, w, \mathcal{R}) = \alpha\sigma(N, v, \mathcal{R}) + \beta.$$

Let $\mathcal{R}$ be a coalition structure for the coalition N and let π be an injection of N into $\mathcal{U}$. Then the coalition structure $\pi(\mathcal{R})$ is defined by

$$\pi(\mathcal{R}) = \{\pi(R) \mid R \in \mathcal{R}\}.$$

Definition 4.1.10. *Let σ be a solution on a set Δ of games with coalition structures. We say that σ is* **anonymous** *(*AN*) if the following condition is satisfied: If $(N, v, \mathcal{R}) \in \Delta$, $\pi : N \to \mathcal{U}$ is an injection, and if $(\pi(N), \pi v, \pi(\mathcal{R})) \in \Delta$, then $\sigma(\pi(N), \pi v, \pi(\mathcal{R})) = \pi(\sigma(N, v, \mathcal{R}))$.*

Note that $\mathcal{PM}$ (on an arbitrary set of games with coalition structures) satisfies COV and AN (see Exercise 4.1.5).

Theorem 4.1.11. *$\mathcal{PM}$ satisfies* RGP *on $\Delta_{\mathcal{U}}$.*

Proof: Let $(N, v, \mathcal{R}) \in \Delta_{\mathcal{U}}$, let $\emptyset \neq S \subseteq N$, and let $x \in \mathcal{PM}(N, v, \mathcal{R})$. Denote $w = v^{\mathcal{R}}_{S,x}$. We have to prove that $x^S \in \mathcal{PM}\left(S, w, \mathcal{R}_{|S}\right)$. Thus, let $k, \ell \in R \in \mathcal{R}_{|S}$, $k \neq \ell$, and let (P, y) be an objection of k against ℓ at x^S with respect to $\left(S, w, \mathcal{R}_{|S}\right)$. Then there exists $P_1 \subseteq N \setminus S$ such that $w(P) = v(P \cup P_1) - x(P_1)$. Let $\widehat{P} = P \cup P_1$ and let $\hat{y} \in \mathbb{R}^{\widehat{P}}$ be defined by $\hat{y} = \left(y, x^{P_1}\right)$. Then $(\widehat{P}, \hat{y})$ is an objection of k against ℓ at x with respect to $(N, v, \mathcal{R})$. As $x \in \mathcal{PM}(N, v, \mathcal{R})$, player ℓ has a counterobjection $(\widehat{Q}, \hat{z})$ to this objection. Let $Q = \widehat{Q} \cap S$ and $z = \hat{z}^S$. Then

$$w(Q) \geq v(\widehat{Q}) - x(\widehat{Q} \setminus Q) \geq v(\widehat{Q}) - \hat{z}(\widehat{Q} \setminus Q) \geq z(Q).$$

We conclude that (Q, z) is a counterobjection to (P, y) with respect to $(S, w, \mathcal{R}_{|S})$.

q.e.d.

Example 4.1.12. Let $g \widehat{=} (4; 1, 1, 1, 1, 1, 1, 1)$ be the seven-player simple majority game and let (N, v) be the corresponding coalitional game. Then $(-1/5, 1/5, \ldots, 1/5) \in \mathcal{PM}(N, v)$. Thus, $\mathcal{PM}$ does *not* satisfy individual rationality. As (N, v) is superadditive, this is a serious drawback. At present, we do not know of a condition that guarantees that every member of the unconstrained bargaining set is individually rational.

Example 4.1.12 leads to the following definition. Let $(N, v, \mathcal{R})$ be a game with coalition structure. We denote

$$I(N, v, \mathcal{R}) = \{x \in X(N, v, \mathcal{R}) \mid x^i \geq v(\{i\}) \text{ for all } i \in N\},$$

which is the set of individually rational payoff vectors in $X(N, v, \mathcal{R})$. Clearly, $I(N, v, \mathcal{R}) \neq \emptyset$ if and only if $v(R) \geq \sum_{i \in R} v(\{i\})$ for every $R \in \mathcal{R}$. In particular, if (N, v) is 0-monotonic (see Exercise 2.1.5), then $I(N, v, \mathcal{R}) \neq \emptyset$ for every coalition structure $\mathcal{R}$ of N.

We are now going to define the bargaining set which is a subsolution of the unconstrained bargaining set. (Let σ and $\widehat{\sigma}$ be solutions on a set Δ of games with coalition structures. Then σ is a *subsolution* of $\widehat{\sigma}$, if $\sigma(N, v, \mathcal{R}) \subseteq \widehat{\sigma}(N, v, \mathcal{R})$ for all $(N, v, \mathcal{R}) \in \Delta$.)

Definition 4.1.13. *Let $(N, v, \mathcal{R})$ be a game with coalition structure. The* **bargaining set** *$\mathcal{M}(N, v, \mathcal{R})$ is defined by*

$$\mathcal{M}(N, v, \mathcal{R}) = I(N, v, \mathcal{R}) \cap \mathcal{PM}(N, v, \mathcal{R}).$$

Thus $\mathcal{M}(N, v, \mathcal{R})$ is the set of all stable members of $I(N, v, \mathcal{R})$.

Remark 4.1.14. Several bargaining sets were introduced by Aumann and Maschler (1964). In Davis and Maschler (1967) the bargaining set $\mathcal{M}(N, v, \mathcal{R})$ is denoted $\mathcal{M}_1^{(i)}(N, v, \mathcal{R})$. The simpler notation is frequently used in the literature (see, e.g., Kahan and Rapoport (1984)).

Remark 4.1.15. Clearly, $\mathcal{C}(N, v, \mathcal{R}) \subseteq \mathcal{M}(N, v, \mathcal{R})$ for every game with coalition structure $(N, v, \mathcal{R})$. Also, $\mathcal{M}$ satisfies COV and AN.

We now proceed to examine whether $\mathcal{M}$ is reasonable (REAB).

Definition 4.1.16. *Let (N, v) be a game. A player $i \in N$ is a* **null player** *(of (N, v)) if $v(S) = v(S \cup \{i\})$ for every $S \subseteq N$.*

Definition 4.1.17. *A solution σ on a set Δ of games with coalition structures satisfies the* **null player property** *(NP) if for every member $(N, v, \mathcal{R}) \in \Delta$, for every $x \in \sigma(N, v, \mathcal{R})$, and for every null player i of (N, v), $x^i = 0$.*

Remark 4.1.18. A solution σ on a set Δ of games with coalition structures satisfies the *dummy property* if for every $(N, v, \mathcal{R}) \in \Delta$, for every $x \in \sigma(N, v, \mathcal{R})$, and for every *dummy* i (that is, a player $i \in N$ satisfying $v(S \cup \{i\}) = v(S) + v(\{i\})$ for every $S \subseteq N \setminus \{i\}$), $x^i = v(\{i\})$. NP is implied by the dummy property, because a null player is a dummy. Conversely, a solution that satisfies NP and COV also satisfies the dummy property, because a dummy of a game with coalition structure $(N, v, \mathcal{R})$ is likewise a dummy of every game with coalition structure which is strategically equivalent to $(N, v, \mathcal{R})$. As our interest is mainly restricted to covariant solutions, it is sufficient to check NP, which is weaker than the dummy property.

Note that RE (see Definition 2.3.9) implies NP and the dummy property. As we do not distinguish between a game (N, v) and $(N, v, \{N\})$ we denote $I(N, v, \{N\}) = I(N, v)$ and $\mathcal{M}(N, v, \{N\}) = \mathcal{M}(N, v)$.

Example 4.1.19. Let (N, v) be the coalitional game associated with the six-person weighted majority game $g \widehat{=} (3; 1, 1, 1, 1, 1, 0)$. Then 6 is a null player. Nevertheless,

$$(1/7, \ldots, 1/7, 2/7) \in \mathcal{M}(N, v).$$

Thus $\mathcal{M}$ does not satisfy NP. Hence it does not satisfy REAB.

Exercises

Exercise 4.1.1. Let (N, v) be a *coalitional simple game*, that is, the coalitional game which is associated with a simple game $(N, \mathcal{W})$. Prove that $\mathcal{C}(N, v) \neq \emptyset$ if and only if the game is weak (see Definition 2.2.7). Describe the core of a weak simple game explicitly.

Exercise 4.1.2. Let $N = \{1, 2\}$ and (N, v_1), (N, v_{-1}), (N, v_0) be the 0-normalized games defined by $v_1(N) = 1$, $v_{-1}(N) = -1$, $v_0(N) = 0$. Determine the unconstrained bargaining sets of (N, v_j), $j = -1, 0, 1$ explicitly.

Exercise 4.1.3. Prove that $\mathcal{PM}(N, v) = \mathcal{C}(N, v)$ for any three-person balanced game (N, v).

Exercise 4.1.4. Let (N, v) be a superadditive three-person game satisfying $\mathcal{C}(N, v) = \emptyset$. Prove that $|\mathcal{PM}(N, v)| = 1$. (Hint: Show that the unique $x \in X(N, v)$ satisfying $v(S) - x(S) = v(T) - x(T)$ for all $S, T \subseteq N$ with $|S| = |T| = 2$ is the unique member of $\mathcal{PM}(N, v)$. See Maschler (1963) for a generalization of this result.)

Exercise 4.1.5. Prove that $\mathcal{PM}$ satisfies COV and AN on any set of games with coalition structures.

Exercise 4.1.6. Prove that (4.1.2) may be replaced by

$$y^i > x^i \text{ for all } i \in P \tag{4.1.8}$$

without changing $\mathcal{PM}$. Prove also that (4.1.3) or (4.1.7) can be replaced by

$$y(P) = v(P) \quad \text{or} \tag{4.1.9}$$
$$z(Q) = v(Q), \tag{4.1.10}$$

respectively, without changing the definition of the bargaining set.

Exercise 4.1.7. Let (N, v) be the game of Example 4.1.19. Find a vector $x \in \mathcal{PM}(N, v)$ such that $x^6 < 0$.

4.2 Existence of the Bargaining Set

Throughout this section let $(N, v, \mathcal{R})$ be a game with coalition structure. We shall prove that $\mathcal{M}(N, v, \mathcal{R})$ is nonempty if $I(N, v, \mathcal{R})$ is nonempty.

The following definition is useful.

Definition 4.2.1. *Let $x \in X(N, v, \mathcal{R})$ and $k, \ell \in R \in \mathcal{R}$, $k \neq \ell$. An objection (P, y) of k against ℓ at x is* **justified** *if ℓ has no counterobjection against (P, y). If k has a justified objection against ℓ at x we say that k is* **stronger than ℓ at** *x and write $k \succ_x^{\mathcal{M}} \ell$.*

Note that $x \in \mathcal{M}(N, v, \mathcal{R})$ if and only if $x \in I(N, v, \mathcal{R})$ and no player has a justified objection against any of his partners at x. We continue with an investigation of the binary relations $\succ_x^{\mathcal{M}}$ and we first prove that $\succ_{\cdot}^{\mathcal{M}}$ is continuous.

Lemma 4.2.2. *The set $\{x \in X(N, v, \mathcal{R}) \,|\, k \succ_x^{\mathcal{M}} \ell\}$ is open relative to $X(N, v, \mathcal{R})$ for all $k, \ell \in R$, $k \neq \ell$, for all $R \in \mathcal{R}$.*

Proof: Let $x \in X(N, v, \mathcal{R})$, $R \in \mathcal{R}$, and $k, \ell \in R$, $k \neq \ell$. Assume that $k \succ_x^{\mathcal{M}} \ell$. Let (P, y) be a justified objection of k against ℓ at x. Here we employ (4.1.8) in the definition of an objection. For every $z \in X(N, v, \mathcal{R})$ define

$$f(z) = \max\{v(Q) - y(Q \cap P) - z(Q \setminus P) \mid Q \in \mathcal{T}_{\ell k}\}.$$

By Exercise 4.1.6, (P, y) is a justified objection of k against ℓ at z if and only if $y^i > z^i$ for all $i \in P$ and $f(z) < 0$. Hence $y^i > x^i$ for all $i \in P$ and $f(x) < 0$. Let $\|\cdot\|$ denote the Euclidean norm of $\mathbb{R}^N$. As f is continuous, there exists $\delta > 0$ such that $f(z) < 0$ for every $z \in X(N, v, \mathcal{R})$ satisfying $\|z - x\| < \delta$. Moreover, if δ is small enough, then $y^i > z^i$ for all $i \in P$. Hence, the pair (P, y) is a justified objection of k against ℓ at z for every $z \in X(N, v, \mathcal{R})$ satisfying $\|z - x\| < \delta$. **q.e.d.**

In order to show that the relations $\succ_x^M$ are acyclic we need the following definition which will also be used in subsequent chapters.

Definition 4.2.3. *Let $S \subseteq N$ and $x \in \mathbb{R}^N$. The* **excess** *of S at x (with respect to (N,v)) is $e(S,x,v) = v(S) - x(S)$.*

Note that a positive excess $e(S,x,v)$ may be interpreted as the *dissatisfaction* of the coalition S when faced with the proposal x. Clearly, $e(\emptyset,x,v) = 0$.

The following lemma shows that a necessary condition for the existence of a justified objection can be expressed with the help of excesses.

Lemma 4.2.4. *Let $x \in X(N,v,\mathcal{R})$, let $k,\ell \in R \in \mathcal{R}$, $k \neq \ell$, and let (P,y) be a justified objection of k against ℓ at x. If $Q \subseteq N$ satisfies $\ell \in Q$ and $e(Q,x,v) \geq e(P,x,v)$, then $k \in Q$.*

Proof: Suppose, on the contrary, that $k \notin Q$. Define $z \in \mathbb{R}^Q$ by

$$z^i = \begin{cases} x^i & \text{, if } i \in Q \setminus P \text{ and } i \neq \ell \\ y^i & \text{, if } i \in Q \cap P \\ v(Q) - z(Q \setminus \{\ell\}) & \text{, if } i = \ell. \end{cases}$$

We shall prove that $z^\ell \geq x^\ell$. Indeed,

$$\begin{aligned} & z^\ell - x^\ell \\ &= v(Q) - z(Q \setminus \{\ell\}) - x^\ell \\ &= v(Q) - y(Q \cap P) - x(Q \setminus P) \\ &\geq v(Q) - \Big(v(P) - y(P \setminus Q)\Big) - x(Q \setminus P) \quad \text{(because } v(P) \geq y(P)\text{)} \\ &\geq v(Q) - x(Q \setminus P) - \Big(v(P) - x(P \setminus Q)\Big) \\ &= e(Q,x,v) - e(P,x,v) \geq 0. \end{aligned}$$

Thus, (Q,z) is a counterobjection to (P,y) and the desired contradiction is obtained. **q.e.d.**

Lemma 4.2.4 leads to the following definition.

Definition 4.2.5. *Let $k,\ell \in N$, $k \neq \ell$, and let $x \in \mathbb{R}^N$. The* **maximum surplus** *of k over ℓ at x (with respect to (N,v)) is*

$$s_{k\ell}(x,v) = \max\{e(S,x,v) \mid S \in \mathcal{T}_{k\ell}\}.$$

If $k,\ell \in R \in \mathcal{R}$ and if $s_{k\ell}(x,v) > s_{\ell k}(x,v)$, then we say that k **outweighs** *ℓ at x and write $k \succ_x^{\mathcal{K}} \ell$.*

If $s_{k\ell}(x,v) \geq 0$, then it is the maximum amount of money that k can use in an objection or counterobjection against ℓ. Thus, in this case $s_{k\ell}(x,v)$ may serve as a measure of the strength of k against ℓ at x. Hence, if k outweighs ℓ at x and $s_{k\ell}(x,v) > 0$, then the strength of k against ℓ is greater than the strength of ℓ against k. In Chapter 5 an interpretation of the surplus, when it is negative, will be presented and the surplus will be used to define a further solution.

The following result is a consequence of Lemma 4.2.4.

Corollary 4.2.6. *Let $x \in X(N,v,\mathcal{R})$ and let $k,\ell \in R \in \mathcal{R}$, $k \neq \ell$. If $s_{k\ell}(x,v) \leq s_{\ell k}(x,v)$, then k has no justified objection against ℓ at x.*

The following lemma enables us to show that the relations $\succ_x^{\mathcal{M}}$ are acyclic.

Lemma 4.2.7. *Let $x \in X(N,v,\mathcal{R})$. Then the relation $\succ_x^{\mathcal{K}}$ is transitive.*

Proof: Let $R \in \mathcal{R}$ and $k,\ell,m \in R$ be distinct players such that

$$s_{k\ell}(x,v) > s_{\ell k}(x,v) \text{ and } s_{\ell m}(x,v) > s_{m\ell}(x,v).$$

It has to be shown that $s_{km}(x,v) > s_{mk}(x,v)$. Let

$$\mathcal{D} = \{D \subseteq N \mid 1 \leq |D \cap \{k,\ell,m\}| \leq 2\}$$

and let $T \in \mathcal{D}$ satisfy

$$e(T,x,v) = \max\{e(D,x,v) \mid D \in \mathcal{D}\}.$$

We claim that $\ell \in T$ implies that $k \in T$. Indeed, assume that $\ell \in T$ and $k \notin T$. Then $s_{\ell k}(x,v) = e(T,x,v) \geq s_{k\ell}(x,v)$, contradicting the assumption that $s_{k\ell}(x,v) > s_{\ell k}(x,v)$. Similarly, it can be shown that $m \in T$ implies that $\ell \in T$. By the definition of $\mathcal{D}$ it now follows that $k \in T$ and $m \notin T$. Hence $s_{km}(x,v) = e(T,x,v) > s_{mk}(x,v)$. **q.e.d.**

We now recall the definition of acyclicity.

Definition 4.2.8. *Let $\succ$ be a binary relation on a set B. The relation $\succ$ is* **acyclic** *if the following condition is satisfied: If $x_1,\ldots,x_k \in B$, $k \geq 2$, and $x_i \succ x_{i+1}$ for all $i = 1,\ldots,k-1$, then $x_k \succ x_1$ does not hold. The relation $\succ$ is* **asymmetric**, *if $x \succ y$ implies that $y \succ x$ is not true.*

Lemma 4.2.9. *For every $x \in X(N,v,\mathcal{R})$ the relation $\succ_x^{\mathcal{M}}$ is acyclic.*

Proof: Let $k,\ell \in R \in \mathcal{R}$, $k \neq \ell$. By Corollary 4.2.6, $k \succ_x^{\mathcal{M}} \ell$ implies that $k \succ_x^{\mathcal{K}} \ell$. Now, $\succ_x^{\mathcal{K}}$ is acyclic, because it is transitive and asymmetric. Hence, the acyclicity of $\succ_x^{\mathcal{M}}$ is implied by that of $\succ_x^{\mathcal{K}}$. **q.e.d.**

Lemma 4.2.7 is due to Davis and Maschler (1965) and Lemma 4.2.9 was proved in Davis and Maschler (1967).

The following notation is useful. For $i \in R \in \mathcal{R}$ let

$$E_i = \{x \in I(N,v,\mathcal{R}) \mid \text{there is no } j \in R \setminus \{i\} \text{ such that } j \succ_x^{\mathcal{M}} i\}.$$

Lemma 4.2.10. *Let $i \in N$ and $R \in \mathcal{R}$. Then E_i is closed and*

$$\{x \in I(N,v,\mathcal{R}) \mid x^i = v(\{i\})\} \subseteq E_i, \quad \bigcup_{j \in R} E_j = I(N,v,\mathcal{R}).$$

Proof: By Lemma 4.2.2, E_i is a closed subset of $I(N,v,\mathcal{R})$. Also, if $x \in I(N,v,\mathcal{R})$, $i \in R \in \mathcal{R}$, and $x^i = v(\{i\})$, then $(\{i\}, v(\{i\}))$ is a counter-objection to any possible objection of a player $j \in R \setminus \{i\}$ against i at x. Thus, no $j \in R \setminus \{i\}$ has a justified objection against i and, therefore, $x \in E_i$. Finally, let $R \in \mathcal{R}$ and let $x \in I(N,v,\mathcal{R})$. As R is finite and in view of Lemma 4.2.9 there exists $j \in R$ such that no $k \in R \setminus \{j\}$ has a justified objection against j at x. Thus, $x \in E_j$ and $I(N,v,\mathcal{R}) \subseteq \bigcup_{j \in R} E_j$. **q.e.d.**

We define the metric $\langle \cdot, \cdot \rangle$ on $\mathbb{R}^N$ by $\langle x, y \rangle = \max\left\{ |x^i - y^i| \;\middle|\; i \in N \right\}$ for all $x, y \in \mathbb{R}^N$. For every nonempty and closed set $A \subseteq \mathbb{R}^N$ and every $x \in \mathbb{R}^N$ we denote

$$\rho(x, A) = \min\{\langle x, y \rangle \mid y \in A\} \tag{4.2.1}$$

and remark that $\rho(\cdot, A) : \mathbb{R}^N \to \mathbb{R}$ is continuous.

Lemma 4.2.11. *Let A_i, $i \in N$, be closed subsets of $I(N,v,\mathcal{R})$ and assume that $I(N,v,\mathcal{R}) \neq \emptyset$. If, for all $i \in N$ and every $R \in \mathcal{R}$,*

$$\{x \in I(N,v,\mathcal{R}) \mid x^i = v(\{i\})\} \subseteq A_i \text{ and } \bigcup_{j \in R} A_j = I(N,v,\mathcal{R}),$$

then $\bigcap_{i \in N} A_i \neq \emptyset$.

Proof: We define a mapping $\eta : I(N,v,\mathcal{R}) \to I(N,v,\mathcal{R})$ as follows. Let $x \in I(N,v,\mathcal{R})$ and $i \in R \in \mathcal{R}$. Then $y = \eta(x)$ is given by

$$y^i = x^i - \rho(x, A_i) + \sum_{j \in R} \frac{\rho(x, A_j)}{|R|}.$$

Then $x^i - \rho(x, A_i) \geq v(\{i\})$, because $\{x \in I(N,v,\mathcal{R}) \mid x^i = v(\{i\})\} \subseteq A_i$. Hence, $y^i \geq v(\{i\})$. Also, $y(R) = x(R)$. Thus, η is well defined. Moreover, η is continuous. By Brouwer's fixed-point theorem η has a fixed point x_0. We shall prove that $x_0 \in \bigcap_{j \in N} A_j$. Assume, on the contrary, that there exists $j \in N$ such that $x_0 \notin A_j$. Then $\rho(x_0, A_j) > 0$, because A_j is closed and nonempty. Let $R \in \mathcal{R}$ contain j. There exists $i \in R$ such that $x_0 \in A_i$, because $\bigcup_{k \in R} A_k = I(N,v,\mathcal{R})$. Thus, $\rho(x_0, A_i) = 0$. Hence

$$x_0^i = x_0^i - \rho(x_0, A_i) + \sum_{k \in R} \frac{\rho(x_0, A_k)}{|R|} \geq x_0^i + \frac{\rho(x_0, A_j)}{|R|}$$

and the desired contradiction is obtained. **q.e.d.**

Now we are able to prove the existence result of the bargaining set.

Theorem 4.2.12. *Let $(N, v, \mathcal{R})$ be a game with coalition structure. Assume that $I(N, v, \mathcal{R}) \neq \emptyset$. Then $\mathcal{M}(N, v, \mathcal{R}) \neq \emptyset$.*

Proof: As $I(N, v, \mathcal{R}) \neq \emptyset$, $v(R) \geq \sum_{j \in R} v(\{j\})$ for every $R \in \mathcal{R}$. By Lemma 4.2.10 the sets $A_i = E_i$, $i \in N$, satisfy all the conditions of Lemma 4.2.11. Hence $\bigcap_{j \in N} E_j \neq \emptyset$. The obvious fact that $\bigcap_{j \in N} E_j = \mathcal{M}(N, v, \mathcal{R})$ finishes the proof.
q.e.d.

Corollary 4.2.13. *Let (N, v) be a game. If $v(S) \geq \sum_{i \in S} v(\{i\})$ for all $\emptyset \neq S \subseteq N$, then $\mathcal{M}(N, v, \mathcal{R}) \neq \emptyset$ for every coalition structure $\mathcal{R}$ of N.*

Lemma 4.2.11 and Theorem 4.2.12 were proved in Peleg (1967b).

Remark 4.2.14. Note that, if $\mathcal{R} = \{N\}$, then Lemma 4.2.11 is equivalent to the KKM Lemma (see, e.g., Border (1985) or Peleg (1967a)). Hence Lemma 4.2.11 may be seen as a generalization of the KKM Lemma to a Cartesian product of finitely many simplices. Also, it should be remarked that Davis and Maschler (1967) used the KKM Lemma to prove Theorem 4.2.12 in the case $\mathcal{R} = \{N\}$.

Exercises

Let $(N, v, \mathcal{R})$ be a game with coalition structure and let $\alpha \in \mathbb{R}^N$. Denote

$$X_\alpha(N, v, \mathcal{R}) = \{x \in X(N, v, \mathcal{R}) \mid x^i \geq \alpha^i \text{ for all } i \in N\}.$$

If $\alpha = 0 \in \mathbb{R}^N$, then $X_\alpha(N, v, \mathcal{R})$ is the set of *pseudo-imputations* of $(N, v, \mathcal{R})$.

Exercise 4.2.1. For $i \in R \in \mathcal{R}$ denote

$$\begin{aligned} E_i &= \{x \in X_\alpha(N, v, \mathcal{R}) \mid \text{there is no } j \in R \setminus \{i\} \text{ such that } j \succ_x^{\mathcal{M}} i\}, \\ X_i &= \{x \in X_\alpha(N, v, \mathcal{R}) \mid x^i = \alpha^i\}. \end{aligned}$$

Show the following modification of Lemma 4.2.10: If $v(\{i\}) \geq \alpha^i$ for all $i \in N$, then

$$\left(i \in N \text{ and } R \in \mathcal{R}\right) \Rightarrow \left(E_i \text{ is closed}, X_i \subseteq E_i, \text{ and } \bigcup_{j \in R} E_j = X_\alpha(N, v, \mathcal{R})\right).$$

Exercise 4.2.2. Using the same notation as in the preceding exercise prove the following modification of Lemma 4.2.11: If $A_i, i \in N$, are closed subsets of $X_\alpha(N, v, \mathcal{R})$ and if $X_\alpha(N, v, \mathcal{R}) \neq \emptyset$, then

$$\left(X_i \subseteq A_i \text{ and } \bigcup_{j \in R} A_j = X_\alpha(N, v, \mathcal{R}) \text{ for all } i \in N,\ R \in \mathcal{R}\right) \Rightarrow \bigcap_{j \in N} A_j \neq \emptyset.$$

Exercise 4.2.3. Prove that $\mathcal{PM}(N, v, \mathcal{R}) \neq \emptyset$ by using Exercises 4.2.1 and 4.2.2.

Exercise 4.2.4. For any game (N, v), any $x \in X(N, v)$, and any $k, \ell \in N$, $k \neq \ell$, we define $k \succeq_x^{\mathcal{K}} \ell$ if $s_{k\ell}(x, v) \geq s_{\ell k}(x, v)$ (compare with Definition 4.2.5). Show that there exist a game (N, v) and $x \in X(N, v)$ such that $\succeq_x^{\mathcal{K}}$ is not transitive.

4.3 Balanced Superadditive Games and the Bargaining Set

In this section we shall prove that there are some interesting classes of balanced superadditive games on which the bargaining set (for the grand coalition) coincides with the core. We start with the following definition.

Definition 4.3.1. *Let (N, v) be a game and let $x \in \mathbb{R}^N$. The* **monotonic cover** *of (N, v) is the game (N, w) defined by*

$$w(S) = \max\{v(R) \mid R \subseteq S\} \text{ for all } S \subseteq N.$$

The **excess game** *of (N, v) at x is the game (N, u) defined by*

$$u(S) = e(S, x, v) \text{ for all } S \subseteq N.$$

The **maximum excess game** *of (N, v) at x is the monotonic cover of the excess game at x, denoted by (N, v_x).*

Clearly, the monotonic cover of a game is a monotonic game. The following result is applied in the sequel.

Theorem 4.3.2 (Solymosi (1999))**.** *Let (N, v) be a superadditive game and let $x \in \mathcal{PM}(N, v)$. Then $x \in \mathcal{C}(N, v)$ if and only if (N, v_x) is balanced.*

Proof: If $x \in \mathcal{C}(N, v)$, then $v_x(S) = 0$ for every $S \subseteq N$. Hence $\mathcal{C}(N, v_x) \neq \emptyset$.

Let, now, $\mathcal{C}(N, v_x) \neq \emptyset$. Assume, on the contrary, that $x \notin \mathcal{C}(N, v)$, so $v_x(N) > 0$. Let $\bar{x} \in \mathcal{C}(N, v_x)$ and denote $P = \{i \in N \mid \bar{x}^i > 0\}$. Then $P \neq \emptyset$, because $\bar{x} \geq 0$ and $\bar{x}(N) = v_x(N)$. Moreover, for every $S \subseteq N$ satisfying $e(S, x, v) = v_x(N)$, $P \subseteq S$. Indeed,

$$e(S, x, v) \leq v_x(S) \leq \bar{x}(S) \leq \bar{x}(N) = v_x(N) = e(S, x, v),$$

so $\bar{x}^i = 0$ for all $i \in N \setminus S$. Let $\widehat{S} \subseteq N$ be maximal (with respect to set inclusion) such that $e(\widehat{S}, x, v) = v_x(N)$. Then $\bar{x}(\widehat{S}) = v_x(\widehat{S})$ and $\emptyset \neq \widehat{S} \neq N$, because x is Pareto optimal. Therefore there exist $k \in P$ and $\ell \in N \setminus \widehat{S}$.

For every $T \subseteq N$ satisfying $T \neq \emptyset = T \cap \widehat{S}$, superadditivity of (N,v) and maximality of $\widehat{S}$ imply that

$$e(\widehat{S},x,v) + e(T,x,v) \leq e(\widehat{S} \cup T,x,v) < e(\widehat{S},x,v).$$

Hence $e(T,x,v) < 0$. Let $y \in \mathbb{R}^{\widehat{S}}$ be defined by

$$y^i = \begin{cases} x^k + \frac{\bar{x}^k}{|\widehat{S}|} & \text{, if } i = k \\ x^i + \bar{x}^i + \frac{\bar{x}^k}{|\widehat{S}|} & \text{, if } i \in \widehat{S} \setminus \{k\}. \end{cases}$$

Then $y^i > x^i$ for all $i \in \widehat{S}$ and $y(\widehat{S}) = v(\widehat{S})$. Hence $(\widehat{S},y)$ is an objection of k against ℓ at x. As $x \in \mathcal{PM}(N,v)$, player ℓ has a counterobjection (Q,z) to $(\widehat{S},y)$. By (4.1.5) and (4.1.7), $e(Q,x,v) \geq 0$. Hence, $Q \cap \widehat{S} \neq \emptyset$. The observation that

$$\begin{aligned} z(Q) &\geq x(Q \setminus \widehat{S}) + y(Q \cap \widehat{S}) \\ &= x(Q \setminus \widehat{S}) + x(Q \cap \widehat{S}) + \bar{x}(Q \cap \widehat{S}) + \tfrac{|Q \cap \widehat{S}|}{|\widehat{S}|}\bar{x}^k \\ &> x(Q) + \bar{x}(Q) \quad \text{(because } \bar{x}(Q \setminus \widehat{S}) = 0 \text{ and } \bar{x}^k > 0) \\ &\geq x(Q) + e(Q,x,v) \quad \text{(because } \bar{x}(Q) \geq v_x(Q) \geq e(Q,x,v)) \\ &= v(Q) \end{aligned}$$

contradicts (4.1.7). **q.e.d.**

Corollary 4.3.3. *Let (N,v) be a superadditive game. Then*
(1) $\mathcal{PM}(N,v) = \mathcal{C}(N,v)$ if and only if $\mathcal{C}(N,v_x) \neq \emptyset$ for every $x \in \mathcal{PM}(N,v)$;
(2) $\mathcal{M}(N,v) = \mathcal{C}(N,v)$ if and only if $\mathcal{C}(N,v_x) \neq \emptyset$ for every $x \in \mathcal{M}(N,v)$.

Proof: Assertion (1) follows from Theorem 4.3.2 and (2) follows from the fact that $\mathcal{M}$ is a subsolution of $\mathcal{PM}$. **q.e.d.**

Corollary 4.3.3 is due to Solymosi (1999) and can be applied to verify the coincidence of the core and the bargaining set in many cases. Two examples are presented.

Theorem 4.3.4. *If (N,v) is a convex game, then $\mathcal{PM}(N,v) = \mathcal{C}(N,v)$.*

Proof: Let (N,v) be a convex game, $x \in \mathcal{PM}(N,v)$, and $S,T \subseteq N$. Then

$$\begin{aligned} v_x(S) + v_x(T) &= \max\{e(P,x,v) + e(Q,x,v) \mid P \subseteq S,\ Q \subseteq T\} \\ &\leq \max\{e(P \cup Q,x,v) + e(P \cap Q,x,v) \mid P \subseteq S,\ Q \subseteq T\} \\ &= v_x(S \cup T) + v_x(S \cap T), \end{aligned}$$

where the inequality is implied by convexity of (N,v). Thus (N,v_x) is a convex game which has a nonempty core by Theorem 3.5.1. Corollary 4.3.3 completes the proof. **q.e.d.**

Theorem 4.3.4 is due to Maschler, Peleg, and Shapley (1972). In fact, a generalization of their proof yields Theorem 4.3.2.

Another class of games, which is closed under taking maximal excess games at imputations, is the class of assignment games (defined in Exercise 3.4.1).

Theorem 4.3.5. *If* (N, v) *is an assignment game, then* $\mathcal{M}(N, v) = \mathcal{C}(N, v)$.

Proof: Let (N, v) be an assignment game with respect to S, B defined by the nonnegative matrix $\left(c_{\{i,j\}}\right)_{i \in S, j \in B}$ of net profits, and let $x \in \mathcal{M}(S \cup B, v)$. For every $i \in S$ and every $j \in B$ denote $\hat{c}_{\{i,j\}} = \left(c_{\{i,j\}} - x_i - x_j\right)_+$. Let (N, w) be the assignment game defined by the matrix $\left(\hat{c}_{\{i,j\}}\right)_{i \in S, j \in B}$. It is left to the reader (Exercise 4.3.1) to prove that

$$w = v_x. \tag{4.3.1}$$

Corollary 4.3.3 completes the proof, because an assignment game is balanced by Theorem 3.4.3 and Exercise 3.4.1. **q.e.d.**

Exercises

Exercise 4.3.1. Prove (4.3.1).

Exercise 4.3.2. A game (N, v) is *veto-controlled*, if there exists $i \in N$ such that $v(S) = 0$ for every $S \subseteq N \setminus \{i\}$. Show that, if (N, v) is a veto-controlled monotonic game, then $\mathcal{M}(N, v) = \mathcal{C}(N, v)$. (See Solymosi (1999).)

4.4 Further Bargaining Sets

In this section first two other bargaining sets, the definitions of which are based on objections and counterobjections, are presented. The definitions differ only inasmuch as "justified objections" are defined differently.

Finally, a bargaining set is described which is based on "global" objections and counterobjections at the given proposal. These "global" objections and counterobjections do not refer to a pair of distinct players as in the "classical" context.

We shall sometimes refer to $\mathcal{M}$ and $\mathcal{PM}$ by calling them the *classical* bargaining or prebargaining set in order to distinguish them from other bargaining sets.

4.4.1 The Reactive and the Semi-reactive Bargaining Set

Let $(N, v, \mathcal{R})$ be a game with coalition structure, $x \in X(N, v, \mathcal{R})$, and $k, \ell \in R \in \mathcal{R}$, $k \neq \ell$. We say that k has an objection against ℓ at x *via* P, if k has an objection (P, y) against ℓ at x. Also, we say that ℓ can counterobject *via* $Q \in \mathcal{T}_{\ell k}(N)$ to an objection (P, y), if there exists a counterobjection (Q, z) to (P, y).

Notation 4.4.1. Let $\succ$ be a binary relation on a set A. Then the negation of $\succ$ is denoted by $\preceq$.

The reactive bargaining set is defined as follows.

Definition 4.4.2. *Let $(N, v, \mathcal{R})$ be a game with coalition structure, let $x \in X(N, v, \mathcal{R})$, and $k, \ell \in R \in \mathcal{R}, k \neq \ell$. Then k has a* **justified objection** *against ℓ* **in the sense of the reactive bargaining set** *at x (abbreviated $k \succ_x^{\mathcal{M}_r} \ell$), if for every $Q \in \mathcal{T}_{\ell k}(N)$ there exists an objection (P, y) of k against ℓ such that ℓ cannot counterobject to (P, y) via Q. The* **reactive prebargaining set** *of $(N, v, \mathcal{R})$ is the set*

$$\mathcal{PM}_r(N, v, \mathcal{R}) = \{x \in X(N, v, \mathcal{R}) \mid k \preceq_x^{\mathcal{M}_r} \ell \text{ for all } k, \ell \in R \in \mathcal{R}, k \neq \ell\}.$$

The expression "reactive" bargaining set refers to the definition of "justified" objections. With respect to the classical bargaining set $\mathcal{M}$ the objector k announces the objection in advance, whereas with respect to the reactive bargaining set the objector is allowed to "react" to the announcement of his partner ℓ, that is, to wait until ℓ announces the coalition which he plans to use to counterobject.

Hence the reactive (pre)bargaining set is a subsolution of the (pre)bargaining set, that is, for every game with coalition structure $(N, v, \mathcal{R})$,

$$\mathcal{PM}_r(N, v, \mathcal{R}) \subseteq \mathcal{PM}(N, v, \mathcal{R}),$$

and using the notation $\mathcal{M}_r(N, v, \mathcal{R}) = \mathcal{PM}_r(N, v, \mathcal{R}) \cap I(N, v, \mathcal{R})$ for the *reactive bargaining set* of the game with coalition structure,

$$\mathcal{M}_r(N, v, \mathcal{R}) \subseteq \mathcal{M}(N, v, \mathcal{R}).$$

Definition 4.4.2 is due to Granot (1994).

Example 4.4.3. Let $N = \{1, \ldots, 7\}$ and $(N, \mathcal{W})$ be the simple game defined by

$$\mathcal{W}^m = \{\{1,2,5\}, \{1,3,4\}, \{1,6,7\}, \{2,3,6\}, \{2,4,7\}, \{3,5,7\}, \{4,5,6\}\}.$$

Let (N, v) be the associated monotonic coalitional game of $(N, \mathcal{W})$ which is the *projective* seven-person game, introduced by von Neumann and Morgenstern (1953). Indeed, the minimal winning coalitions are the members of the lines

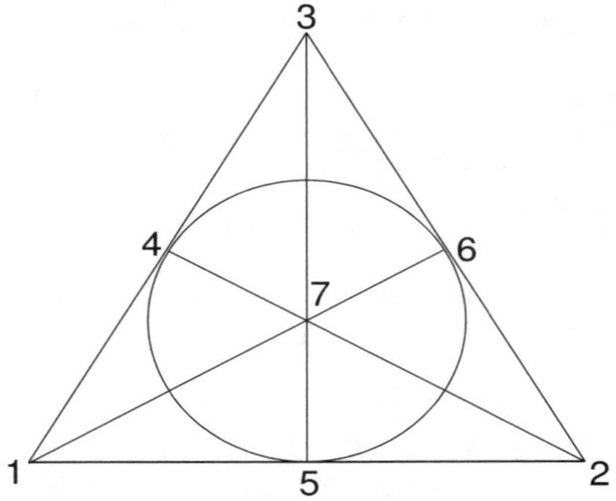

Fig. 4.4.1. The Projective Seven-Person Game

in Figure 4.4.1. In Chapter 5 we shall show that $\mathcal{M}_r(N,v)$ is star-shaped. In fact (see Granot and Maschler (1997)),

$$\mathcal{M}_r(N,v) = \bigcup_{S \in \mathcal{W}^m} \text{convh} \left\{ \frac{1}{7}\chi_N, \frac{1}{3}\chi_S \right\}. \qquad (4.4.1)$$

The center $(1/7,\ldots,1/7)$ of this star is the symmetric imputation of the game and, in any element $1/3\chi_S$, $S \in \mathcal{W}^m$, the members of S share $v(S)$ equally. Granot and Maschler (1997) show that convh $\mathcal{M}_r(N,v) \subseteq \mathcal{M}(N,v)$.

The following definition shows that there exists a "natural" solution concept between the reactive and the classical bargaining set.

Definition 4.4.4. *Let $(N,v,\mathcal{R})$ be a game with coalition structure, let $x \in X(N,v,\mathcal{R})$, and $k,\ell \in R \in \mathcal{R}, k \neq \ell$. Then k has a* **justified objection** *against ℓ* **in the sense of the semi-reactive bargaining set** *at x (abbreviated $k \succ_x^{\mathcal{M}_{sr}} \ell$), if there exists $P \in \mathcal{T}_{k\ell}(N)$ satisfying the following properties: (1) Player k has an objection via P and (2) for every $Q \in \mathcal{T}_{\ell k}(N)$ there exists an objection (P,y) of k against ℓ at x such that ℓ cannot counter via Q. The* **semi-reactive prebargaining set** *of $(N,v,\mathcal{R})$ is the set*

$$\mathcal{PM}_{sr}(N,v,\mathcal{R}) = \{x \in X(N,v,\mathcal{R}) \mid k \preceq_x^{\mathcal{M}_{sr}} \ell \text{ for all } k,\ell \in R \in \mathcal{R}, k \neq \ell\}.$$

Let $\mathcal{M}_{sr}(N,v,\mathcal{R}) = I(N,v,\mathcal{R}) \cap \mathcal{PM}_{sr}(N,v,\mathcal{R})$ denote the *semi-reactive bargaining set* of $(N,v,\mathcal{R})$.

Hence, k has a justified objection against ℓ with respect to the semi-reactive bargaining set, if he is able to announce in advance a coalition P which he can use to object against ℓ and then is allowed to wait until the coalition which ℓ plans to use to counterobject is announced. Hence, the reactive (pre)bargaining set is a subsolution of the semi-reactive (pre)bargaining set and the semi-reactive (pre)bargaining set is a subsolution of the classical (pre)bargaining set. Definition 4.4.4 is due to Sudhölter and Potters (2001).

In order to compare the definitions of the classical, the reactive, and the semi-reactive (pre)bargaining set formally, it is useful to describe the relations $\preceq_x^{\mathcal{M}}$, $\preceq_x^{\mathcal{M}_r}$, and $\preceq_x^{\mathcal{M}_{sr}}$ (that is, the negations of $\succ_x^{\mathcal{M}}$, $\succ_x^{\mathcal{M}_r}$, and $\succ_x^{\mathcal{M}_{sr}}$) in detail. The following notation is useful.

Notation 4.4.5. Let N be a coalition and $x, y \in \mathbb{R}^N$. Then we use the following abbreviations:

$$\begin{aligned} x \geq y &\Leftrightarrow x^i \geq y^i \text{ for all } i \in N; \\ x > y &\Leftrightarrow x \geq y \text{ and } x \neq y; \\ x \gg y &\Leftrightarrow x^i > y^i \text{ for all } i \in N. \end{aligned}$$

Let $(N, v, \mathcal{R})$ be a game with coalition structure, let $x \in X(N, v, \mathcal{R})$, let $R \in \mathcal{R}$ and $k, \ell \in R, k \neq \ell$. Then

(1) $k \preceq_x^{\mathcal{M}} \ell$, if

$$\begin{array}{l} \text{for all } P \in \mathcal{T}_{k\ell} \text{ with } e(P, x, v) > 0 \\ \text{for all } y \in \mathbb{R}^P \text{ with } y(P) = v(P),\ y \gg x^P \\ \text{there exists } Q \in \mathcal{T}_{\ell k} \text{ such that} \\ \text{there exists } z \in \mathbb{R}^Q \text{ with } z(Q) = v(Q), z \geq x^Q, z^{P \cap Q} \geq y^{P \cap Q}; \end{array} \tag{4.4.2}$$

(2) $k \preceq_x^{\mathcal{M}_r} \ell$, if

$$\begin{array}{l} \text{there exists } Q \in \mathcal{T}_{\ell k} \text{ such that} \\ \text{for all } P \in \mathcal{T}_{k\ell} \text{ with } e(P, x, v) > 0 \\ \text{for all } y \in \mathbb{R}^P \text{ with } y(P) = v(P),\ y \gg x^P \\ \text{there exists } z \in \mathbb{R}^Q \text{ with } z(Q) = v(Q), z \geq x^Q, z^{P \cap Q} \geq y^{P \cap Q}; \end{array} \tag{4.4.3}$$

(3) $k \preceq_x^{\mathcal{M}_{sr}} \ell$, if

$$\begin{array}{l} \text{for all } P \in \mathcal{T}_{k\ell} \text{ with } e(P, x, v) > 0 \\ \text{there exists } Q \in \mathcal{T}_{\ell k} \text{ such that} \\ \text{for all } y \in \mathbb{R}^P \text{ with } y(P) = v(P),\ y \gg x^P \\ \text{there exists } z \in \mathbb{R}^Q \text{ with } z(Q) = v(Q), z \geq x^Q, z^{P \cap Q} \geq y^{P \cap Q}. \end{array} \tag{4.4.4}$$

Note that (4.4.2), (4.4.3), and (4.4.4) differ from each other only inasmuch as the row "there exists $Q \in \mathcal{T}_{\ell k}$ such that" is the third, first, and second row, respectively. Hence

$$k \preceq_x^{\mathcal{M}_r} \ell \Rightarrow k \preceq_x^{\mathcal{M}_{sr}} \ell \Rightarrow k \preceq_x^{\mathcal{M}} \ell.$$

Therefore, the reactive prebargaining set is a subsolution of the semi-reactive prebargaining set and the latter is a subsolution of the prebargaining set:

$$\begin{array}{ccccc} \mathcal{PM}_r(N,v,\mathcal{R}) & \subseteq & \mathcal{PM}_{sr}(N,v,\mathcal{R}) & \subseteq & \mathcal{PM}(N,v,\mathcal{R}) \\ \mathcal{M}_r(N,v,\mathcal{R}) & \subseteq & \mathcal{M}_{sr}(N,v,\mathcal{R}) & \subseteq & \mathcal{M}(N,v,\mathcal{R}). \end{array} \qquad (4.4.5)$$

Example 4.1.19, Theorem 4.4.8, and Exercise 4.4.7 show that the inclusions of (4.4.5) may be strict.

Remark 4.4.6. Remark 4.1.7 remains valid for the reactive bargaining set, that is, the core is a subsolution of the reactive bargaining set. Also, the (semi-)reactive (pre)bargaining set satisfies COV and AN.

Remark 4.4.7. Both $\mathcal{PM}_r$ and $\mathcal{PM}_{sr}$ satisfy RGP on $\Delta_{\mathcal{U}}$.

The proof of Remark 4.4.7 is left to the reader (see Exercise 4.4.8).

The main advantage of the (semi-)reactive (pre)bargaining set over the classical (pre)bargaining set may be seen in the following result.

Theorem 4.4.8. *Let Γ be a set of superadditive games. The semi-reactive prebargaining set satisfies* IR *on Γ.*

Proof: Let (N,v) be a superadditive game and $x \in \mathcal{PM}_{sr}(N,v)$. We have to show that $x^i \geq v(\{i\})$ for all $i \in N$. Assume, on the contrary, that $x^k < v(\{k\})$ for some player k. Denote $\mu = \max_{S \subseteq N} e(S,x,v)$. Then $\mu > 0$, because $e(\{k\},x,v) > 0$. Also, by superadditivity of (N,v),

$$(S \subseteq N,\ e(S,x,v) = \mu) \Rightarrow k \in S.$$

Let $P \subseteq N$ be maximal such that $e(P,x,v) = \mu$. By Pareto optimality, $P \neq N$. Take $\ell \in N \setminus P$. Then (see Exercise 4.4.3) there exists $Q \in \mathcal{T}_{\ell k}$ satisfying

$$\Big(Q \cap P = \emptyset \text{ and } e(Q,x,v) \geq 0\Big) \text{ or } e(Q,x,v) \geq e(P,x,v). \qquad (4.4.6)$$

However, $e(Q,x,v) < \mu$, because $k \notin Q$. Hence $Q \cap P = \emptyset$ and $e(Q,x,v) \geq 0$. By superadditivity,

$$e(P \cup Q,x,v) \geq e(P,x,v) + e(Q,x,v) \geq \mu,$$

which is impossible by maximality of P. **q.e.d.**

Theorem 4.4.9. *Let Δ be a set of superadditive games with coalition structures. The semi-reactive prebargaining set satisfies* REAB *on Δ.*

Proof: Assume, on the contrary, that

$$x^\ell > b^\ell_{\max} = \max_{S\subseteq N\setminus\{\ell\}} \Big(v(S\cup\{\ell\}) - v(S)\Big) \text{ for some } \ell \in R \in \mathcal{R}.$$

Denote $\mu = \max_{S\subseteq N} e(S,x,v)$. Then $\mu > 0$, because $e(R\setminus\{\ell\},x,v) > e(R,x,v) = 0$. Also,

$$\Big(S \subseteq N, e(S,x,v) = \mu\Big) \Rightarrow \ell \notin S.$$

Let $P \subseteq N$ be maximal such that $e(P,x,v) = \mu$. Then $P \cap R \neq \emptyset$, because by superadditivity

$$\begin{aligned} e(P\cap(N\setminus R),x,v) &< e(P\cap(N\setminus R),x,v) + e(R\setminus\{\ell\},x,v) \\ &\leq e(P\cup(R\setminus\{\ell\}),x,v). \end{aligned}$$

Thus there exists $k \in P \cap R$. Hence there exists $Q \in \mathcal{T}_{\ell k}$ such that every objection of k against ℓ via P can be countered via Q. However, $e(Q,x,v) < \mu$, because $\ell \in Q$, and $P \cap Q \neq \emptyset$, because P is maximal. Exercise 4.4.3 yields the desired contradiction. **q.e.d.**

Corollary 4.4.10. *On every set of superadditive games both $\mathcal{PM}_{sr}$ and $\mathcal{M}_{sr}$ satisfy reasonableness.*

In the next chapter an example is presented (Example 5.5.13) which shows that Theorem 4.4.8 is not valid for all sets of superadditive games with coalition structures.

Corollary 4.4.10 is due to Sudhölter and Potters (2001).

4.4.2 The Mas-Colell Bargaining Set

The variant of the bargaining set discussed in this subsection was only defined for the coalition structure $\{N\}$; thus we restrict our attention to this case and assume throughout that (N,v) is a game and that $x \in X(N,v)$. For $R,S \subseteq N$ satisfying $R \cap S = \emptyset$ and $a \in \mathbb{R}^R, b \in \mathbb{R}^S$, we define $c = (a,b) \in \mathbb{R}^{R\cup S}$ by $c^i = a^i$ for all $i \in R$ and $c^j = b^j$ for all $j \in S$.

Definition 4.4.11. *A pair (P,y) is an* **objection** *(in the sense of the Mas-Colell bargaining set), if $\emptyset \neq P \subseteq N$, $y(P) = v(P)$, and $y > x^P$. A pair (Q,z) is a* **counterobjection** *to the objection (P,y) if $\emptyset \neq Q \subseteq N$ and*

$$z(Q) = v(Q) \text{ and } z > \left(y^{P\cap Q}, x^{Q\setminus P}\right). \tag{4.4.7}$$

An objection at x is **justified**, *if it has no counterobjection. The* **Mas-Colell prebargaining set** *of (N,v) is the set*

$$\mathcal{PMB}(N,v) = \{x \in X(N,v) \mid \text{There is no justified objection at } x\}$$

and

$$\mathcal{MB}(N,v) = \mathcal{PMB}(N,v) \cap I(N,v)$$

is the **Mas-Colell bargaining set** *of* (N,v).

Hence an objection (P,y) and a counterobjection (Q,z) in the sense of the Mas-Colell bargaining set do not explicitly refer to some players k and ℓ. Though (P,y) is also an objection of any $k \in P$ satisfying $y^k > x^k$ against any $\ell \in N \setminus P$ at x in the classical sense, neither $Q \setminus P \neq \emptyset$ nor $P \setminus Q \neq \emptyset$ is required from (Q,z). Thus, $(P,y),(Q,z)$ may not establish a pair consisting of an objection and counterobjection of any pair (k,ℓ) of distinct players.

Clearly, $\mathcal{C}(N,v) \subseteq \mathcal{MB}(N,v)$. Definition 4.4.11 is due to Mas-Colell (1989).

Remark 4.4.12. In view of the fact that an objection (P,y) in the sense of the Mas-Colell bargaining set cannot be countered via P, we may additionally assume that $Q \neq P$ in the definition of a counterobjection. If, now, the strict inequality (4.4.7) is replaced by the weak inequality, that is, by

$$z \geq \left(y^{P\cap Q}, x^{Q\setminus P}\right), \tag{4.4.8}$$

then the resulting (pre)bargaining set contains the classical (pre)bargaining set. Indeed, if x is in the (pre)bargaining set of (N,v) and if (P,y) is an objection at x in the sense of the Mas-Colell bargaining set, then there exists $k \in P$ with $y^k > x^k$ and $\ell \in N \setminus P$, because $v(P) = y(P) > x(P)$ and $x(N) = v(N)$. Hence (P,y) is an objection of k against ℓ. Any counterobjection (Q,z) in the sense of the bargaining set satisfies (4.4.8).

Remark 4.4.12 leads to the question whether the classical prebargaining set is contained in the Mas-Colell prebargaining set. Theorem 4.4.13 shows that the answer is affirmative for superadditive games.

Theorem 4.4.13. *If* (N,v) *is a superadditive game, then*

$$\mathcal{PM}(N,v) \subseteq \mathcal{PMB}(N,v).$$

Proof: Let (N,v) be a superadditive game and assume that there exists $x \in \mathcal{PM}(N,v) \setminus \mathcal{PMB}(N,v)$. Let $P \subseteq N$ be a maximal coalition such that there exists a justified objection (P,y) in the sense of $\mathcal{MB}$. By Pareto optimality of x, $P \neq N$. Let $\ell \in N \setminus P$. Also, let $k \in P$ satisfy $y^k > x^k$. Then (P,y) is an objection of k against ℓ in the sense of $\mathcal{M}$. Hence there is a counterobjection (Q,z) of ℓ to the objection (P,y) of k.

We claim that for every $R \subseteq N \setminus P$, $e(R,x,v) < 0$. Assume the contrary. By superadditivity,

$$e(P \cup R, x, v) \geq e(P,x,v) + e(R,x,v) \geq e(P,x,v).$$

Therefore there exists $\widehat{y} \in \mathbb{R}^{P\cup R}$ such that $\widehat{y}(P\cup R) = v(P\cup R)$, $\widehat{y}^i = y^i$ for all $i \in P$ and $\widehat{y}^j \geq y^j$ for all $j \in R$. Hence $(P \cup R, \widehat{y})$ is an objection in the sense of $\mathcal{MB}$. By maximality of P the objection $(P \cup R, \widehat{y})$ has a counterobjection. Clearly this counterobjection is also a counterobjection to (P, y).

As $e(Q, x, v) \geq 0$, our claim implies that $P \cap Q \neq \emptyset$. Choose $\varepsilon > 0$ satisfying $y^k - \varepsilon > x^k$ and define $\tilde{y} \in \mathbb{R}^P$ by

$$\tilde{y}^k = y^k - \varepsilon \text{ and } \tilde{y}^i = y^i + \frac{\varepsilon}{|P| - 1} \text{ for all } i \in P \setminus \{k\}.$$

Then $(P, \tilde{y})$ is an objection of k against ℓ as well. Let $(\widehat{Q}, \widehat{z})$ be a counterobjection of ℓ to the objection $(P, \tilde{y})$ of k. Then, again by the claim, there exists $j \in \widehat{Q} \cap P$. The observations that

$$\widehat{z}(\widehat{Q}) = v(\widehat{Q}),\ \widehat{z}^i \geq x^i \text{ for all } i \in \widehat{Q} \setminus P,\ \widehat{z}^j \geq \tilde{y}^j > y^j \text{ for all } j \in \widehat{Q} \cap P$$

directly imply that $(\widehat{Q}, \widehat{z})$ is a counterobjection to (P, y), which was excluded by our assumption.

q.e.d.

A different proof of Theorem 4.4.13 is contained in Holzman (2000). This paper also shows that the closure of the Mas-Colell prebargaining set of a game contains the classical prebargaining set of that game.

Exercises

Let $(N, v, \mathcal{R})$ be a game with coalition structure, $x \in X(N, v, \mathcal{R})$, and $k, \ell \in R \in \mathcal{R}, k \neq \ell$.

Exercise 4.4.1. Verify that $k \succ_x^{\mathcal{M}_r} \ell$ if and only if for every $Q \in \mathcal{T}_{\ell k}$ the following conditions are satisfied:

$$e(Q, x, v) < 0 \Rightarrow \text{there exists } P \in \mathcal{T}_{k\ell} \text{ with } e(P, x, v) > 0;$$
$$e(Q, x, v) \geq 0 \Rightarrow \text{there exists } P \in \mathcal{T}_{k\ell} \text{ with } \left\{\begin{matrix} e(P, x, v) > e(Q, x, v) \\ \text{and } P \cap Q \neq \emptyset \end{matrix}\right\}.$$

(See Granot and Maschler (1997).)

Exercise 4.4.2. Let (N, v) be the seven-person projective game and let

$$x = \frac{1}{3}(1, 1, 0, 0, 1, 0, 0),\ y = \frac{1}{3}(1, 0, 1, 1, 0, 0, 0),\ z = \frac{1}{2}(x + y).$$

Show that $x, y \in \mathcal{M}(N, v)$ by first showing that Corollary 4.2.6 is also valid in the context of $\mathcal{M}_r$. Also prove that $z \notin \mathcal{M}_r(N, v)$ by verifying that 6 has a justified objection against 1.

Exercise 4.4.3. Verify that $k \preceq_x^{\mathcal{M}_{sr}} \ell$ if and only if for every $P \in \mathcal{T}_{k\ell}$ with $e(P, x, v) > 0$ there exists $Q \in \mathcal{T}_{\ell k}$ such that (4.4.6) is satisfied.

Exercise 4.4.4. Prove that $\succ_{\cdot}^{\mathcal{M}_r}$ and $\succ_{\cdot}^{\mathcal{M}_{sr}}$ are continuous. (See the proof of Lemma 4.2.2.)

Exercise 4.4.5. Show that Corollary 4.2.6 remains valid in the context of the reactive and of the semi-reactive bargaining set. Deduce that $\succ_x^{\mathcal{M}_r}$ and $\succ_x^{\mathcal{M}_{sr}}$ are acyclic.

Exercise 4.4.6. Use Exercises 4.4.4 and 4.4.5 to show that

$$I(N,v,\mathcal{R}) \neq \emptyset \Rightarrow \mathcal{M}_r(N,v,\mathcal{R}) \neq \emptyset \neq \mathcal{M}_{sr}(N,v,\mathcal{R}).$$

Exercise 4.4.7. Let $N = \{1,2,3,4\}$ and let $v(S)$, $S \subseteq N$, be defined by the following formula:

$$v(S) = \begin{cases} 8 \text{ , if } S = N \\ 6 \text{ , if } |S| = 3 \text{ or } \Big(|S| = 2 \text{ and } 1 \in S\Big) \\ 5 \text{ , if } |S| = 2 \text{ and } 1 \notin S \\ 0 \text{ , if } |S| \leq 1. \end{cases}$$

Show that $(2,2,2,2) \in \mathcal{M}_{sr}(N,v) \setminus \mathcal{M}_r(N,v)$. (See Sudhölter and Potters (2001).)

Exercise 4.4.8. Show Remark 4.4.7. (The proof of the assertion concerning the semi-reactive prebargaining set is similar to the proof of Theorem 4.1.11. For a proof of the other assertion see Granot and Maschler (1997).)

Exercise 4.4.9. Show that the Mas-Colell bargaining set and the bargaining set of the simple majority three-person game (N,v) (defined by $N = \{1,2,3\}$, $v(S) = 1$, if $|S| \geq 2$, and $v(S) = 0$, otherwise) are

$$\mathcal{MB}(N,v) = \left\{x \in X(N,v) \,\middle|\, x^i < \frac{1}{2}\right\} \text{ and } \mathcal{M}(N,v) = \left\{\left(\frac{1}{3},\frac{1}{3},\frac{1}{3}\right)\right\}.$$

(See Holzman (2000).)

Exercise 4.4.10. Show that $\mathcal{PMB}$ does not satisfy RGP on $\Gamma_{\mathcal{U}}$, provided $|\mathcal{U}| \geq 3$.

4.5 Non-monotonicity of Bargaining Sets

In this section an example is presented which shows that none of the bargaining sets based on *individual* objections and counterobjections, satisfies strong aggregate monotonicity. See Definition 2.3.23 for the definition of SAM. Also, Example 4.1.19 is used to describe a phenomenon which may be seen as the "dummy paradox" of the classical bargaining set.

The following example in fact shows that the reactive and the semi-reactive (pre)bargaining sets do not even satisfy the following relaxed version of SAM.

Definition 4.5.1. *A solution σ on a set Γ of games satisfies* **aggregate monotonicity** *(AM) if for all games $(N,u),(N,v) \in \Gamma$ satisfying $u(N) \geq v(N)$ and $u(S) = v(S)$ for all $S \subsetneqq N$, and for all $x \in \sigma(N,v)$, there exists $y \in \sigma(N,u)$ such that $y \geq x$.*

The next remark is useful. Its proof is straightforward and left to the reader (see Exercise 4.5.1).

Remark 4.5.2. Let (N,v) be a game, $x \in \mathcal{PM}(N,v)$, and $k,\ell \in N,\ k \neq \ell$. Denote $\mathcal{S}_{\ell k}(x,v) = \{S \in \mathcal{T}_{\ell k} \mid e(S,x,v) \geq 0\}$. If $P \in \mathcal{T}_{k\ell}$ satisfies

$$e(P,x,v) > 0 \text{ and } P \cap \bigcap_{Q \in \mathcal{S}_{\ell k}(x,v)} Q \neq \emptyset,$$

then $e(P,x,v) \leq \max_{Q \in \mathcal{S}_{\ell k}(x,v)} e(Q,x,v)$.

In the next chapter the following example will also be used to show that two other remarkable solutions do not satisfy AM.

Example 4.5.3. Let $N = \{1,\ldots,6\}$ and $S_i,\ i \in N$, be defined by $S_1 = \{1,2,4,6\},\ S_2 = \{1,2\},\ S_3 = \{1,3,5,6\},\ S_4 = \{1,4,5\},\ S_5 = N \setminus \{1\}$, and $S_6 = \{2,3,4\}$. Hence the characteristic vectors $\chi_{S_i},\ i \in N$, can be represented in the square 6×6 matrix D with the entries $D_{ij} = \chi^j_{S_i},\ i,j \in N$:

$$D = \begin{pmatrix} 1\,1\,0\,1\,0\,1 \\ 1\,1\,0\,0\,0\,0 \\ 1\,0\,1\,0\,1\,1 \\ 1\,0\,0\,1\,1\,0 \\ 0\,1\,1\,1\,1\,1 \\ 0\,1\,1\,1\,0\,0 \end{pmatrix}.$$

Let $v(S),\ S \subseteq N$, be defined by $v(S_i) = 2$ for all $i \in N$, $v(\emptyset) = v(N) = 0$, and $v(S) = -1$, otherwise. Also, denote $\bar{x} = 0 \in \mathbb{R}^N$. Then $s_{k\ell}(\bar{x},v) = 2$ for all $k,\ell \in N,\ k \neq \ell$. By Corollary 4.2.6 and Exercise 4.4.5, $\bar{x} \in \mathcal{PM}_r(N,v)$. Clearly, $\bar{x}$ is individually rational, so $\bar{x} \in \mathcal{M}_r(N,v)$.

Let (N,u) be a game which differs from (N,v) only inasmuch as $0 < u(N) \leq 1$. The following lemma shows that none of the variants of the prebargaining set satisfies AM.

Lemma 4.5.4. $x \in X(N,u),\ x > \bar{x} \Rightarrow x \notin \mathcal{PM}(N,u)$.

Proof: Assume, on the contrary, that $x \in \mathcal{PM}(N,u)$. Note that $e(S_i,x,u) \geq 1$ for all $i \in N$, $e(N,x,u) = e(\emptyset,x,u) = 0$, and $e(S,x,u) \leq -1$, otherwise.

Let $\alpha^i = e(S_i,x,u)$ for all $i \in N$ and let $\alpha = \alpha^4$. We shall now employ Remark 4.5.2 several times to show that $\alpha^i = \alpha$ for all $i \in N$. We use the abbreviation

$\mathcal{S}_{k\ell} = \mathcal{S}_{k\ell}(x,u)$ for all distinct $k, \ell \in N$. A careful inspection of the matrix D shows that

$$S_k \cap S_\ell \neq \emptyset \text{ for all } k, \ell \in N \tag{4.5.1}$$

and that the following equations are valid:

$$\{S_2\} = \mathcal{S}_{24}; \tag{4.5.2}$$
$$\{S_4\} = \mathcal{S}_{42} = \mathcal{S}_{53}; \tag{4.5.3}$$
$$\{S_6\} = \mathcal{S}_{35} = \mathcal{S}_{36}; \tag{4.5.4}$$
$$\{S_1\} = \mathcal{S}_{63}; \tag{4.5.5}$$
$$\{S_5\} = \mathcal{S}_{51}, \ \{S_1, S_2\} = \mathcal{S}_{15}. \tag{4.5.6}$$

Remark 4.5.2 and (4.5.1) together with (4.5.2) and (4.5.3) imply that $\alpha^2 = \alpha$. Analogously, (4.5.3) and (4.5.4) or (4.5.4) and (4.5.5), respectively, imply that $\alpha^6 = \alpha = \alpha^1$. Also, by (4.5.6), $\alpha^5 = \alpha$, because $S_1 \cap S_2 \cap S_5 \neq \emptyset$.

As $x \geq 0$ and $S_1 \setminus S_2 = \{4,6\}$, $S_5 \setminus S_6 = \{5,6\}$, $x_4 = x_5 = x_6 = 0$. Therefore,

$$\alpha = \alpha^1 = 2 - x(S_1) = 2 - x^1 - x^2 = \alpha^4 = 2 - x^1$$

implies that $x^2 = 0$. We conclude that $\alpha^3 = 2 - x^1 - x^3 \leq \alpha^1 = \alpha$. On the other hand, $\mathcal{S}_{32} = \{S_3\}$ and $S_1 \in \mathcal{S}_{23}$. Hence $\alpha^3 \geq \alpha$ by Remark 4.5.2. Thus, $x^2 = \cdots = x^6 = 0$. But $\alpha^1 = \alpha^6$ yields $x^1 = 0$ as well, which is impossible by Pareto optimality of x. **q.e.d.**

Remark 4.5.5. Lemma 4.5.4 remains valid if the coalition function v is replaced by a coalition function v' satisfying $v'(S) \leq v(S)$ for all $S \subseteq N$ and $v'(S_i) = v(S_i)$ for all $i \in N$ and $v'(N) = v(N)$. Of course the coalition function u has to be replaced by u', which coincides with v' except that $0 < u'(N) \leq 1$. The proof can be literally copied. Also, $\bar{x} \in \mathcal{M}(N, v')$ remains valid. Especially $v'(S)$, $S \subseteq N$, defined by

$$v'(S) = \max\left\{v(T) - 3|S \setminus T| \,\middle|\, T \in \{\emptyset, N, S_1, \ldots, S_6\},\ T \subseteq S\right\},$$

has the aforementioned property. Clearly (N, v') and the corresponding game (N, u') are superadditive.

Corollary 4.5.6. *Let* $|\mathcal{U}| \geq 6$ *and let* Γ *be a set of games containing all superadditive games. None of the solutions* $\mathcal{M}$, $\mathcal{PM}$, $\mathcal{M}_{sr}$, $\mathcal{PM}_{sr}$, $\mathcal{M}_r$, $\mathcal{PM}_r$ *satisfies* AM *on* Γ.

Remark 4.5.7. An example which shows that the bargaining set does not satisfy AM is due to Megiddo (1974). He uses a nine-person game which is not superadditive. We presented a different game to get our stronger result.

Now, we consider the five-person simple majority game with one additional dummy (N, v), defined in Example 4.1.19, and show that, if the worth of the grand coalition is increased by any small amount δ, then, with respect to every member of the bargaining set of the new game, the former null player receives less than he receives by some member of the bargaining set of the initial game. In view of the fact that the dummy (player 6) of (N, v) is not a dummy of the new game, because he contributes δ when joining the coalition of the five other players, the phenomenon of decreasing the null player's payoff may be seen as the "dummy paradox of the bargaining set".

Recall that

$$\bar{x} = \left(\frac{1}{7}, \ldots, \frac{1}{7}, \frac{2}{7}\right) \in \mathcal{M}(N, v).$$

Let (N, u) be a game satisfying $u(S) = v(S)$ for all $S \subsetneqq N$ and $u(N) = v(N) + \delta$ for some $\delta \in \mathbb{R}$ satisfying $0 < \delta < \frac{2}{7}$.

Lemma 4.5.8. *If $x \in \mathcal{M}(N, u)$, then $x^6 < \bar{x}^6$.*

Proof: Let $x \in I(N, u)$ satisfy $x^6 \geq 2/7$. It remains to show that $x \notin \mathcal{M}(N, u)$. Without loss of generality we may assume that

$$x^1 \leq \cdots \leq x^5, \tag{4.5.7}$$

because $\mathcal{M}$ satisfies AN. In what follows we shall construct a justified objection of 1 against 6 via the coalition $P = \{1, 2, 3\}$. Indeed, P can be used to object, because

$$e(P, x, u) = 1 - x(P) = 1 - x(N) + x(\{4, 5\}) + x^6 \geq -\delta + x^6 > 0. \tag{4.5.8}$$

Let $Q_{\{i\}}, i = 2, 3$, and $Q_{\{2,3\}}$ be the members of $\mathcal{T}_{61}$ defined by

$$Q_{\{i\}} = \{i, 4, 5, 6\},\ i = 2, 3, \text{ and } Q_{\{2,3\}} = \{2, 3, 4, 6\}.$$

Then

$$Q \in \mathcal{T}_{61},\ e(Q, x, u) \geq 0 \Rightarrow u(Q) = 1, \tag{4.5.9}$$

because $x \geq 0$ and $x^6 > 0$. Also, we have

$$Q \in \mathcal{T}_{61},\ u(Q) = 1 \Rightarrow e(Q, x, u) \leq e(Q_{Q \cap \{2,3\}}, x, u). \tag{4.5.10}$$

Indeed, every $Q \in \mathcal{T}_{61}$ satisfying $u(Q) = 1$ intersects $\{2, 3\}$, so $Q_{Q \cap \{2,3\}}$ is defined. The inequality follows from (4.5.7). Also, $x \geq 0$, $x^6 > 0$, (4.5.7) - (4.5.10) imply that

$$e(P, x, u) > (e(Q, x, u))_+ \text{ for all } Q \in \mathcal{T}_{61}. \tag{4.5.11}$$

We claim that

$$e(P, x, u) > (e(Q_{\{2\}}, x, u))_+ + (e(Q_{\{3\}}, x, u))_+. \tag{4.5.12}$$

By (4.5.11) it suffices to show that

$$e(P,x,u) > e(Q_{\{2\}},x,u) + (e(Q_{\{3\}},x,u), \tag{4.5.13}$$

which is equivalent to

$$1 - x(P) > 1 - x(Q_{\{2\}}) + 1 - x(Q_{\{3\}})$$

and, thus, to $-1 - x^1 + 2x(\{4,5,6\}) > 0$. By the observation that

$$\begin{aligned} & -1 - x^1 + 2x(\{4,5,6\}) \\ = & -1 + x(N) - 2x^1 - x(\{2,3\}) + x(\{4,5,6\}) \\ \geq & \; \delta + x^6 - 2x^1 \end{aligned}$$

it suffices to show that $\delta + x^6 - 2x^1 > 0$. By (4.5.7), $5x^1 + x^6 \leq 1 + \delta$, so

$$\delta + x^6 - 2x^1 \geq \frac{3\delta + 7x^6 - 2}{5} > 0.$$

The last inequality is implied by the assumption that $x^6 \geq 2/7$.

Now the proof can be finished. By (4.5.8), (4.5.11) and (4.5.12) there exists $t \in \mathbb{R}^P$ satisfying $t \gg 0$, $t(P) = e(P,x,u)$, $t(\{2,3\}) > e(Q_{\{2,3\}},x,u)$, and $t^i > (e(Q_{\{i\}},x,u))_+$ for every $i \in \{2,3\}$. Let $y \in \mathbb{R}^P$ be defined by $y = x^P + t$. Then (P,y) is a justified objection of 1 against 6 at x. **q.e.d.**

Remark 4.5.9. By Corollary 4.4.10 the reactive and the semi-reactive (pre)-bargaining sets do not satisfy a dummy paradox when the attention is restricted to superadditive games. On the contrary, if the game is modified by only increasing the worth of the grand coalition, then a former dummy cannot receive less than before with respect to every distribution of the semi-reactive prebargaining set.

Exercises

Exercise 4.5.1. Prove Remark 4.5.2.

Exercise 4.5.2. Let Γ be a set of two-person games. Show that, on Γ, $\mathcal{M}$ satisfies SAM and $\mathcal{PM}$ does not satisfy SAM.

Exercise 4.5.3. Show that $\mathcal{M}$ does not satisfy SAM on the set of three-person games. (Let v be a coalition function which is positive for one two-person coalition and 0 otherwise. Show that $0 \in \mathcal{M}(N,v)$.)

Exercise 4.5.4. Show that $\mathcal{PM}$ satisfies SAM on the set of superadditive three-person games.

4.6 The Bargaining Set and Syndication: An Example

In this section we examine the effect of syndication on the core and the bargaining set $\mathcal{M}$ of the following five-person market: Let

$$P = \{1, 2\},\ Q = \{3, 4, 5\},\ N = P \cup Q,\ m = 2,\ a^i = (1, 0),\ a^j = (0, 1/2)$$

for all $i \in P$, $j \in Q$, and $w(x_1, x_2) = w^k(x_1, x_2) = \min\{x_1, x_2\}$ for all $k \in N$. Then $\left(N, \mathbb{R}^2_+, (a^i)_{i \in N}, (w)_{i \in N}\right)$ is a market. The corresponding market game (N, v) is given by

$$v(S) = \min\left\{|S \cap P|, \frac{|S \cap Q|}{2}\right\} \text{ for all } S \subseteq N.$$

The straightforward proof that

$$\mathcal{C}(N, v) = \left\{\left(0, 0, \frac{1}{2}, \frac{1}{2}, \frac{1}{2}\right)\right\} \tag{4.6.1}$$

is the content of Exercise 4.6.1.

The core outcome $(0, 0, 1/2, 1/2, 1/2)$ seems to give the members of Q more than they deserve in the game. Although there is an oversupply of the first commodity, there is **no** oversupply of traders of the first type. Indeed, each member of P contributes at least $1/2$ to $v(N)$. For example, $v(\{2, 3, 4, 5\}) = 1$, and, therefore, trader 1 may demand a positive payoff if he joins the rest of the players to obtain the total payoff of $3/2 = v(N)$. The fact that the bargaining power of the member of P is non-null is reflected in the bargaining set of the game.

Maschler (1976) showed that

$$\mathcal{M}(N, v) = \left\{\left(\frac{t}{2}, \frac{t}{2}, \frac{3-2t}{6}, \frac{3-2t}{6}, \frac{3-2t}{6}\right) \,\middle|\, 0 \le t \le \frac{3}{2}\right\}. \tag{4.6.2}$$

It can also be shown (see Exercise 4.6.3) that

$$\mathcal{M}(N, v) = \mathcal{PM}_{sr}(N, v) = \mathcal{M}_{sr}(N, v) = \mathcal{PM}_r(N, v) = \mathcal{M}_r(N, v). \tag{4.6.3}$$

Thus, the bargaining sets contain points outside the core that represent "reasonable" payoff distributions. For example, $(1/4, 1/4, 1/3, 1/3, 1/3)$ is a stable outcome, that is, a member of the bargaining set, that seems to reflect the bargaining power of the players better than the core payoff. Also, $(3/8, 3/8, 2/8, 2/8, 2/8)$ is stable and it distributes the same aggregate amount to P and Q which may be regarded as reasonable (see Chapter 6, Example 6.6.13).

Now suppose that the members of Q have decided to form a syndicate, that is, to operate as if they were a single player. Let $N_1 = \{1, 2, Q\}$ be the new set of agents. The new market is $\left(N_1, \mathbb{R}^2_+, (a^i_*)_{i \in N_1}, (w)_{i \in N_1}\right)$, where $a^i_* = a^i$ for all $i \in P$ and $a^Q_* = (0, 3/2)$. The corresponding market game (N_1, v_1) is given by

$$v_1(S) = \begin{cases} 0 \ , & \text{if } S \subseteq P \text{ or } S = Q, \\ 1 \ , & \text{if } S = \{1, Q\} \text{ or } S = \{2, Q\}, \\ 3/2 \ , & \text{if } S = N_1. \end{cases}$$

The proof that $\mathcal{C}(N_1, v_1)$ is given by

$$\mathcal{C}(N_1, v_1) = \text{convh}\left\{\left(0, 0, \frac{3}{2}\right), \left(\frac{1}{2}, 0, 1\right), \left(0, \frac{1}{2}, 1\right), \left(\frac{1}{2}, \frac{1}{2}, \frac{1}{2}\right)\right\} \tag{4.6.4}$$

is left to the reader (see Exercise 4.6.4). Here "convh" is the abbreviation of "convex hull of".

We conclude from (4.6.1) and (4.6.4) that syndication is disadvantageous for Q, if the concept of the core is used to solve the games. Indeed, the syndicate Q can get in the core at most 3/2 and it may get a smaller payoff, whereas without syndication Q receives exactly 3/2.

The foregoing example is taken from Postlewaite and Rosenthal (1974). These authors claim that the core in this example is intuitively acceptable and their arguments for justifying the core also explain the foregoing disadvantage of syndication. However, following Maschler (1976), we have argued that the core of the foregoing market game (N, v) is too small. Indeed, this is also the reason that generates the syndication paradox of the core. Thus, we reach the same conclusion as Aumann (1973) and Maschler (1976); namely, that the core is not the proper solution concept for studying syndication.

Now we consider the effect of syndication on the bargaining set $\mathcal{M}$. The pair (N_1, v_1) is a balanced three-person game. Hence, by Exercise 4.1.2, $\mathcal{M}(N_1, v_1) = \mathcal{C}(N_1, v_1)$. Thus, Q may get in $\mathcal{M}(N, v)$ any payoff $x(Q)$ satisfying $0 \leq x(Q) \leq 3/2$, whereas in $\mathcal{M}(N_1, v_1)$ the "single player" Q may get any payoff x^Q satisfying $1/2 \leq x^Q \leq 3/2$. Therefore, if $\mathcal{M}$ is applied in both situations to solve the games (N, v) and (N_1, v_1), then syndication is profitable.

Exercises

Exercise 4.6.1. Prove (4.6.1).

Exercise 4.6.2. Verify (4.6.2) by, e.g., consulting Maschler (1976). Verify, for every $x \in \mathcal{M}(N, v)$ and every pair $k, \ell \in N,\ k \neq \ell$, that $s_{k\ell}(x, v) = s_{\ell k}(x, v)$.

Exercise 4.6.3. Assume that (4.6.2) is true. Show (4.6.3) by applying Theorem 4.4.8.

Exercise 4.6.4. Prove (4.6.4).

4.7 Notes and Comments

(1) Aumann and Maschler (1964) proposed several variants of Definition 4.1.13. Additional bargaining sets were defined by other authors (see, e.g., (8) below). The reader may find a survey of various bargaining sets in Kahan and Rapoport (1984). We emphasize that most of the theoretical work has been devoted to $\mathcal{M}$. Also, some work has been done on a bargaining set which is based on a notion of coalitional stability and has no existence theorem (see Maschler (1964) and Peleg (1964)).

(2) In the next chapter we shall introduce the prekernel $\mathcal{PK}$ and we shall prove that $\mathcal{PK}(N, v, \mathcal{R}) \neq \emptyset$ for every game with coalition structure $(N, v, \mathcal{R})$. The proof will be constructive in the following way: It will only use the theory of linear inequalities in finite-dimensional real vector spaces. As a corollary we shall obtain an existence theorem for $\mathcal{PM}$ and $\mathcal{PM}_r$.

(3) A proof of Brouwer's fixed-point theorem is contained in Klein (1973) and Franklin (1980).

(4) In the next chapter we shall provide an elementary proof of Theorem 4.2.12 (and of the claim in Exercise 4.4.6). Nevertheless, we have chosen to give the foregoing proof for the following reasons. Firstly, the present proof may be modified to yield existence theorems for bargaining sets of cooperative NTU games, discussed in Part II. In particular, Lemma 4.2.11 is essential for such a generalization. Secondly, the results that were obtained on the binary relation $\succ_x^{\mathcal{M}}$ are interesting and instructive; they deserve a detailed presentation.

(5) Aumann and Maschler (1964) have shown that for each game with coalition structure $(N, v, \mathcal{R})$, $\mathcal{M}(N, v, \mathcal{R})$ is a finite union of closed convex polyhedra. Also, $\mathcal{M}_r(N, v, \mathcal{R})$ and $\mathcal{M}_{sr}(N, v, \mathcal{R})$ are unions of closed convex polyhedra. The same is true for the corresponding prebargaining sets (see Granot and Maschler (1997) and Sudhölter and Potters (2001)). In Chapter 9 we shall show that the classical prebargaining set of a game is bounded. Thus the prebargaining sets (the classical, the reactive and the semi-reactive) of a game with coalition structure are finite unions of polytopes. In Maschler (1966) the inequalities that determine $\mathcal{M}(N, v, \mathcal{R})$ are given in explicit form. The Mas-Colell bargaining set of a game is not necessarily closed (see Exercise 4.4.9).

(6) Sudhölter and Potters (2001) show that the reactive and the semi-reactive prebargaining set coincide for many games. Also, they show that the semi-reactive prebargaining set on $\Gamma_{\mathcal{U}}$ can be characterized by NE, PO, RGP, a variant of CRGP, and the requirement that no player can justifiably object against any other player ℓ via $N \setminus \{\ell\}$. The "same" characterization (only the meaning of "justified objections" differs) applies to the classical prebargaining set.

(7) Solymosi (1999) contains many examples of classes of games for which the bargaining set coincides with the core.

(8) A different kind of bargaining set was introduced by Zhou (1994). Let (N, v) be a game and $x \in \mathbb{R}^N$. Objections are objections in the sense of the Mas-Colell bargaining set. A counterobjection (Q, z) to an objection (P, y) has to satisfy (4.4.8) and $Q \setminus P,\ P \setminus Q,\ P \cap Q \neq \emptyset$. Now, the bargaining set $\mathcal{ZB}(N, v)$ is defined to be the set of all $(x, \mathcal{R})$ such that $\mathcal{R}$ is a coalition structure of N, $x \in X^*(N, v, \mathcal{R})$, and every objection at x has a counterobjection. Hence this bargaining set may not contain any element of $\{(x, \mathcal{R}) \mid x \in X^*(N, v, \mathcal{R})\}$ for some given $\mathcal{R}$, even if $I(N, v, \mathcal{R}) \neq \emptyset$. In such a case $\mathcal{R}$ is *not compatible* with the bargaining set. Fortunately, Zhou proved the existence of a compatible coalition structure. This bargaining set satisfies IR, that is, if $(x, \mathcal{R}) \in \mathcal{ZB}(N, v)$, then $x^i \geq v(\{i\})$, and efficiency, that is, $x \in X(N, v, \mathcal{R})$. If (N, v) is superadditive, then $\{N\}$ is the unique compatible coalition structure. Hence, on superadditive games $\mathcal{ZB}$ can be regarded as a solution in the sense of Definition 2.3.1. Example 4.1.19 can be used to show that the bargaining set introduced by Zhou does not satisfy REAB or NP on superadditive games.

(9) Further investigations of $\mathcal{PMB}$, $\mathcal{ZB}$, and of some related subsolutions are contained in Shimomura (1997).

5

The Prekernel, Kernel, and Nucleolus

This chapter is devoted to a systematic study of the prekernel. However, many results on the kernel and nucleolus, two related solutions, are obtained. In the first section we present the definition of the prekernel and show that this solution is contained in the reactive prebargaining set. Also, we introduce a very general version of the nucleolus and prove existence and single-valuedness results under very mild conditions. As a corollary we deduce that the prenucleolus of every game consists of a single point. Then we show that the prenucleolus belongs to the prekernel. Thereby we obtain existence theorems for the prekernel and the unconstrained bargaining sets $\mathcal{PM}$, $\mathcal{PM}_{sr}$, and $\mathcal{PM}_r$. In Section 5.2 we prove that the prekernel satisfies RGP and CRGP, and that the prenucleolus satisfies RGP.

The desirability relation which is derived from a coalitional game is introduced in Section 5.3. Also, it is proved that the prekernel preserves the desirability relation of every game and, therefore, has the equal treatment property. An axiomatization of the prekernel is given in Section 5.4. The following six independent axioms characterize the prekernel: NE, PO, COV, ETP, RGP, and CRGP.

The kernel of a coalitional game is introduced in Section 5.5. Then it is proved that the prekernel of a 0-monotonic game is equal to the kernel. Also, the relation of the semi-reactive bargaining set, the kernel, and the core, when restricted to superadditive simple games, is discussed. The reasonableness of the prekernel and of the kernel is investigated in the next section. Section 5.7 is devoted to the kernel of convex games. It is proved that the kernel (for the grand coalition) of a convex game coincides with the nucleolus of the game. In the next section some examples of prekernels are discussed.

Finally, several notes and comments are given in Section 5.9.

5.1 The Nucleolus and the Prenucleolus

We start with the definition of the prekernel of a game with coalition structure.

Definition 5.1.1. *Let $(N, v, \mathcal{R})$ be a game with coalition structure and $\mathcal{P} = \mathcal{P}(\mathcal{R})$ (see (3.8.1)). The* **prekernel** *of $(N, v, \mathcal{R})$, $\mathcal{PK}(N, v, \mathcal{R})$, is defined by*

$$\mathcal{PK}(N, v, \mathcal{R}) = \{x \in X(N, v, \mathcal{R}) \mid s_{k\ell}(x, v) = s_{\ell k}(x, v) \text{ for all } \{k, \ell\} \in \mathcal{P}\}.$$

The **prekernel** *of (N, v), $\mathcal{PK}(N, v)$ is $\mathcal{PK}(N, v, \{N\})$.*

Remark 5.1.2. Let $(N, v, \mathcal{R})$ be a game with coalition structure. By Corollary 4.2.6 and Exercise 4.4.5, $\mathcal{PK}(N, v, \mathcal{R}) \subseteq \mathcal{PM}_r(N, v, \mathcal{R})$.

The next example shows that the prekernel can be strictly contained in the reactive prebargaining set even when the core is empty.

Example 5.1.3. Let $N = \{1, \ldots, 4\}$ and let $v(S)$, $S \subseteq N$, be defined by

$$v(S) = \begin{cases} 1, & \text{if } S = N \text{ or } S \in \{\{1, 2\}, \{3, 4\}\}, \\ 0, & \text{otherwise.} \end{cases}$$

Let $x = (1/4, \ldots, 1/4) \in \mathbb{R}^N$. Then $\mathcal{PK}(N, v) = \{x\}$. Indeed, clearly $x \in \mathcal{PK}(N, v)$. Let $y \in \mathcal{PK}(N, v)$. Then $y^1 = y^2$ and $y^3 = y^4$ can easily be checked (see Section 5.3). Also, $y^1 < y^3$ or $y^1 > y^3$ is not possible. However, $(0, 0, 1/2, 1/2) \in \mathcal{M}_r(N, v)$.

It is the aim of this section to show that $\mathcal{PK}$ satisfies NE. The existence of the prekernel can be verified in the same manner as the existence of the prebargaining set (see Section 4.2) by applying Lemma 4.2.7, the modification of Lemma 4.2.11 suggested by Exercise 4.2.2, and the fact that the relation $\succ^{\mathcal{K}}$ (see Definition 4.2.5) is continuous. However, in this section we shall present an elementary proof of nonemptiness of the prekernel. We shall describe a method to construct a preimputation of the prekernel of a game with coalition structure.

Let N be a coalition of players, let $X \subseteq \mathbb{R}^N$, and let $H = (h_i)_{i \in D}$ be a finite sequence of real-valued functions defined on X. Denote by $d \in \mathbb{N}$ the cardinality of D. For $x \in X$ let $\theta(x) = (\theta^1(x), \ldots, \theta^d(x))$ be the vector in $\mathbb{R}^d$, the Euclidean space of dimension d, whose components are the numbers $(h_i(x))_{i \in D}$ arranged in non-increasing order, that is,

$$\theta : X \to \mathbb{R}^d,\ \theta^t(x) = \max_{T \subseteq D, |T| = t} \min_{i \in T} h_i(x) \text{ for all } t = 1, \ldots, d. \tag{5.1.1}$$

Let $\geq_{lex}$ denote the lexicographical ordering of $\mathbb{R}^d$; that is, $x \geq_{lex} y$, where $x, y \in \mathbb{R}^d$, if either $x = y$ or there is $1 \leq t \leq d$ such that $x^j = y^j$ for $1 \leq j < t$ and $x^t > y^t$.

Definition 5.1.4. *The* **nucleolus** *of H with respect to X, $\mathcal{N}(H,X)$, is defined by*

$$\mathcal{N}(H,X) = \{x \in X \mid \theta(y) \geq_{lex} \theta(x) \text{ for all } y \in X\}.$$

If $h_i(x)$, $i \in D$, measures the dissatisfaction of some subcoalition of N at the outcome x, then the vector $\theta(x)$ orders the "complaints" of the various groups according to their magnitude, the highest complaint first, the second-highest second, and so forth. At the points of the nucleolus the "total dissatisfaction" $\theta(x)$ is minimized.

The following example is highly important.

Definition 5.1.5. *Let (N,v) be a game, let $X \subseteq \mathbb{R}^N$, and let*

$$H = (e(S,\cdot,v))_{S \in 2^N}.$$

Then $\mathcal{N}(H,X)$ is the **nucleolus of** (N,v) **with respect to** X *and it is denoted by $\mathcal{N}(N,v,X)$.*

Obviously, $e(S,x,v)$ is a natural measure of the dissatisfaction of the coalition S at x.

Theorem 5.1.6. *If X is nonempty and compact and if h_i, $i \in D$, are continuous, then $\mathcal{N}(H,X) \neq \emptyset$.*

The following lemma is used to prove Theorem 5.1.6.

Lemma 5.1.7. *If all h_i, $i \in D$, are continuous, then θ is continuous.*

Proof: Assume that all h_i, $i \in D$, are continuous. Then θ^t is continuous for every $t = 1,\ldots,d$ by (5.1.1). **q.e.d.**

Proof of Theorem 5.1.6: Define $X_0 = X$ and

$$X_t = \{x \in X_{t-1} \mid \theta^t(y) \geq \theta^t(x) \text{ for all } y \in X_{t-1}\},\ t = 1,\ldots,d.$$

By Lemma 5.1.7, X_t is a nonempty and compact subset of X_{t-1}, for $t = 1,\ldots,d-1$. Hence, in particular, $X_d \neq \emptyset$. As the reader may easily verify, $\mathcal{N}(X,H) = X_d$. **q.e.d.**

Theorem 5.1.8. *Assume that X is convex and all h_i, $i \in D$, are convex. Then $\mathcal{N}(H,X)$ is convex. Furthermore, if $x,y \in \mathcal{N}(H,X)$, then $h_i(x) = h_i(y)$ for all $i \in D$.*

The following Lemma is used in the proof of the foregoing theorem.

Lemma 5.1.9. *Let X be convex and all h_i, $i \in D$, be convex. If $x,y \in X$ and $0 \leq \alpha \leq 1$, then*

$$\alpha\theta(x) + (1-\alpha)\theta(y) \geq_{lex} \theta(\alpha x + (1-\alpha)y). \tag{5.1.2}$$

Proof: Let $i_1, \ldots, i_d$ be an ordering of D such that

$$\theta(\alpha x + (1-\alpha)y) = (h_{i_1}(\alpha x + (1-\alpha)y), \ldots, h_{i_d}(\alpha x + (1-\alpha)y)).$$

Denote $a = (h_{i_1}(x), \ldots, h_{i_d}(x))$ and $b = (h_{i_1}(y), \ldots, h_{i_d}(y))$. As h_i is convex for every $i \in D$,

$$\alpha h_{i_t}(x) + (1-\alpha)h_{i_t}(y) \geq h_{i_t}(\alpha x + (1-\alpha)y), \; t = 1, \ldots, d.$$

Thus, $\alpha a_i + (1-\alpha)b_i \geq \theta^i(\alpha x + (1-\alpha)y), \; i = 1, \ldots, d$. Hence

$$\alpha a + (1-\alpha)b \geq_{lex} \theta(\alpha x + (1-\alpha)y).$$

Therefore (5.1.2) holds, because $\theta(x) \geq_{lex} a$ and $\theta(y) \geq_{lex} b$. **q.e.d.**

Proof of Theorem 5.1.8: Let $x, y \in \mathcal{N}(H, X)$, $x \neq y$, and $0 < \alpha < 1$. Then $\theta(x) = \theta(y)$. By Lemma 5.1.9

$$\theta(\alpha x + (1-\alpha)y) = \alpha\theta(x) + (1-\alpha)\theta(y).$$

Hence $\theta(\alpha x + (1-\alpha)y) = \theta(x)$. Thus, $\alpha x + (1-\alpha)y \in \mathcal{N}(H, X)$. Furthermore, using the proof of Lemma 5.1.9, we conclude that $a = \theta(x)$ and $b = \theta(y)$. Therefore, $h_i(x) = h_i(y)$ for all $i \in D$. **q.e.d.**

Theorems 5.1.6 and 5.1.8 yield the following corollary.

Corollary 5.1.10. *Let (N, v) be a game and $X \subseteq \mathbb{R}^N$. If X is nonempty and compact, then $\mathcal{N}(N, v, X) \neq \emptyset$. If, in addition, X is convex, then $\mathcal{N}(N, v, X)$ consists of a single point.*

Proof: By Theorem 5.1.6, $\mathcal{N}(N, v, X) \neq \emptyset$. Now assume that X is also convex. If $x, y \in \mathcal{N}(N, v, X)$, then, by Theorem 5.1.8, $e(S, x, v) = e(S, y, v)$ for all $S \subseteq N$. Hence $x = y$. **q.e.d.**

Definition 5.1.11. *Let $(N, v, \mathcal{R})$ be a game with coalition structure. Then $\mathcal{N}(N, v, I(N, v, \mathcal{R}))$ is called the* **nucleolus** *of $(N, v, \mathcal{R})$ and is denoted by $\mathcal{N}(N, v, \mathcal{R})$. Also, $\mathcal{N}(N, v, \{N\}) = \mathcal{N}(N, v)$ is the* **nucleolus** *of (N, v).*

Remark 5.1.12. If $I(N, v, \mathcal{R}) \neq \emptyset$, then $\mathcal{N}(N, v, \mathcal{R})$ contains exactly one payoff vector (see Corollary 5.1.10).

Definition 5.1.13. *Let $(N, v, \mathcal{R})$ be a game with coalition structure. The* **prenucleolus** *of $(N, v, \mathcal{R})$, $\mathcal{PN}(N, v, \mathcal{R})$, is defined by*

$$\mathcal{PN}(N, v, \mathcal{R}) = \mathcal{N}(N, v, X^*(N, v, \mathcal{R})).$$

The **prenucleolus** *of (N, v), $\mathcal{PN}(N, v)$, is $\mathcal{PN}(N, v, \{N\})$.*

Theorem 5.1.14. *If $(N, v, \mathcal{R})$ is a game with coalition structure, then the prenucleolus $\mathcal{PN}(N, v, \mathcal{R})$ consists of a single point.*

Proof: Let $y \in X^*(N, v, \mathcal{R})$ and let $\mu = \max_{S \subseteq N} e(S, y, v)$. Define

$$X = \{x \in X^*(N, v, \mathcal{R}) \mid e(S, x, v) \leq \mu \text{ for all } S \subseteq N\}.$$

Then X is nonempty, convex, and compact. By Corollary 5.1.10, the set $\mathcal{N}(N, v, X)$ contains exactly one outcome. Clearly, $\mathcal{N}(N, v, X) = \mathcal{PN}(N, v, \mathcal{R})$.
q.e.d.

The unique element of the prenucleolus $\mathcal{PN}(N, v, \mathcal{R})$ of a game with coalition structure $(N, v, \mathcal{R})$ is, again, called the *prenucleolus* (point) and it is denoted by $\nu(N, v, \mathcal{R})$. Also, $\nu(N, v) = \nu(N, v, \{N\})$.

Remark 5.1.15. Let $G = (N, v, \mathcal{R})$ be a game with coalition structure. The prenucleolus of G can be obtained by solving a sequence of linear programs. Indeed, $\nu(N, v, \mathcal{R})$ is reached by first minimizing the maximal excess, then minimizing the number of coalitions attaining the maximal excess, then minimizing the second-highest excess and the number of coalitions attaining this excess, and so on. This procedure is suggested by the proof of Theorem 5.1.6, where $X_0 = X(N, v, \mathcal{R})$.

Theorem 5.1.16. *If $(N, v, \mathcal{R})$ is a game with coalition structure such that $\mathcal{C}(N, v, \mathcal{R}) \neq \emptyset$, then $\mathcal{PN}(N, v, \mathcal{R}) = \mathcal{N}(N, v, \mathcal{R})$ and $\nu(N, v, \mathcal{R}) \in \mathcal{C}(N, v, \mathcal{R})$.*

Proof: Definitions 3.8.2, 5.1.11, and 5.1.13 and Corollary 5.1.10 show this theorem.
q.e.d.

Theorem 5.1.17. *If $(N, v, \mathcal{R})$ is a game with coalition structure, then $\nu(N, v, \mathcal{R}) \in \mathcal{PK}(N, v, \mathcal{R})$.*

Proof: Let $x = \nu(N, v, \mathcal{R})$ and assume that $x \notin \mathcal{PK}(N, v, \mathcal{R})$. Then there are two distinct players k and ℓ in some $R \in \mathcal{R}$ such that $s_{k\ell}(x, v) > s_{\ell k}(x, v)$. Let $\delta = \frac{s_{k\ell}(x,v) - s_{\ell k}(x,v)}{2}$ and let $y \in \mathbb{R}^N$ be given by $y^i = x^i$, if $i \in N \setminus \{k, \ell\}$, $y^k = x^k + \delta$, and $y^\ell = x^\ell - \delta$. Then $y \in X(N, v, \mathcal{R})$, because $x \in X(N, v, \mathcal{R})$. Denote

$$\mathcal{S} = \{S \in 2^N \setminus \mathcal{T}_{k\ell} \mid e(S, x, v) \geq s_{k\ell}(x, v)\} \text{ and } s = |\mathcal{S}|.$$

Clearly, $\theta^{s+1}(x) = s_{k\ell}(x)$. Now, if $S \in 2^N \setminus (\mathcal{T}_{k\ell} \cup \mathcal{T}_{\ell k})$, then $e(S, x, v) = e(S, y, v)$. If $S \in \mathcal{T}_{k\ell}$, then $e(S, y, v) = e(S, x, v) - \delta$. Finally, if $S \in \mathcal{T}_{\ell k}$, then

$$e(S, y, v) = e(S, x, v) + \delta \leq s_{\ell k}(x, v) + \delta = s_{k\ell}(x, v) - \delta.$$

Thus, we may conclude that $\theta^t(y) = \theta^t(x)$ for all $t \leq s$ and $\theta^{s+1}(y) < s_{k\ell}(x, v) = \theta^{s+1}(x)$. Hence $\theta(x) >_{lex} \theta(y)$ and the desired contradiction has been obtained.
q.e.d.

Corollary 5.1.18. *If $(N, v, \mathcal{R})$ is a game with coalition structure, then $\mathcal{PK}(N, v, \mathcal{R}) \neq \emptyset$ and $\mathcal{PM}_r(N, v, \mathcal{R}) \neq \emptyset$.*

Definition 5.1.1 is due to Maschler, Peleg, and Shapley (1972). Definitions 5.1.5 and 5.1.11, Corollary 5.1.10, Remark 5.1.12, and Theorem 5.1.16 are due to Schmeidler (1969). Definition 5.1.4 and Theorems 5.1.6 and 5.1.8 are taken from Justman (1977). Note that a nucleolus $\mathcal{N}(H, X)$ is a *general nucleolus* in Maschler, Potters, and Tijs (1992).

Exercises

Exercise 5.1.1. Let (N, v) be a symmetric game and let $x \in \mathbb{R}^N$ be defined by $x^i = v(N)/|N|$, $i \in N$. Show that $\nu(N, v) = x$ and $\mathcal{PK}(N, v) = \{x\}$.

Exercise 5.1.2. Prove that the prenucleolus satisfies AN and COV.

Exercise 5.1.3. Prove that $\mathcal{PK}(N, v) \subseteq \mathcal{PMB}(N, v)$ for every game (N, v).

Let $(N, v, \mathcal{R})$ be a game with coalition structure and $\mathcal{P} = \mathcal{P}(\mathcal{R})$. The *positive prekernel* of $(N, v, \mathcal{R})$, $\mathcal{PK}_+(N, v, \mathcal{R})$, is defined by

$$\begin{aligned}&\mathcal{PK}_+(N, v, \mathcal{R}) = \\ &\{x \in X(N, v, \mathcal{R}) \mid (s_{k\ell}(x, v))_+ = (s_{\ell k}(x, v))_+ \text{ for all } \{k, \ell\} \in \mathcal{P}\}.\end{aligned} \tag{5.1.3}$$

The *positive prekernel* of (N, v), $\mathcal{PK}_+(N, v)$, is $\mathcal{PK}_+(N, v, \{N\})$. This solution is discussed in Sudhölter and Peleg (2000). Note that

$$\mathcal{C}(N, v, \mathcal{R}) \cup \mathcal{PK}(N, v, \mathcal{R}) \subseteq \mathcal{PK}_+(N, v, \mathcal{R}).$$

Exercise 5.1.4. Show that

$$\mathcal{PK}(N, v, \mathcal{R}) \subseteq \mathcal{PK}_+(N, v, \mathcal{R}) \subseteq \mathcal{PM}_r(N, v, \mathcal{R}).$$

Exercise 5.1.5. Let $x \in \mathcal{PK}_+(N, v, \mathcal{R})$ and let

$$X = \{y \in X(N, v, \mathcal{R}) \mid e(S, y, v)_+ = e(S, x, v)_+ \text{ for all } S \subseteq N\}.$$

Show that $\mathcal{N}(N, v, X)$ is a singleton contained in $\mathcal{PK}(N, v, \mathcal{R})$.

5.2 The Reduced Game Property

In this section we shall prove that the prekernel and the prenucleolus satisfy RGP (see Definition 3.8.9). Let $\mathcal{U}$ be a set of players and let $\Delta_{\mathcal{U}}$ be the set of all games with coalition structures.

Lemma 5.2.1. *The prekernel satisfies* RGP *on* $\Delta_{\mathcal{U}}$.

Proof: Let $(N,v,\mathcal{R}) \in \Delta_{\mathcal{U}}$, $x \in X(N,v,\mathcal{R})$, $\emptyset \neq S \subseteq N$, and let $k,\ell \in R \in \mathcal{R}_{|S}$, $k \neq \ell$. We compute

$$\begin{aligned}
&s_{k\ell}(x^S, v^{\mathcal{R}}_{S,x}) \\
&= \max\{v^{\mathcal{R}}_{S,x}(T) - x(T) \mid T \in \mathcal{T}_{k\ell}(S)\} \\
&= \max\{\max\{v(T \cup Q) - x(T \cup Q) \mid Q \subseteq N \setminus S\} \mid T \in \mathcal{T}_{k\ell}(S)\} \\
&= \max\{v(P) - x(P) \mid P \in \mathcal{T}_{k\ell}(N)\} = s_{k\ell}(x,v).
\end{aligned}$$

Hence, if $x \in \mathcal{PK}(N,v,\mathcal{R})$, then

$$s_{k\ell}(x^S, v^{\mathcal{R}}_{S,x}) = s_{k\ell}(x,v) = s_{\ell k}(x,v) = s_{\ell k}(x^S, v^{\mathcal{R}}_{S,x})$$

and the proof is complete. **q.e.d.**

Let $\Gamma_{\mathcal{U}} = \{(N,v) \mid N \subseteq \mathcal{U}\}$ be the set of all games.

Corollary 5.2.2. *The prekernel satisfies* RGP *on $\Gamma_{\mathcal{U}}$.*

Lemma 5.2.3. *The prekernel satisfies* CRGP *on $\Delta_{\mathcal{U}}$ (see Definition 3.8.11).*

The proof of Lemma 5.2.3 is similar to the proof of Lemma 5.2.1; hence it is omitted.

Corollary 5.2.4. *The prekernel satisfies* CRGP *on $\Gamma_{\mathcal{U}}$.*

Now we prove Kohlberg's (1971) characterization of the prenucleolus, which will enable us to deduce RGP. (Actually, Kohlberg (1971) deals only with the nucleolus.) We start with the following definition.

Definition 5.2.5. *Let $(N,v) \in \Gamma_{\mathcal{U}}$ and let $\mathcal{R}$ be a coalition structure of N. For every $x \in \mathbb{R}^N$ and $\alpha \in \mathbb{R}$ denote*

$$\mathcal{D}(\alpha,x,v) = \{S \subseteq N \mid e(S,x,v) \geq \alpha\}.$$

A vector $x \in X(N,v,\mathcal{R})$ has **Property I** *with respect to $(N,v,\mathcal{R})$ if the following condition is satisfied for all $\alpha \in \mathbb{R}$ such that $\mathcal{D}(\alpha,x,v) \neq \emptyset$: If $y \in \mathbb{R}^N$ satisfies $y(R) = 0$ for all $R \in \mathcal{R}$ and $y(S) \geq 0$ for all $S \in \mathcal{D}(\alpha,x,v)$, then $y(S) = 0$ for all $S \in \mathcal{D}(\alpha,x,v)$.*

Theorem 5.2.6. *Let $(N,v,\mathcal{R}) \in \Delta_{\mathcal{U}}$ and $x \in X(N,v,\mathcal{R})$. Then $x = \nu(N,v,\mathcal{R})$ if and only if x has property* I.

Proof: Necessity. Assume that $x = \nu(N,v,\mathcal{R})$. Let $\alpha \in \mathbb{R}$ satisfy $\mathcal{D}(\alpha,x,v) \neq \emptyset$ and let $y \in \mathbb{R}^N$ satisfy $y(R) = 0$ for all $R \in \mathcal{R}$ and $y(S) \geq 0$ for all $S \in \mathcal{D}(\alpha,x,v)$. Define $z_\varepsilon = x + \varepsilon y$ for every $\varepsilon > 0$. Then $z_\varepsilon \in X(N,v,\mathcal{R})$. Choose $\varepsilon^* > 0$ such that, for all $S \in \mathcal{D}(\alpha,x,v)$ and all $T \in 2^N \setminus \mathcal{D}(\alpha,x,v)$,

$$e(S, z_{\varepsilon^*}, v) > e(T, z_{\varepsilon^*}, v). \tag{5.2.1}$$

For every $S \in \mathcal{D}(\alpha,x,v)$,

$$\begin{aligned} e(S, z_{\varepsilon^*}, v) &= v(S) - (x(S) + \varepsilon^* y(S)) \\ &= e(S, x, v) - \varepsilon^* y(S) \qquad \leq e(S, x, v). \end{aligned} \tag{5.2.2}$$

Now assume, on the contrary, that there is $S \in \mathcal{D}(\alpha, x, v)$ such that $y(S) > 0$. By (5.2.1) and (5.2.2) we obtain $\theta(x) >_{lex} \theta(z_{\varepsilon^*})$, which is a contradiction.

Sufficiency. Let $x \in X(N, v, \mathcal{R})$ have Property I and let $z = \nu(N, v, \mathcal{R})$. Denote

$$\{e(S, x, v) \mid S \subseteq N\} = \{\alpha_1, \ldots, \alpha_p\},$$

where $\alpha_1 > \cdots > \alpha_p$. Define $y = z - x$. Then $y(R) = 0$ for all $R \in \mathcal{R}$. Also, as $\theta(x) \geq_{lex} \theta(z)$, if $S \in \mathcal{D}(\alpha_1, x, v)$, then $e(S, x, v) = \alpha_1 \geq e(S, z, v)$. Hence

$$e(S, x, v) - e(S, z, v) = (z - x)(S) = y(S) \geq 0.$$

Therefore, by our assumption, $y(S) = 0$ for all $S \in \mathcal{D}(\alpha_1, x, v)$.

Assume now that $y(S) = 0$ for all $S \in \mathcal{D}(\alpha_t, x, v)$ for some $1 \leq t < p$. Then, as $\theta(x) \geq_{lex} \theta(z)$,

$$e(S, x, v) = \alpha_{t+1} \geq e(S, z, v) \text{ for all } S \in \mathcal{D}(\alpha_{t+1}, x, v) \setminus \mathcal{D}(\alpha_t, x, v).$$

Hence $y(S) \geq 0$ for all $S \in \mathcal{D}(\alpha_{t+1}, x, v)$. Again, by our assumption, $y(S) = 0$ for all $S \in \mathcal{D}(\alpha_{t+1}, x, v)$. We conclude that $y(S) = 0$ for all $S \subseteq N$. Hence, $y = 0$ and $x = z$. **q.e.d.**

Theorem 5.2.7. *The prenucleolus satisfies* RGP *on* $\Delta_{\mathcal{U}}$.

Proof: Let $(N, v, \mathcal{R}) \in \Delta_{\mathcal{U}}$, $x = \nu(N, v, \mathcal{R})$, and let $\emptyset \neq S \subseteq N$. We shall prove that x^S has Property I with respect to the game with coalition structure $(S, v^{\mathcal{R}}_{S,x})$. Indeed, let $\alpha \in \mathbb{R}$ satisfy $\mathcal{D}(\alpha, x^S, v^{\mathcal{R}}_{S,x}) \neq \emptyset$ and let $y^S \in \mathbb{R}^S$ satisfy $y^S(R) = 0$ for all $T \in \mathcal{R}_{|S}$ and $y^S(Q) \geq 0$ for all $Q \in \mathcal{D}(\alpha, x^S, v^{\mathcal{R}}_{S,x})$. Then

$$\begin{aligned} &\{T \cap S \mid T \in \mathcal{D}(\alpha, x, v) \text{ and } \emptyset \neq T \cap S \notin \mathcal{R}_{|S}\} \\ &= \mathcal{D}(\alpha, x^S, v^{\mathcal{R}}_{S,x}) \setminus \left(\mathcal{R}_{|S} \cup \{\emptyset\}\right). \end{aligned} \tag{5.2.3}$$

Let $y = (y^S, 0^{N \setminus S})$. Then $y \in \mathbb{R}^N$, $y(R) = 0$ for all $R \in \mathcal{R}$, and, by (5.2.3), $y(Q) \geq 0$ for all $Q \in \mathcal{D}(\alpha, x, v)$. By Theorem 5.2.6, x has Property I. Hence $y(Q) = 0$ for all $Q \in \mathcal{D}(\alpha, x, v)$. Therefore (5.2.3) implies that $y^S(Q) = 0$ for all $Q \in \mathcal{D}(\alpha, x^S, v^{\mathcal{R}}_{S,x})$. **q.e.d.**

Corollary 5.2.8. *The prenucleolus satisfies* RGP *on* $\Gamma_{\mathcal{U}}$.

Example 5.2.9. Let $\Gamma^{\mathcal{C}}_{\mathcal{U}} = \{(N, v) \in \Gamma_{\mathcal{U}} \mid \mathcal{C}(N, v) \neq \emptyset\}$ and let $\sigma = \mathcal{PN}$ on $\Gamma^{\mathcal{C}}_{\mathcal{U}}$. By Theorem 5.1.14, $\sigma(N, v)$ is a singleton for every $(N, v) \in \Gamma^{\mathcal{C}}_{\mathcal{U}}$. Also, by Theorem 5.1.16, $\sigma(N, v) \subseteq \mathcal{C}(N, v)$ for all $(N, v) \in \Gamma^{\mathcal{C}}_{\mathcal{U}}$. Thus, σ satisfies NE and IR. Moreover, by Lemma 2.3.16 and Corollary 5.2.8, σ satisfies RGP on $\Gamma^{\mathcal{C}}_{\mathcal{U}}$. By Theorem 3.6.1, σ does not satisfy SUPA, if $\mathcal{U}$ contains at least three players. Thus, SUPA is independent of NE, IR, and RGP on $\Gamma^{\mathcal{C}}_{\mathcal{U}}$ (see Remark 3.6.8).

Example 5.2.10. Let σ be the solution on the set $\Gamma_{\mathcal{U}}^{tb}$ of totally balanced games defined by $\sigma(N,v) = \mathcal{C}(N,v) \cap \mathcal{PK}(N,v)$, $(N,v) \in \Gamma_{\mathcal{U}}^{tb}$. Thus, σ satisfies IR. Theorems 5.1.16 and 5.1.17 imply that σ satisfies NE. By Theorem 3.7.1 and Corollaries 5.2.2 and 5.2.4, σ satisfies WRGP and CRGP. Thus, SUPA is independent of NE, IR, RGP, and CRGP in Theorem 3.7.1 when $|\mathcal{U}| \geq 4$.

Exercises

Exercise 5.2.1. Show that there exist a game (N,v), $|N| = 3$, a vector $x \in X(N,v)$, and a sufficient set $\mathcal{Q} \subseteq \mathcal{P}(N)$ (see Exercise 2.3.2) such that $x^S \in \mathcal{PK}(S, v_{S,x})$ for all $S \in \mathcal{Q}$ and $x \notin \mathcal{PK}(N,v)$. (Exercise 4.2.4 is useful.)

Exercise 5.2.2. Prove that the positive prekernel (defined by (5.1.3)) satisfies RGP and CRGP on $\Delta_{\mathcal{U}}$.

Exercise 5.2.3. Show by means of an example that none of the following solutions satisfies RGP^M (the Moulin reduced game property; see Exercise 2.3.3) on $\Gamma_{\mathcal{U}}$, provided that $|\mathcal{U}| \geq 3$: $\mathcal{PN}$, $\mathcal{PK}$, $\mathcal{PM}_r$, $\mathcal{PM}_{sr}$, $\mathcal{PM}$, and $\mathcal{PMB}$.

Exercise 5.2.4. Let $N = \{1,\ldots,5\}$. Define $v(S)$, $S \subseteq N$, by

$$v(S) = \begin{cases} 0 & \text{, if } S \in \{\emptyset, N, \{1,2,5\}, \{1,4,5\}, \{2,3\}, \{3,4\}\}, \\ -1 & \text{, if } |S| = 1, \\ -4 & \text{, otherwise.} \end{cases}$$

Verify that $\nu(N,v) = 0 \in \mathbb{R}^N$.

Exercise 5.2.5. Show by an example that the prekernel for the grand coalition does not satisfy RCP (see Definition 2.3.18) on $\Gamma_{\mathcal{U}}$, if $|\mathcal{U}| \geq 5$. (Hint: Consider the game (N,v) of Exercise 5.2.4 and the prekernel of the reduced game with respect to the prenucleolus and $\{1,\ldots,4\}$.)

Remark 5.2.11. Note that Chang and Kan (1992) present an eight-person game with a coalition structure which shows that the prekernel does not satisfy RCP.

5.3 Desirability, Equal Treatment, and the Prekernel

Let (N,v) be a game. We now introduce a partial ordering of the players in their ability to produce payoffs.

Definition 5.3.1. *Let (N,v) be a game. A player $k \in N$ is said to be* **at least as desirable as** *a player $\ell \in N$ with respect to (N,v), and this is denoted by $k \succsim_v \ell$, if*

$$v(S \cup \{k\}) \geq v(S \cup \{\ell\}) \text{ for all } S \subseteq N \setminus \{k,\ell\}.$$

Intuitively, if $k \succsim_v \ell$, and $k, \ell \notin S$, then S will prefer k to ℓ as an additional partner.

Remark 5.3.2. If $k \succsim_v \ell$ and $\ell \succsim_v k$, then we write $k \sim_v \ell$. Clearly, $k \sim_v \ell$ if and only if k and ℓ are substitutes (see Definition 3.8.5) with respect to the game (N, v).

Theorem 5.3.3. *For every game (N, v) the desirability relation $\succsim_v$ is reflexive and transitive.*

Proof: Obviously $\succsim_v$ is a reflexive relation. To prove transitivity, let $k, \ell, m \in N$ satisfy $k \succsim_v \ell$ and $\ell \succsim_v m$. We may assume that $k \neq m$. Let $S \subseteq N \setminus \{k, m\}$. If $\ell \notin S$, then $v(S \cup \{k\}) \geq v(S \cup \{\ell\}) \geq v(S \cup \{m\})$. If $\ell \in S$ let $T = S \setminus \{\ell\}$. Then

$$\begin{aligned} v(S \cup \{k\}) = v(T \cup \{k\} \cup \{\ell\}) &\geq v(T \cup \{k\} \cup \{m\}) \\ &\geq v(T \cup \{\ell\} \cup \{m\}) = v(S \cup \{m\}). \end{aligned}$$

Thus, $k \succsim_v \ell$. **q.e.d.**

Corollary 5.3.4. *For every game (N, v), $\sim_v$ is an equivalence relation.*

Theorem 5.3.5. *Let $(N, v, \mathcal{R})$ be a game with coalition structure and let $x \in \mathcal{PK}(N, v, \mathcal{R})$. If $k, \ell \in R \in \mathcal{R}$ satisfy $k \succsim_v \ell$, then $x^k \geq x^\ell$.*

Proof: Assume, on the contrary, that $x^\ell > x^k$. Choose $S \in \mathcal{T}_{\ell k}$ such that $s_{\ell k}(x, v) = e(S, x, v)$. Let $T = (S \setminus \{\ell\}) \cup \{k\}$. Then $v(T) \geq v(S)$, because $k \succsim_v \ell$. Therefore, $e(T, x, v) > e(S, x, v)$. Thus,

$$s_{k\ell}(x, v) \geq e(T, x, v) > e(S, x, v) = s_{\ell k}(x, v),$$

which contradicts the assumption that $x \in \mathcal{PK}(N, v, \mathcal{R})$. **q.e.d.**

Corollary 5.3.6. *Let $(N, v, \mathcal{R})$ be a game with coalition structure and let $x \in \mathcal{PK}(N, v, \mathcal{R})$. If k and ℓ are partners in $\mathcal{R}$ and $k \sim_v \ell$, then $x^k = x^\ell$.*

Corollary 5.3.7. *Let (N, v) be a symmetric game and let $\mathcal{R}$ be a coalition structure of N. Then $\mathcal{PK}(N, v, \mathcal{R})$ consists of the unique member of $X(N, v, \mathcal{R})$ in which the members of every coalition $R \in \mathcal{R}$ receive equal payments.*

Theorem 5.3.5 motivates the following definition. Let $\mathcal{U}$ be a set of players and Γ be a set of games.

Definition 5.3.8. *Let $(N, v) \in \Gamma$ and let σ be a solution on Γ. Then σ* **preserves the desirability relation** *of (N, v) if $x^k \geq x^\ell$ for every $x \in \sigma(N, v)$ and all players $k, \ell \in N$ satisfying $k \succsim_v \ell$.*

By Theorem 5.3.5 the prekernel preserves the desirability relation of every game.

Definition 5.3.9. *A solution σ on Γ has the* **equal treatment property** *(*ETP*) if the following condition is satisfied: If $(N, v) \in \Gamma$, $x \in \sigma(N, v)$, and if $k, \ell \in N$ satisfy $k \sim_v \ell$, then $x^k = x^\ell$.*

Remark 5.3.10. By Corollary 5.3.6, the prekernel satisfies ETP on every set of games.

Let Δ be a set of games with coalition structures.

Definition 5.3.11. *A solution σ on Δ has the* **restricted** *equal treatment property (*RETP*) if, for all $(N, v, \mathcal{R}) \in \Delta$, $x \in \sigma(N, v, \mathcal{R})$, the following condition is satisfied: If $\{k, \ell\} \in \mathcal{P}(\mathcal{R})$ satisfies $k \sim_v \ell$, then $x^k = x^\ell$.*

Remark 5.3.12. The prekernel satisfies RETP on every set Δ of games with coalition structures.

Definition 5.3.13. *Let (N, v) be a game and let $k, \ell \in N$. Player k is* **more** *desirable than player ℓ, and this is denoted by $k \succ_v \ell$, if $k \succsim_v \ell$ but not $\ell \succsim_v k$.*

Remark 5.3.14. The prekernel may not preserve the strong desirability relation $\succ_v$, as shown in Exercise 5.3.1.

The results of this section are due to Maschler and Peleg (1966).

Exercises

Exercise 5.3.1. Let $N = \{1, 2, 3\}$ and $v(S)$, $S \subseteq N$, be defined by $v(\{1, 2\}) = 1$, $v(N) = 6$, and $v(S) = 0$ otherwise. Prove that $\mathcal{PK}(N, v) = \{(2, 2, 2)\}$ and that $1 \succ_v 3$.

Exercise 5.3.2. Let (N, v) be the coalitional game corresponding to a weighted majority game (see Definition 2.2.11). Show that $\succsim_v$ is a complete relation.

5.4 An Axiomatization of the Prekernel

Let (N, v) be a game. We now recall the definition of the prekernel of (N, v),

$$\mathcal{PK}(N, v) = \{x \in X(N, v) \mid s_{ij}(x, v) = s_{ji}(x, v) \text{ for all } i \neq j\}. \tag{5.4.1}$$

A direct interpretation of (5.4.1) has always been problematic. For example, Maschler, Peleg, and Shapley (1979) write: "Any attempt at providing an intuitive interpretation to the definition of the prekernel seems to rely on *interpersonal comparison of utilities*. The quantity $s_{ij}(x)$, which measures i's "strength" against j, would there be interpreted as, essentially, the maximum gain (or, if negative, the minimal loss) that i would obtain by "bribing" some players other than j to depart from x, giving each of them a very small bonus. If we compare $s_{ij}(x)$ with $s_{ji}(x)$, we in effect compare i's utility units with j's utility units and implicitly assume that the "intensity of feeling" of i towards i's utility units is, in some sense, equal to the "intensity of feeling" of j towards j's utility units. Since no clear meaning of "intensity of feeling" – interpersonally comparable – is known at present, the prekernel was never considered a satisfactory solution concept, from the intuitive point of view."

In this section we shall provide an axiomatization of the prekernel, which avoids any reference to interpersonal comparison of utilities.

Let $\mathcal{U}$ be a set of players and

$$\Gamma_{\mathcal{U}} = \{(N, v) \mid (N, v) \text{ is a game and } N \subseteq \mathcal{U}\}$$

be the set of all games. First, we need the following simple result. The easy proof is left to the reader (see Exercise 5.4.1).

Remark 5.4.1. The prekernel satisfies COV on every set of games.

Now we are ready for the axiomatization of the prekernel.

Theorem 5.4.2. *There is a unique solution on* $\Gamma_{\mathcal{U}}$ *that satisfies* NE, PO, COV, ETP, RGP, *and* CRGP, *and it is the prekernel.*

We postpone the proof of Theorem 5.4.2 and show the following lemma.

Lemma 5.4.3. *Let* σ *be a solution on* $\Gamma_{\mathcal{U}}$ *that satisfies* NE, PO, COV, *and* ETP *on the class of all two-person games. Then for every two-person game* $(N, v) \in \Gamma_{\mathcal{U}}$, $\sigma(N, v)$ *is the* **standard solution** *of* (N, v). *That is,* $\sigma(N, v) = \{x\}$, *where*

$$x^i = \frac{v(N) - \sum_{j \in N} v(\{j\})}{2} + v(\{i\}) \text{ for all } i \in N. \tag{5.4.2}$$

Proof: Let (N, v) be a two-person game and let (N, w) be the 0-normalized game defined by $w(\emptyset) = w(\{i\}) = 0$ for $i \in N$ and $w(N) = v(N) - \sum_{i \in N} v(\{i\})$. Let $y, z \in \mathbb{R}^N$ be defined by $y^i = w(N)/2$ and $z^i = v(\{i\})$ for all $i \in N$. By NE, ETP, and PO, $\sigma(N, w) = \{y\}$. Also, $v = w + z$, so COV completes the proof. **q.e.d.**

Proof of Theorem 5.4.2: By Corollary 5.1.18, Definition 5.1.1, Remarks 5.4.1 and 5.3.10, and Corollaries 5.2.2 and 5.2.4, the prekernel satisfies the foregoing six axioms. Thus, it remains to prove the uniqueness part of the theorem. Let σ be a solution on $\Gamma_{\mathcal{U}}$ that satisfies the foregoing six axioms and let $(N, v) \in \Gamma_{\mathcal{U}}$ be an n-person game. If $n = 1$, then $\sigma(N, v) = \mathcal{PK}(N, v)$ by NE and PO. For the case $n = 2$ Lemma 5.4.3 shows that $\sigma(N, v) = \mathcal{PK}(N, v)$.

Now, we assume that $n \geq 3$. If $x \in \sigma(N,v)$, then, by RGP, $x^S \in \sigma(S, v_{S,x})$ for every $S \in \mathcal{P}(N)$. Hence $x^S \in \mathcal{PK}(S, v_{S,x})$ for all $S \in \mathcal{P}(N)$. As $\mathcal{PK}$ satisfies CRGP, $x \in \mathcal{PK}(N,v)$. Conversely, let $x \in \mathcal{PK}(N,v)$. Then $x^S \in \mathcal{PK}(S, v_{S,x})$ for every $S \in \mathcal{P}(N)$, because $\mathcal{PK}$ satisfies RGP. Hence, $x^S \in \sigma(S, v_{S,x})$ for every $S \in \mathcal{P}(N)$. Thus, by CRGP, $x \in \sigma(N,v)$. **q.e.d.**

The following examples show that the six properties that characterize the prekernel are logically independent.

Example 5.4.4 (The empty solution). Let $\sigma(N,v) = \emptyset$ for every game $(N,v) \in \Gamma_{\mathcal{U}}$. Then σ satisfies all the foregoing properties except NE.

Example 5.4.5. For $(N,v) \in \Gamma_{\mathcal{U}}$ let

$$\sigma(N,v) = \{x \in X^*(N,v) \mid s_{ij}(x,v) = s_{ji}(x,v) \text{ for all } i \neq j\}.$$

Then, as the reader can easily verify, σ satisfies NE, COV, ETP, RGP, and CRGP. Clearly, σ violates PO.

Example 5.4.6 (The equal split solution). For every game $(N,v) \in \Gamma_{\mathcal{U}}$ define $\sigma(N,v) = \{x\}$, where $x^i = v(N)/|N|$ for all $i \in N$. Then σ satisfies all axioms of Theorem 5.4.2 except COV.

Example 5.4.7. The solution $X(\cdot)$ on $\Gamma_{\mathcal{U}}$ satisfies all axioms except ETP.

Example 5.4.8. Let (N,v) be a game and $i,j \in N$. Players i and j are *equivalent*, written $i \cong_v j$, if

$$v(S \cup \{i\}) - v(\{i\}) = v(S \cup \{j\}) - v(\{j\}) \text{ for all } S \subseteq N \setminus \{i,j\}.$$

Now, let

$$\sigma(N,v) = \{x \in X(N,v) \mid x^i - v(\{i\}) = x^j - v(\{j\}), \text{ if } i \cong_v j,\ i,j \in N\}.$$

The reader can easily verify that σ satisfies NE, PO, COV, and ETP. Thus, by Lemma 5.4.3, $\sigma(N,v) = \mathcal{PK}(N,v)$ in the case $|N| = 2$. In order to prove that σ satisfies CRGP, let $x \in X(N,v)$ such that $x^S \in \sigma(S, v_{S,x})$ for all $S \in \mathcal{P}(N)$. By the foregoing remarks, $x^S \in \mathcal{PK}(S, v_{S,x})$ for every $S \in \mathcal{P}(N)$. As $\mathcal{PK}$ satisfies CRGP, $x \in \mathcal{PK}(N,v)$. Now, $\mathcal{PK}$ satisfies ETP and COV. Hence $\mathcal{PK}(N,v) \subseteq \sigma(N,v)$. Thus, $x \in \sigma(N,v)$. Finally, if $\mathcal{U}$ contains at least three players, then $\sigma \neq \mathcal{PK}$; therefore σ does not satisfy RGP.

Example 5.4.9. The prenucleolus $\mathcal{PN}$ on $\Gamma_{\mathcal{U}}$ satisfies NE, PO, COV, ETP, and RGP. If $|\mathcal{U}| \geq 4$, then $\mathcal{PN} \neq \mathcal{PK}$ (see Exercise 5.4.3). Hence $\mathcal{PN}$ does not satisfy CRGP.

Now we generalize Theorem 5.4.2 to games with coalition structures. Let, as usual, $\Delta_{\mathcal{U}}$ denote the set of all games with coalition structures. First we need the following definition.

Definition 5.4.10. *A solution σ on a set Δ of games with coalition structures is* **efficient** (EFF) *if, for every $(N, v, \mathcal{R}) \in \Delta$, $\sigma(N, v, \mathcal{R}) \subseteq X(N, v, \mathcal{R})$.*

Clearly, EFF is a generalization of PO to games with coalition structures.

Theorem 5.4.11. *There is a unique solution on $\Delta_{\mathcal{U}}$ that satisfies* NE, EFF, COV, RETP, RGP, *and* CRGP, *and it is the prekernel.*

The proof of Theorem 5.4.11 is similar to that of Theorem 5.4.2. Hence it is omitted.

Theorems 5.4.2 and 5.4.11 are due to Peleg (1986).

Exercises

Exercise 5.4.1. Prove Remark 5.4.1.

Exercise 5.4.2. Show that $|\mathcal{PK}(N, v)| = 1$ for every three-person game (N, v).

Exercise 5.4.3. Let (M, u) be given by (3.7.1). Show that $\mathcal{PK}(M, u) = \mathcal{C}(M, u)$ (see Lemma 3.7.3).

5.5 Individual Rationality and the Kernel

We shall now prove that the prekernel satisfies IR on the class of weakly superadditive (i.e., 0-monotonic) games. This will enable us to explore the relationship between the prekernel and the kernel, an individually rational variant of the prekernel. Moreover, we shall show that the semi-reactive bargaining set coincides with the union of the core and the kernel for every superadditive coalitional simple game. We start with the following definition.

Definition 5.5.1. *Let N be a coalition and $\mathcal{D} \subseteq 2^N$. The collection $\mathcal{D}$ is* **separating** *if the following condition is satisfied. If $k, \ell \in N$, $k \neq \ell$, and $\mathcal{D} \cap \mathcal{T}_{k\ell}(N) \neq \emptyset$ then $\mathcal{D} \cap \mathcal{T}_{\ell k}(N) \neq \emptyset$.*

Notation 5.5.2. For every game (N, v) and every $x \in \mathbb{R}^N$ let $\mathcal{D}(x, v)$ be defined by

$$\mathcal{D}(x, v) = \\ \{S \in 2^N \setminus \{N, \emptyset\} \mid e(S, x, v) \geq e(T, x, v) \text{ if } T \in 2^N \setminus \{N, \emptyset\}\}. \tag{5.5.1}$$

Lemma 5.5.3. *Let (N, v) be a game and $x \in \mathcal{PK}(N, v)$. Then $\mathcal{D}(x, v)$ is separating.*

The proof of Lemma 5.5.3 is straightforward.

Corollary 5.5.4. *Under the assumptions of Lemma 5.5.3,*

$$\bigcap_{S \in \mathcal{D}(x,v)} S = \emptyset.$$

Let $\mathcal{U}$ be a set of players. We shall show that the prekernel satisfies REBE (see Definition 2.3.9 (2)) on every set of games.

Theorem 5.5.5. *The prekernel satisfies* REBE *on every set* $\Gamma \subseteq \Gamma_{\mathcal{U}}$ *of games.*

Proof: Let $(N,v) \in \Gamma$ and $x \in \mathcal{PK}(N,v)$. We have to prove that

$$x^i \geq b^i_{\min}(N,v) = \min_{S \subseteq N \setminus \{i\}} (v(S \cup \{i\}) - v(S)) \text{ for all } i \in N.$$

Assume, on the contrary, that there exists a player $k \in N$ such that

$$x^k < b^k_{\min}(N,v). \tag{5.5.2}$$

By (5.5.2), for all $S \subseteq N \setminus \{k\}$, we obtain

$$\begin{aligned} e(S \cup \{k\}, x, v) &= v(S \cup \{k\}) - x(S) - x^k \\ &> v(S) - x(S) \qquad\qquad = e(S, x, v). \end{aligned} \tag{5.5.3}$$

In particular, $e(\{k\}, x, v) > e(\emptyset, x, v) = 0$ so that $N \setminus \{k\} \notin \mathcal{D}(x,v)$. Hence $k \in \bigcap_{S \in \mathcal{D}(x,v)} S$, which contradicts Corollary 5.5.4. **q.e.d.**

Let $\Gamma^w_{\mathcal{U}}$ be the set of all weakly superadditive games.

Theorem 5.5.6. *The prekernel satisfies* IR *on* $\Gamma^w_{\mathcal{U}}$.

Proof: Note that $b^i_{\min}(N,v) = v(\{i\})$ for $i \in N$ and every 0-monotonic game (N,v). Thus the proof is completed by Theorem 5.5.5. **q.e.d.**

The kernel is defined in the following way.

Definition 5.5.7. *Let $(N,v,\mathcal{R})$ be a game with coalition structure. An imputation $x \in I(N,v,\mathcal{R})$ belongs to the* **kernel** *$\mathcal{K}(N,v,\mathcal{R})$ of $(N,v,\mathcal{R})$ if the following condition is satisfied for all $R \in \mathcal{R}$: If $k, \ell \in R$, $k \neq \ell$, then $s_{k\ell}(x,v) \geq s_{\ell k}(x,v)$ or $x^k = v(\{k\})$.*

Remark 5.5.8. If $(N,v,\mathcal{R})$ is a game with coalition structure, then $\mathcal{PK}(N,v) \cap I(N,v,\mathcal{R}) \subseteq \mathcal{K}(N,v,\mathcal{R})$.

Remark 5.5.9. The kernel satisfies COV and AN.

Remark 5.5.8 is an immediate consequence of Definitions 5.1.1 and 5.5.7. The straightforward proof of Remark 5.5.9 is left to the reader (see Exercise 5.5.1).

Definition 5.5.7 is due to Davis and Maschler (1965). Its rationale is very simple: The inequality $s_{\ell k}(x,v) > s_{k\ell}(x,v)$ calls for a transfer of money from k to ℓ unless it is prevented by individual rationality, that is, by the fact that $x^k = v(\{k\})$.

Theorem 5.5.10. *$\mathcal{PK}(N,v) = \mathcal{K}(N,v)$ for every $(N,v) \in \Gamma_{\mathcal{U}}^w$.*

Proof: Let $(N,v) \in \Gamma_{\mathcal{U}}^w$. By Remarks 5.4.1 and 5.5.9 we may assume that (N,v) is 0-normalized. By Theorem 5.5.6 and Remark 5.5.8, $\mathcal{PK}(N,v) \subseteq \mathcal{K}(N,v)$. Thus, it remains to prove that $\mathcal{K}(N,v) \subseteq \mathcal{PK}(N,v)$. We may assume that $|N| \geq 2$. Let $x \in \mathcal{K}(N,v)$. We claim that

$$H = \bigcap_{S \in \mathcal{D}(x,v)} S = \emptyset. \tag{5.5.4}$$

Assume, on the contrary, that there exists $k \in H$. Then $s_{k\ell}(x,v) > s_{\ell k}(x,v)$ for every $\ell \in N \setminus H$. Hence $x^\ell = 0$ for all $\ell \in N \setminus H$. Thus, if $S \in \mathcal{D}(x,v)$, then

$$e(S,x,v) = v(S) - x(S) = v(S) - x(H) = v(S) - x(N) \leq v(N) - x(N) = 0.$$

Now, for every $\ell \in N \setminus H$, $e(\{\ell\},x,v) = 0$, because $x^\ell = 0$. Hence, $\{\ell\} \in \mathcal{D}(x,v)$ and the desired contradiction has been obtained. Therefore (5.5.4) holds.

To complete the proof assume, on the contrary, that there exist $k, \ell \in N$, $k \neq \ell$, such that $s_{k\ell}(x,v) > s_{\ell k}(x,v)$. As $x \in \mathcal{K}(N,v)$, $x^\ell = 0$. By (5.5.4) there exists $S \in \mathcal{D}(x,v)$ such that $k \notin S$. Let $T = S \cup \{\ell\}$. Then

$$e(T,x,v) = v(T) - x(T) = v(T) - x(S) \geq v(S) - x(S) = e(S,x,v),$$

so $T \in \mathcal{D}(x,v)$. Therefore $s_{\ell k}(x,v) \geq e(T,x,v) \geq s_{k\ell}(x,v)$, and the desired contradiction has been obtained. **q.e.d.**

Corollary 5.5.11. *If $(N,v) \in \Gamma_{\mathcal{U}}^w$, then $\mathcal{PN}(N,v) = \mathcal{N}(N,v)$.*

The following examples show that Theorems 5.5.6 and 5.5.10 cannot be extended to all games which are not 0-monotonic or to games with non-trivial coalition structures.

Example 5.5.12. Let $N = \{1,2,3\}$, $v(\{1,2\}) = 10$, $v(N) = 2$, and $v(S) = 0$ otherwise. Then $\mathcal{PK}(N,v) = \{(3,3,-4)\}$ is not individually rational. Clearly, (N,v) is not 0-monotonic. Note that $\mathcal{K}(N,v) = \{(1,1,0)\}$.

Example 5.5.13. Let $N = \{1,2,3\}$, $v(\{1,2\}) = 2$, $v(\{1,3\}) = v(N) = 4$, and $v(S) = 0$ otherwise. Moreover, let $\mathcal{R} = \{\{1,2\},\{3\}\}$. An easy computation shows that $\mathcal{PK}(N,v,\mathcal{R}) = \{(3,-1,0)\}$ is not individually rational. In this example (N,v) is superadditive and $\mathcal{K}(N,v,\mathcal{R}) = \{(2,0,0)\}$.

Theorems 5.5.6 and 5.5.10 are due to Maschler, Peleg, and Shapley (1972).

The following remark provides an alternative proof of the existence theorem of the bargaining set $\mathcal{M}$.

Remark 5.5.14. Let $(M, v, \mathcal{R})$ be a game with coalition structure and let $I(N, v, \mathcal{R}) \neq \emptyset$. Then

$$\mathcal{N}(N, v, \mathcal{R}) \subseteq \mathcal{K}(N, v, \mathcal{R}) \text{ (Schmeidler (1969));} \tag{5.5.5}$$
$$\mathcal{K}(N, v, \mathcal{R}) \subseteq \mathcal{M}(N, v, \mathcal{R}) \text{ (Davis and Maschler (1965)).} \tag{5.5.6}$$

The proof of this remark is left to the reader (see Exercises 5.5.3 and 5.5.5). It should be noted that Example 5.5.13 also shows that Theorem 4.4.8 cannot be extended to games with coalition structures. Indeed, the prekernel is a subsolution even of the reactive prebargaining set (see Exercise 5.5.5).

Now we shall discuss the relation of the semi-reactive bargaining set and the kernel and the core when restricted to superadditive coalitional simple games (see Exercise 4.1.1).

Theorem 5.5.15. *If (N, v) is a superadditive coalitional simple game, then $\mathcal{PM}_{sr}(N, v) = \mathcal{C}(N, v) \cup \mathcal{PK}(N, v)$.*

Proof: Let (N, v) be a superadditive coalitional simple game. In view of Theorems 4.4.8, 4.4.9, and 5.5.10, and of Remarks 4.4.6 and 5.1.2 we have to show that $\mathcal{M}_{sr}(N, v) \subseteq \mathcal{C}(N, v) \cup \mathcal{K}(N, v)$. We distinguish the following cases.

(1) $\mathcal{C}(N, v) \neq \emptyset$. Let $x \in I(N, v) \setminus \mathcal{C}(N, v)$. It remains to show that $x \notin \mathcal{M}_{sr}(N, v)$. In view of the fact that x does not belong to the core of (N, v), there is some player $\ell \in N \setminus V$ satisfying $x^\ell > v(\{\ell\}) = 0$, where V denotes the set of veto players. With $P = N \setminus \{\ell\}$ we come up with $e(P, x, v) = 1 - x(P) = 1 - x(N) + x^\ell = x^\ell$. Moreover, $e(Q, x, v) = -x(Q) \leq -x^\ell < 0$ holds true for any $Q \subseteq N$ satisfying $\ell \in Q$ and $V \setminus Q \neq \emptyset$. These observations directly show that every player in V has a justified objection against ℓ via P in the sense of the semi-reactive bargaining set.

(2) $\mathcal{C}(N, v) = \emptyset$. Let $x \in I(N, v) \setminus \mathcal{K}(N, v)$. It remains to show that $x \notin \mathcal{M}_{\mathrm{sr}}(N, v)$. Indeed, there are distinct players k and ℓ such that $s_{k\ell}(x, v) > s_{\ell k}(x, v)$ and $x^\ell > v(\{\ell\})$. By the absence of veto players (see Exercise 4.1.1) we have $e(N \setminus \{k\}, x, v) = x^k \geq 0$, so $s_{k\ell}(x, v) > s_{\ell k}(x, v) \geq 0$. Let $P \in \mathcal{T}_{k\ell}$ be a maximal coalition with $e(P, x, v) = s_{k\ell}(x, v)$. For every coalition $Q \in \mathcal{T}_{\ell k}$ we have $e(Q, x, v) < e(P, x, v)$ and, therefore, Q can only be used in a counterobjection if $Q \cap P = \emptyset$ and $e(Q, x, v) \geq 0$. In this case Q must be a winning coalition, because $x^\ell > 0$. However, disjoint winning coalitions do not exist in a superadditive simple game. We conclude that k has a justified objection against ℓ via P. **q.e.d.**

Remark 5.5.16. Theorem 5.5.15 shows, e.g., that the kernel of the seven-person projective game (see 4.4.3) coincides with its (semi-)reactive prebargaining set. Maschler and Peleg (1967) computed the explicit shape of the (pre)kernel (see (4.4.1)).

Remark 5.5.17. Neither the (pre)nucleolus nor the (pre)kernel satisfies aggregate monotonicity. Indeed, in order to apply Example 4.5.3 we first repeat that the games (N, v') and (N, u') defined in Remark 4.5.5 are superadditive. Hence, by Theorem 5.5.10 and Corollary 5.5.11, we only have to show that $\bar{x} = 0 \in \mathbb{R}^N$ is the prenucleolus point of (N, v'), which can easily be done by applying Corollary 6.1.3.

Exercises

Exercise 5.5.1. Prove Remark 5.5.9.

Let $(N, v, \mathcal{R})$ be a game with coalition structure.

Exercise 5.5.2. Let $x \in \mathcal{K}(N, v, \mathcal{R})$ and let $k, \ell \in R \in \mathcal{R},\ k \neq \ell$. Prove that if $k \succsim_v \ell$, then $x^k \geq x^\ell$.

Exercise 5.5.3. Prove (5.5.5).

The *positive kernel* of $(N, v, \mathcal{R})$, $\mathcal{K}_+(N, v, \mathcal{R})$, is defined by

$$\mathcal{K}_+(N, v, \mathcal{R}) = \left\{ x \in I(N, v, \mathcal{R}) \,\middle|\, \begin{array}{l} (s_{k\ell}(x, v))_+ \geq s_{\ell k}(x, v) \text{ or } x^k = v(\{k\}) \\ \text{for all } \{k, \ell\} \in \mathcal{P}(\mathcal{R}) \end{array} \right\}$$

and the *positive kernel* of (N, v) is $\mathcal{K}_+(N, v) = \mathcal{K}_+(N, v, \{N\})$. Note that

$$\mathcal{C}(N, v, \mathcal{R}) \cup \mathcal{K}(N, v, \mathcal{R}) \subseteq \mathcal{K}_+(N, v, \mathcal{R}).$$

Exercise 5.5.4. Show that $\mathcal{K}_+(N, v) = \mathcal{PK}_+(N, v)$ for every 0-monotonic game (N, v). (See Sudhölter and Peleg (2000).)

Exercise 5.5.5. Show that $\mathcal{K}_+(N, v, \mathcal{R}) \subseteq \mathcal{M}_r(N, v, \mathcal{R})$.

5.6 Reasonableness of the Prekernel and the Kernel

Let $\mathcal{U}$ be a set of players and $\Gamma_{\mathcal{U}}$ be the set of all games with players in $\mathcal{U}$.

Theorem 5.6.1. *The prekernel is reasonable on $\Gamma_{\mathcal{U}}$.*

Proof: Let $(N, v) \in \Gamma_{\mathcal{U}}$ and $x \in \mathcal{PK}(N, v)$. By Theorem 5.5.5 it remains to show that

$$x^i \leq b^i_{\max}(N, v) = \max_{S \subseteq N \setminus \{i\}} (v(S \cup \{i\}) - v(S)) \text{ for all } i \in N.$$

Assume, on the contrary, that there exists $k \in N$ such that

$$x^k > b^k_{\max}(N, v). \tag{5.6.1}$$

By (5.6.1), for all $S \subseteq N$ satisfying $k \in S$, we obtain

$$\begin{aligned} e(S \setminus \{k\}, x, v) &= v(S \setminus \{k\}) - x(S) + x^k \\ &> v(S) - x(S) \qquad = e(S, x, v). \end{aligned} \tag{5.6.2}$$

In particular, $e(N \setminus \{k\}, x, v) > e(N, x, v) = 0$ by Pareto optimality of x. Hence $k \notin \bigcup_{S \in \mathcal{D}(x,v)} S$, which contradicts Lemma 5.5.3. **q.e.d.**

Remark 5.6.2. The kernel is reasonable on every set of games.

The easy proof of Remark 5.6.2 is left to the reader (see Exercise 5.6.1). The following example shows that Theorem 5.6.1 cannot be extended to games with coalition structures.

Example 5.6.3. Let (N_1, v_1) and (N, v) be defined by $N_1 = \{1, \ldots, 4\}$, $v_1(\{1, 2\}) = 1$, $v_1(\{3, 4\}) = 2$, and $v_1(S) = 0$ otherwise, and $N = N_1 \cup \{5\}$, $v(T) = v_1(T \cap N_1)$ for all $T \subseteq N$. Then 5 is a null player (see Definition 4.1.16) of (N, v). Furthermore, let $\mathcal{R} = \{\{1, 2, 5\}, \{3\}, \{4\}\}$. Then $\mathcal{PK}(N, v, \mathcal{R}) = \{(0, 0, 0, 0, 1)\} = \mathcal{K}(N, v, \mathcal{R})$. Thus, both the prekernel and the kernel of $(N, v, \mathcal{R})$ do not satisfy REAB.

Notwithstanding Example 5.6.3 the following corollary is true. Let

$$\Delta^s_{\mathcal{U}} = \{(N, v, \mathcal{R}) \in \Delta_{\mathcal{U}} \mid (N, v) \text{ is superadditive}\},$$

where $\Delta_{\mathcal{U}}$ denotes, as usual, the set of all games with coalition structures $(N, v, \mathcal{R})$ such that $N \subseteq \mathcal{U}$.

Corollary 5.6.4. *The prekernel and the kernel satisfy* REAB *on every subset of* $\Delta^s_{\mathcal{U}}$.

Proof: By Exercises 5.1.4 and 5.5.5 the prekernel is a subsolution of the semi-reactive prebargaining set and the kernel is a subsolution of the semi-reactive bargaining set. Hence, Theorem 4.4.9 completes the proof. **q.e.d.**

Note that Corollary 5.6.4 can also be proved similarly to Theorem 5.6.1. Theorem 5.6.1 and Corollary 5.6.4 are due to Wesley (1971).

Exercises

Exercise 5.6.1. Show Remark 5.6.2.

Exercise 5.6.2. Show that the positive prekernel and the positive kernel satisfy RE on every set of games.

5.7 The Prekernel of a Convex Game

The purpose of this section is to show that the prekernel of a convex game (for the grand coalition) consists of a single point; hence it coincides with the prenucleolus. We start with the following lemmata.

Lemma 5.7.1. *Let (N,v) be a convex game, let $x \in \mathcal{C}(N,v)$, and let $\emptyset \neq S \subseteq N$. Then the reduced game $(S, v_{S,x})$ is convex.*

Proof: Let $P, T \subseteq S$. We shall show that

$$v_{S,x}(P) + v_{S,x}(T) \leq v_{S,x}(P \cap T) + v_{S,x}(P \cup T). \tag{5.7.1}$$

Clearly, (5.7.1) holds if $P \subseteq T$ or $T \subseteq P$. Therefore we may assume that $P \neq S \neq T$ and $P \neq \emptyset \neq T$. By the definition of $v_{S,x}$ there exist $Q_1, Q_2 \subseteq N \setminus S$ such that

$$v_{S,x}(P) = v(P \cup Q_1) - x(Q_1) \text{ and } v_{S,x}(T) = v(T \cup Q_2) - x(Q_2).$$

Hence

$$\begin{aligned} & v_{S,x}(P) + v_{S,x}(T) \\ & = v(P \cup Q_1) + v(T \cup Q_2) - x(Q_1) - x(Q_2) \\ & \leq v(P \cup T \cup Q_1 \cup Q_2) - x(Q_1 \cup Q_2) + \\ & \qquad v((P \cap T) \cup (Q_1 \cap Q_2)) - x(Q_1 \cap Q_2) \\ & \leq \max_{Q \subseteq N \setminus S}(v(P \cup T \cup Q) - x(Q)) + \\ & \qquad \max_{Q \subseteq N \setminus S}(v((P \cap T) \cup Q) - x(Q)). \end{aligned}$$

Therefore it suffices to show that

$$\max_{Q \subseteq N \setminus S}(v(P \cup T \cup Q) - x(Q)) = v_{S,x}(P \cup T); \text{ and} \tag{5.7.2}$$

$$\max_{Q \subseteq N \setminus S}(v((P \cap T) \cup Q) - x(Q)) = v_{S,x}(P \cap T). \tag{5.7.3}$$

If $P \cup T \neq S$, then (5.7.2) holds by definition of $v_{S,x}$. If $P \cup T = S$, then, as $x \in \mathcal{C}(N,v)$,

$$v(P \cup T \cup Q) - x(Q) \leq x(S) = v_{S,x}(S) \text{ for every } Q \subseteq N \setminus S,$$

so (5.7.2) holds in any case. In order to prove (5.7.3) we proceed similarly. If $P \cap T \neq \emptyset$, then (5.7.3) holds by definition of $v_{S,x}$. If $P \cap T = \emptyset$, then, as $x \in \mathcal{C}(N,v)$,

$$v((P \cap T) \cup Q) - x(Q) = v(Q) - x(Q) \leq 0 = v(P \cap T),$$

so (5.7.3) holds in this case. **q.e.d.**

Lemma 5.7.2. *Let (N,v) be a convex game, let $x \in X(N,v)$, and let $S \in \mathcal{D}(x,v)$ (see (5.5.1)). Then, for every $P \subseteq S$ and every $T \subseteq N \setminus S$,*

$$\max_{Q \subseteq N \setminus S} e(P \cup Q, x, v) = \max\{e(P,x,v), e(P \cup (N \setminus S), x, v)\}; \tag{5.7.4}$$

$$\max_{R \subseteq S} e(T \cup R, x, v) = \max\{e(T,x,v), e(T \cup S, x, v)\}. \tag{5.7.5}$$

Proof: Let $P, \widehat{R} \subseteq S$ and $T, \widehat{Q} \subseteq N \setminus S$ such that

$$\max_{Q \subseteq N \setminus S} e(P \cup Q, x, v) = e(P \cup \widehat{Q}, x, v), \quad \max_{R \subseteq S} e(T \cup R, x, v) = e(T \cup \widehat{R}, x, v).$$

As (N, v) is convex, we obtain

$$e(P \cup \widehat{Q}, x, v) + e(S, x, v) \leq e(P, x, v) + e(S \cup \widehat{Q}, x, v); \tag{5.7.6}$$
$$e(T \cup \widehat{R}, x, v) + e(S, x, v) \leq e(T \cup S, x, v) + e(\widehat{R}, x, v). \tag{5.7.7}$$

If $\widehat{Q} \neq N \setminus S$, then $e(S, x, v) \geq e(S \cup \widehat{Q}, x, v)$, so $e(P \cup \widehat{Q}, x, v) \leq e(P, x, v)$ by (5.7.6). Hence (5.7.4) holds in any case. If $\widehat{R} \neq \emptyset$, then $e(S, x, v) \geq e(\widehat{R}, x, v)$, so $e(T \cup \widehat{R}, x, v) \leq e(T \cup S, x, v)$ by (5.7.7). Hence (5.7.5) holds in any case.

q.e.d.

Corollary 5.7.3. *Let (N, v) be a convex game, let $x \in X(N, v)$, and let $S \in \mathcal{D}(x, v)$. Then the coalition functions $v_{S,x}$ and $v_{N \setminus S, x}$ are given by*

$$v_{S,x}(P) = \begin{cases} x(P) & , \text{ if } P = S, \emptyset \\ \max\{v(P), v(P \cup (N \setminus S)) - x(N \setminus S)\} & , \text{ otherwise} \end{cases}$$

and

$$v_{N \setminus S, x}(T) = \begin{cases} x(T) & , \text{ if } T = N \setminus S, \emptyset \\ \max\{v(T), v(T \cup S) - x(S)\} & , \text{ otherwise.} \end{cases}$$

Definition 5.7.4. *A collection $\mathcal{D}$ of subsets of N, $\emptyset \neq N$ finite, is called a* **near-ring** *of N if*

$$S, T \in \mathcal{D} \Rightarrow \begin{cases} S \cup T = N \text{ or} \\ S \cap T = \emptyset \text{ or} \\ \text{both } S \cap T \in \mathcal{D} \text{ and } S \cup T \in \mathcal{D}. \end{cases}$$

Lemma 5.7.5. *Let (N, v) be a convex game and let $x \in X(N, v)$. Then $\mathcal{D}(x, v)$ is a near-ring.*

Proof: Let $S, T \in \mathcal{D}(x, v)$. As (N, v) is convex, we obtain

$$e(S, x, v) + e(T, x, v) \leq e(S \cap T, x, v) + e(S \cup T, x, v). \tag{5.7.8}$$

If both $S \cap T \neq \emptyset$ and $S \cup T \neq N$, then $e(S, x, v) \geq e(S \cap T, x, v)$ and $e(T, x, v) \geq e(S \cup T, x, v)$. Hence, by (5.7.8), both $S \cap T$ and $S \cup T$ are in $\mathcal{D}(x, v)$.

q.e.d.

Lemma 5.7.6. *Every separating near-ring $\mathcal{D}$ of a set N of cardinality $n \in \mathbb{N}$, with the exception of $\{\emptyset\}$, contains a balanced collection of subsets of N.*

Proof: For every $k \in N$ let $\widehat{k} = \{\ell \in N \setminus \{k\} \mid \mathcal{T}_{k\ell}(N) \cap \mathcal{D} = \emptyset\} \cup \{k\}$ and define, for every $S \subseteq N$, $\widehat{S} = \{\widehat{k} \mid k \in S\}$. Then $\widehat{\mathcal{D}} = \{\widehat{S} \mid S \in \mathcal{D}\}$ is a separating near-ring of $\widehat{N}$ (see Definition 5.5.1). In fact, $\widehat{\mathcal{D}}$ is *completely* separating, that is, $\widehat{\mathcal{D}} \cap \mathcal{T}_{ij}(\widehat{N}) \neq \emptyset$ for all $i, j \in \widehat{N}$, $i \neq j$. Also, if $\widehat{\mathcal{D}}_* \subseteq \widehat{\mathcal{D}}$ is a balanced collection of $\widehat{N}$ and $\mathcal{D}_* = \{S \in 2^N \mid \widehat{S} \in \widehat{\mathcal{D}}_*\}$, then $\mathcal{D}_* \subseteq \mathcal{D}$ and $\mathcal{D}_*$ is a balanced collection of N.

Hence we may assume that $\mathcal{D}$ is completely separating. Clearly, the lemma holds for $n = 1$. Thus, let $n \geq 2$. For every $i \in N$ let $\mathcal{D}_{-i} = \{S \in \mathcal{D} \mid i \notin S\}$ and let $\mathcal{D}^*_{-i}$ be the set of members of $\mathcal{D}_{-i}$ which are maximal under inclusion. As $\mathcal{D}$ is completely separating, each member of $N \setminus \{i\}$ belongs to at least one element of $\mathcal{D}^*_{-i}$. Furthermore, by the near-ring property, the elements of $\mathcal{D}^*_{-i}$ are disjoint. Thus, for every player $i \in N$, $\mathcal{D}^*_{-i}$ is a partition of $N \setminus \{i\}$. We now observe that $\mathcal{D}^* = \bigcup_{i \in N} \mathcal{D}^*_{-i}$ is balanced. Indeed, for $S \in \mathcal{D}^*$ let $c(S) = |\{i \in N \mid S \in \mathcal{D}^*_{-i}\}|$. Then $(c(S)/(n-1))_{S \in \mathcal{D}^*}$ is a system of balancing weights for $\mathcal{D}^*$. Clearly $\mathcal{D}^* \subseteq \mathcal{D}$, so the proof is complete. **q.e.d.**

Theorem 5.7.7. *The prekernel (for the grand coalition) of a convex game consists of a single point.*

Proof: Let (N, v) be an n-person convex game. We shall proceed by induction on n. The case $n = 1$ is obvious. Thus, let $n \geq 2$. For $x \in X(N, v)$ let $\mu(x) = e(S, x, v)$ for $S \in \mathcal{D}(x, v)$. Now, let $x, y \in \mathcal{PK}(N, v)$. Without loss of generality we may assume that

$$\mu(x) \leq \mu(y). \tag{5.7.9}$$

By Lemmata 5.5.3 and 5.7.5, $\mathcal{D}(y, v)$ is a separating near-ring. Hence, by Lemma 5.7.6, $\mathcal{D}(y, v)$ contains a balanced collection $\mathcal{B}$. Let $T \in \mathcal{B}$. By (5.7.9), $e(T, x, v) \leq \mu(x) \leq \mu(y) = e(T, y, v)$. Hence, $x(T) \geq y(T)$ for all $T \in \mathcal{B}$. Multiplying these inequalities by the balancing weights and summing over all coalitions of $\mathcal{B}$, we obtain $x(N) \geq y(N)$, with equality occurring only if $\mathcal{B} \subseteq \mathcal{D}(x, v)$ and $\mu(x) = \mu(y)$. By Pareto optimality of x and y equality must occur. We conclude from the foregoing observation that there exists $S \in \mathcal{D}(x, v) \cap \mathcal{D}(y, v)$ such that

$$x(S) = y(S) \text{ and } x(N \setminus S) = y(N \setminus S). \tag{5.7.10}$$

We now conclude from (5.7.10) and Corollary 5.7.3 that $v_{S,x} = v_{S,y}$ and $v_{N \setminus S, x} = v_{N \setminus S, y}$. By Remark 5.1.2, $x, y \in \mathcal{PM}_r(N, v)$. By Theorem 4.3.4, $\mathcal{PM}(N, v) = \mathcal{C}(N, v)$. Thus $x, y \in \mathcal{C}(N, v)$. Therefore, by Lemma 5.7.1, the games $(S, v_{S,x})$ and $(N \setminus S, v_{N \setminus S, x})$ are convex. Invoking now the induction hypothesis we deduce that the prekernel of each of the reduced games $(S, v_{S,x}) = ((S, v_{S,y}))$ and $(N \setminus S, v_{N \setminus S, x}) = ((N \setminus S, v_{N \setminus S, y}))$ consists of a single point. But, by Corollary 5.2.2, $x^S \in \mathcal{PK}(S, v_{S,x})$ and $y^S \in \mathcal{PK}(S, v_{S,x})$. Hence $x^S = y^S$. Similarly, $x^{N \setminus S} = y^{N \setminus S}$, and the proof is complete. **q.e.d.**

Corollary 5.7.8. *The kernel (for the grand coalition) of a convex game consists of a single point.*

Corollary 5.7.9. *The kernel of a convex game coincides with the nucleolus of the game.*

Theorem 5.7.7 is due to Maschler, Peleg, and Shapley (1972).

5.8 The Prekernel and Syndication

We shall consider the effect of syndication on the prekernel of the following one-parameter family of five-person markets. Let $P = \{1, 2\}$, $Q = \{3, 4, 5\}$, $N = P \cup Q$, $m = 2$, $a^i = (1, 0)$, $a^j = (0, a)$ for all $i \in P$, $j \in Q$, and $w^k(x_1, x_2) = \min\{x_1, x_2\} = w(x_1, x_2)$. Then

$$\left(N, \mathbb{R}^2_+, (a^i)_{i\in N}, (w)_{i\in N}\right)$$

is a market. The corresponding market game (N, v) is given by

$$v(S) = \min\{|S \cap P|, a|S \cap Q|\} \text{ for all } S \subseteq N. \tag{5.8.1}$$

The case $a = 1/2$ was considered in Section 4.6 in relation to the bargaining set. We shall compute $\mathcal{PK}(N, v)$ for $0 \leq a < \infty$. By Remark 5.3.10 and Theorem 5.5.6 the prekernel satisfies ETP and IR. Thus, if $x \in \mathcal{PK}(N, v)$, then $x = (\alpha, \alpha, \beta, \beta, \beta)$ for some α, β satisfying

$$\alpha, \beta \geq 0 \text{ and } 2\alpha + 3\beta = v(N). \tag{5.8.2}$$

The following result is very useful.

Lemma 5.8.1. *Let $x = (\alpha, \alpha, \beta, \beta, \beta)$ for some $\alpha, \beta \geq 0$ satisfying (5.8.2). Then $x \in \mathcal{PK}(N, v)$ if and only if $\mathcal{D}(x, v) \cap \mathcal{T}_{pq} \neq \emptyset$ and $\mathcal{D}(x, v) \cap \mathcal{T}_{qp} \neq \emptyset$ for all $p \in P$ and all $q \in Q$.*

Proof: If $\mathcal{D}(x, v) \cap \mathcal{T}_{pq} \neq \emptyset$ and $\mathcal{D}(x, v) \cap \mathcal{T}_{qp} \neq \emptyset$ for all $p \in P$ and all $q \in Q$, then $s_{pq}(x, v) = s_{qp}(x, v)$ for all $p \in P$ and $q \in Q$, so $s_{k\ell}(x, v) = s_{\ell k}(x, v)$ for all $k \in N$ and all $\ell \in N \setminus \{k\}$. Thus $x \in \mathcal{PK}(N, v)$ in this case. Conversely, let $x \in \mathcal{PK}(N, v)$, and let $p \in P$ and $q \in Q$. As $\mathcal{D}(x, v)$ is separating, there exists $S \in \mathcal{D}(x, v)$ such that $p \in S$ and $Q \setminus S \neq \emptyset$. As all members of Q are substitutes, we may assume that $q \notin S$. Hence $S \in \mathcal{T}_{pq}$. **q.e.d.**

Using Lemma 5.8.1 the reader may easily verify that the prekernel is given by the following formulae:

$$0 \leq a \leq \frac{1}{3} \Rightarrow \mathcal{PK}(N, v) = \{(0, 0, a, a, a)\} \tag{5.8.3}$$

$$\frac{1}{3} < a \leq \frac{1}{2} \Rightarrow \mathcal{PK}(N,v) = \left\{ (\alpha,\alpha,\beta,\beta,\beta) \,\middle|\, \begin{array}{c} \alpha \leq \frac{9a-3}{2} \\ \alpha,\beta \geq 0, \\ 2\alpha + 3\beta = 3a \end{array} \right\} \tag{5.8.4}$$

$$\frac{1}{2} < a \leq \frac{2}{3} \Rightarrow \mathcal{PK}(N,v) = \left\{ (\alpha,\alpha,\beta,\beta,\beta) \,\middle|\, \begin{array}{c} \alpha \leq \frac{3a}{2}, \\ \alpha,\beta \geq 0, \\ 2\alpha + 3\beta = 3a \end{array} \right\} \tag{5.8.5}$$

$$\frac{2}{3} < a \leq 1 \Rightarrow \mathcal{PK}(N,v) = \left\{ (\alpha,\alpha,\beta,\beta,\beta) \,\middle|\, \begin{array}{c} \beta \leq 2(1-a), \\ \beta,\alpha \geq 0, \\ 2\alpha + 3\beta = 2 \end{array} \right\} \tag{5.8.6}$$

$$1 < a \Rightarrow \mathcal{PK}(N,v) = \{(1,1,0,0,0)\}. \tag{5.8.7}$$

Now suppose that the members of Q have decided to form a syndicate. Then $N_1 = \{1,2,Q\}$ is the new set of agents and the coalition function v_1 of the new game is given by

$$v_1(S) = \begin{cases} 0 & \text{, if } S \subseteq N_1 \text{ and } Q \notin S, \\ \min\{|S \cap P|, 3a\} & \text{, if } S \subseteq N_1 \text{ and } Q \in S. \end{cases}$$

Therefore, $\mathcal{PK}(N_1, v_1)$ can easily be computed as

$$\mathcal{PK}(N_1,v_1) = \begin{cases} \{(0,0,3a)\} & \text{, if } 0 \leq a \leq \frac{1}{3}, \\ \{(\frac{3a-1}{2}, \frac{3a-1}{2}, 1)\} & \text{, if } \frac{1}{3} < a \leq \frac{2}{3}, \\ \{(\frac{1}{2}, \frac{1}{2}, 1)\} & \text{, if } \frac{2}{3} < a. \end{cases}$$

Thus, for $a \geq 5/6$ syndication is advantageous. For $0 \leq a \leq 1/3$ there is no change in the payoff as a result of syndication, whereas for $1/3 < a < 5/6$ the payoff to the syndicate is in the "upper part" of the range of the previous payoffs. So, we may conclude that syndication is advantageous according to the prekernel.

Remark 5.8.2. For $a = 1/2$ (the example of Section 4.6)

$$\mathcal{PK}(N,v) = \text{convh}\left\{ \left(0,0,\frac{1}{2},\frac{1}{2},\frac{1}{2}\right), \left(\frac{3}{4},\frac{3}{4},0,0,0\right) \right\}$$

and $\mathcal{PK}(N_1, v_1) = \{(\frac{1}{4}, \frac{1}{4}, 1)\}$. Thus, syndication seems advantageous. However, if we consider the nucleolus, then syndication is disadvantageous. Indeed,

$$\nu(N,v) = \left(0,0,\frac{1}{2},\frac{1}{2},\frac{1}{2}\right) \text{ and } \nu(N_1,v_1) = \left(\frac{1}{4},\frac{1}{4},1\right). \tag{5.8.8}$$

Maschler (1976) has shown that, according to the bargaining set $\mathcal{M}$, syndication is advantageous. Legros (1987) investigates the relation between syndication and the nucleolus. Granot and Maschler (1997) show that the bargaining set of (N, v) coincides with the reactive bargaining set; thus it coincides with the semi-reactive (pre)bargaining set as well.

Exercises

Exercise 5.8.1. Compute the prekernel when P forms a syndicate.

Exercise 5.8.2. Show (5.8.8).

5.9 Notes and Comments

(1) In view of Exercise 5.2.3 the Moulin reduced game property may not be used to characterize the prekernel or the prenucleolus. In order to axiomatize these solutions we shall employ RGP, the Davis-Maschler reduced game property, because RGP is satisfied by the core and all of the "core-based" solutions of Exercise 5.2.3 except $\mathcal{PMB}$. However, we shall also present a characterization of the core using RGP^M (see Theorem 12.5.1).

(2) Let (N, v) be a strong simple game (see Definition 2.2.7) without null players and let $n = |N|$. Then Isbell (1956) shows (a) that $|\mathcal{W}^m| \geq n$ and (b) that a strong simple game without null players has exactly as many minimal winning coalitions as players if and only if it is a certain homogeneous weighted majority game called a *partition* game **or** if it is the seven-person projective game. Peleg (1966) showed that the kernels of partition games are star-shaped. Hence, by Remark 5.5.16, the kernel of each strong simple game without null players which has a minimal number of minimal winning coalitions, is star-shaped.

(3) Sudhölter (1996b) generalized Peleg's (1966) result by showing that the prekernel of every homogeneous weighted majority game is star-shaped. In fact, the kernels of partition games are singletons. Also, a center of this star can be selected by taking the member of the prekernel which is most egalitarian with respect to some weighted Gini index.

(4) We have seen in this chapter that the prekernel has many nice properties. Nevertheless it sometimes yields counterintuitive results. The following example is due to Davis and Maschler (1965).

Example 5.9.1. Let (N, v) be the coalitional game associated with the homogeneous weighted majority game which has $(4; 3, 1, 1, 1, 1)$ as a representation. Let $\mathcal{R} = \{\{1, 2\}, \{3\}, \{4\}, \{5\}\}$. Then

$$\mathcal{PK}(N, v, \mathcal{R}) = \mathcal{K}(N, v, \mathcal{R}) = \mathcal{N}(N, v, \mathcal{R}) = \left\{\left(\frac{1}{2}, \frac{1}{2}, 0, 0, 0\right)\right\}.$$

Thus, although 1 is much stronger than 2 in the game (N, v), the prekernel with respect to the foregoing coalition structure dictates an equal split.

The computation of the bargaining set of the foregoing example is instructive. Indeed,

$$\mathcal{M}(N, v, \mathcal{R}) = \text{convh}\left\{\left(\frac{1}{2}, \frac{1}{2}, 0, 0, 0\right), \left(\frac{3}{4}, \frac{1}{4}, 0, 0, 0\right)\right\}.$$

If $\mathcal{R}_0 = \{\{1\}, \{2, 3, 4, 5\}\}$, then

$$\mathcal{M}(N, v, \mathcal{R}_0) = \mathcal{PK}(N, v, \mathcal{R}_0) = \left\{\left(0, \frac{1}{4}, \frac{1}{4}, \frac{1}{4}, \frac{1}{4}\right)\right\}.$$

Thus 1 cannot demand more than $\frac{3}{4}$, because 2 receives $\frac{1}{4}$ in $\mathcal{M}(n, v, \mathcal{R}_0)$. Also, 2 cannot demand more than $\frac{1}{2}$, because 1 is more desirable than 2. So the bargaining set tells us that 1 has to offer 2 a payoff t satisfying $\frac{1}{4} \leq t \leq \frac{1}{2}$ in order to induce him to form the coalition $\{1, 2\}$. Unfortunately, the prekernel picks the "questionable" endpoint of the foregoing interval, namely $t = \frac{1}{2}$.

The reader is referred to Davis and Maschler (1965) for a discussion of the foregoing example which includes letters by several game theorists.

(5) Let (N, v) be a game and let $X \subseteq X(N, v)$. The kernel of (N, v) *with respect to* X, $\mathcal{K}(N, v, X)$, is the set of all $x \in X$ that satisfy the following condition: If $k, \ell \in N$, $k \neq \ell$, and $s_{k\ell}(x, v) < s_{\ell k}(x, v)$, then for every $\varepsilon > 0$ there exists $0 < \delta \leq \varepsilon$ such that $x - \delta\chi_{\{k\}} + \delta\chi_{\{\ell\}} \notin X$. Kikuta (1997) proves that, if

$$X = \{x \in X(N, v) \mid b^i_{\min}(N, v) \leq x^i \leq b^i_{\max}(N, v) \text{ for all } i \in N\},$$

then $\mathcal{K}(N, v, X) = \mathcal{PK}(N, v)$. By Remark 5.6.2 this result implies Theorem 5.5.10.

6

The Prenucleolus

This chapter is devoted to the following three topics: (1) an axiomatization of the prenucleolus, (2) an investigation of the prenucleolus on weighted majority games, and (3) an investigation of the modiclus of a game which is the restriction of the prenucleolus of a certain "replicated" game. We start with Kohlberg's characterization of the prenucleolus by balanced collections. In Sections 6.2 and 6.3 we present Sobolev's characterization of the prenucleolus. More precisely, we prove that single-valuedness, covariance, anonymity, and the reduced game property, completely determine the prenucleolus (on the set of all games). We introduce a new interesting solution in order to show that the anonymity is independent of the remaining axioms. An outline of a generalization of Sobolev's characterization to games with coalition structures is given in Section 6.4.

In Section 6.5 we study the prenucleolus of strong weighted majority games and prove that the prenucleolus induces a representation. Also, under suitable additional conditions, minimal and minimum integral representations can be derived from the prenucleolus.

In Section 6.6 we introduce a further interesting solution. The modiclus of a game with coalition structure is obtained by lexicographically minimizing the differences of excesses. It is shown that the modiclus has many properties in common with the prenucleolus. All results of Section 6.5 apply as well to the modiclus, because this solution is the prenucleolus on constant-sum games. Also, some of the results can be generalized to arbitrary weighted majority games.

6.1 A Combinatorial Characterization of the Prenucleolus

In this section we prove Kohlberg's (1971) result on the characterization of the prenucleolus by balanced collections.

Theorem 6.1.1. *Let (N, v) be a game and let $x \in X(N, v)$. Then $x = \nu(N, v)$, if and only if, for every $\alpha \in \mathbb{R}$, $\mathcal{D}(\alpha, x, v) \neq \emptyset$ implies that $\mathcal{D}(\alpha, x, v)$ is a balanced collection over N.*

Proof: Sufficiency. Let $\alpha \in \mathbb{R}$ be such that $\mathcal{D}(\alpha, x, v) \neq \emptyset$. Also, let $y \in \mathbb{R}^N$ satisfy $y(N) = 0$ and $y(S) \geq 0$ for all $S \in \mathcal{D}(\alpha, x, v)$. By our assumption $\mathcal{D}(\alpha, x, v)$ is balanced. Hence, there exist $\delta_S > 0$, $S \in \mathcal{D}(\alpha, x, v)$, such that

$$\sum_{S \in \mathcal{D}(\alpha, x, v)} \delta_S \chi_S = \chi_N. \tag{6.1.1}$$

Taking the scalar product of both sides of (6.1.1) with y we obtain

$$\sum_{S \in \mathcal{D}(\alpha, x, v)} \delta_S y(S) = y(N) = 0.$$

Therefore $y(S) = 0$ for every $S \in \mathcal{D}(\alpha, x, v)$. Thus, x has Property I (see Definition 5.2.5). By Theorem 5.2.6, $x = \nu(N, v)$.

Necessity. Assume that $x = \nu(N, v)$. By Theorem 5.2.6, x has Property I. Let $\alpha \in \mathbb{R}$ such that $\mathcal{D}(\alpha, x, v) \neq \emptyset$. Consider the linear programming problem

$$\begin{cases} \max \sum_{S \in \mathcal{D}(\alpha, x, v)} y(S) \\ \text{subject to } \; -y(S) \leq 0, \;\; S \in \mathcal{D}(\alpha, x, v) \text{ and } y(N) = 0. \end{cases} \tag{6.1.2}$$

The linear program (6.1.2) is feasible and its value is 0. Hence, its dual is feasible, that is, there exist $\beta_S \geq 0$, $S \in \mathcal{D}(\alpha, x, v)$ and $\beta_N \in \mathbb{R}$ such that

$$-\sum_{S \in \mathcal{D}(\alpha, x, v)} \beta_S \chi_S + \beta_N \chi_N = \sum_{S \in \mathcal{D}(\alpha, x, v)} \chi_S.$$

Hence $\beta_N \chi_N = \sum_{S \in \mathcal{D}(\alpha, x, v)} (1 + \beta_S) \chi_S$. As $1 + \beta_S > 0$ for $S \in \mathcal{D}(\alpha, x, v)$, $\beta_N > 0$ and $\mathcal{D}(\alpha, x, v)$ is balanced. **q.e.d.**

The following lemma is useful. If $X \subseteq Y$ for some linear space Y, then $\langle X \rangle$ denotes the linear span of X.

Lemma 6.1.2. *Let $\emptyset \neq N$ be a finite set and let $\mathcal{D}$ be a balanced collection over N. If $T \subseteq N$ satisfies $\chi_T \in \langle \{\chi_S \mid S \in \mathcal{D}\} \rangle$, then $\mathcal{D} \cup \{T\}$ is balanced.*

The straightforward proof of Lemma 6.1.2 is left to the reader (see Exercise 6.1.2). Theorem 6.1.1 and Lemma 6.1.2 imply the following result.

Corollary 6.1.3. *Let (N, v) be a game, let $x \in X(N, v)$, and $\mu_1 > \cdots > \mu_p$ be given by $\{e(S, x, v) \mid S \subseteq N\} = \{\mu_1, \ldots, \mu_p\}$. Let $1 \leq q \leq p$ be such that $\langle\{\chi_S \mid S \in \mathcal{D}(\mu_q, x, v)\}\rangle = \mathbb{R}^N$. Then $x = \nu(N, v)$ if and only if, for every $i = 1, \ldots, q$, $\mathcal{D}(\mu_i, x, v)$ is balanced.*

Exercises

Exercise 6.1.1. Let (N, v) be a game satisfying $I(N, v) \neq \emptyset$, and let $x \in I(N, v)$. Then $\{x\} = \mathcal{N}(N, v)$ (see Remark 5.1.12) if and only if the following condition is valid for every $\alpha \in \mathbb{R}$: If $\mathcal{D}(\alpha, x, v) \neq \emptyset$, then there exists $\mathcal{E}(\alpha, x, v) \subseteq \{\{j\} \mid j \in N$ and $x^j = v(\{j\}\}$ such that $\mathcal{D}(\alpha, x, v) \cup \mathcal{E}(\alpha, x, v)$ is balanced (see Kohlberg (1971)).

Exercise 6.1.2. Prove Lemma 6.1.2.

Exercise 6.1.3. Let (N, v') be the game defined in Remark 4.5.5. Show that $\nu(N, v') = 0 \in \mathbb{R}^N$.

Exercise 6.1.4. Let (N, v) be a game such that $v(S) \in \mathbb{Q}$ for every $S \subseteq N$. Show that every coordinate of $x = \nu(N, v)$ is also rational. (Verify that every balanced collection over N has a system of rational balancing weights and show that $x(S) \in \mathbb{Q}$ for all $S \subseteq N$ by applying Theorem 6.1.1 recursively to the different $\mathcal{D}(\alpha, x, v)$.)

6.2 Preliminary Results

We shall now prove some combinatorial lemmata which will be used in the characterization of the prenucleolus. The axiomatization itself will be given in the next section.

Definition 6.2.1. *A* **coalitional family** *is a pair $(N, (\mathcal{B}_\ell)_{\ell \in L})$ where (i) N and L are finite nonempty sets, (ii) $\mathcal{B}_\ell \subseteq 2^N$ for every $\ell \in L$, and (iii) $\mathcal{B}_\ell \cap \mathcal{B}_{\ell^*} = \emptyset$ if $\ell \neq \ell^*$, $\ell, \ell^* \in L$.*

Definition 6.2.2. *Let $\mathcal{H} = (N, (\mathcal{B}_\ell)_{\ell \in L})$ be a coalitional family. A permutation π of N is a* **symmetry** *of $\mathcal{H}$ if for every $\ell \in L$ and every $S \in \mathcal{B}_\ell$, $\pi(S) \in \mathcal{B}_\ell$. $\mathcal{H}$ is* **transitive** *if for every pair $(i, j) \in N \times N$ there exists a symmetry π of $\mathcal{H}$ such that $\pi(i) = j$.*

Remark 6.2.3. A coalitional family is transitive if and only if its symmetry group is transitive.

Notation 6.2.4. If N is a finite set, $i \in N$, and $\mathcal{B} \subseteq 2^N$, then we denote $\mathcal{B}^i = \{S \in \mathcal{B} \mid i \in S\}$.

Lemma 6.2.5. *Let $(N, \mathcal{B})$ be a coalitional family. If there exists $k \in \mathbb{N}$ such that*

$$|\mathcal{B}^i| = k \text{ for every } i \in N, \tag{6.2.1}$$

$$\mathcal{B}^i \neq \mathcal{B}^j \text{ if } i \neq j,\ i,j \in N, \text{ and} \tag{6.2.2}$$

$$n = \binom{t}{k} \text{ where } n = |N| \text{ and } t = |\mathcal{B}|, \tag{6.2.3}$$

then $(N, \mathcal{B})$ is transitive.

Proof: Let $\widehat{\mathcal{B}} = \{\mathcal{B}^* \subseteq \mathcal{B} \mid |\mathcal{B}^*| = k\}$. By (6.2.3), $|\widehat{\mathcal{B}}| = n$. Hence, by (6.2.1) and (6.2.2), $\widehat{\mathcal{B}} = \{\mathcal{B}^i \mid i \in N\}$. Now, if $\pi^* : \mathcal{B} \to \mathcal{B}$ is a permutation of $\mathcal{B}$, then π^* has the following properties:

(1) If $\mathcal{B}^* \in \widehat{\mathcal{B}}$, then $\pi^*(\mathcal{B}^*) \in \widehat{\mathcal{B}}$.

(2) If $\mathcal{B}_1, \mathcal{B}_2 \in \widehat{\mathcal{B}}$ and $\mathcal{B}_1 \neq \mathcal{B}_2$, then $\pi^*(\mathcal{B}_1) \neq \pi^*(\mathcal{B}_2)$.

Therefore, π^* induces a permutation on $\widehat{\mathcal{B}}$ which, in turn, induces a permutation π on N. More precisely, π is defined by

$$\pi(i) = j \text{ if } \pi^*(\mathcal{B}^i) = \mathcal{B}^j,\ i,j \in N. \tag{6.2.4}$$

We shall now prove that π is a symmetry of $(N, \mathcal{B})$, that is, $S \in \mathcal{B}$ implies that $\pi(S) \in \mathcal{B}$. Let $S \in \mathcal{B}$ and let $\widehat{\mathcal{B}}^S$ be defined by

$$\widehat{\mathcal{B}}^S = \{\mathcal{B}^* \in \widehat{\mathcal{B}} \mid S \in \mathcal{B}^*\}.$$

By the definition of $\widehat{\mathcal{B}}$, $|\widehat{\mathcal{B}}^S| = \binom{t-1}{k-1}$. We now prove that $|S| = |\widehat{\mathcal{B}}^S|$. Indeed, if $i \in S$ then $S \in \mathcal{B}^i$ and, therefore, $\mathcal{B}^i \in \widehat{\mathcal{B}}^S$. Conversely, if $\mathcal{B}^i \in \widehat{\mathcal{B}}^S$, then $S \in \mathcal{B}^i$ and, thus, $i \in S$. Hence the mapping $S \to \widehat{\mathcal{B}}^S$, $i \mapsto \mathcal{B}^i$ is a bijection.

We conclude that all the members of $\mathcal{B}$ have the same number $\binom{t-1}{k-1}$ of elements. Therefore $\{S\} = \bigcap_{i \in S} \mathcal{B}^i$. Hence, by (6.2.4),

$$\bigcap_{j \in \pi(S)} \mathcal{B}^j = \bigcap_{i \in S} \pi^*(\mathcal{B}^i) = \pi^*\left(\bigcap_{i \in S} \mathcal{B}^i\right) = \{\pi^*(S)\}. \tag{6.2.5}$$

By (6.2.5), $\pi(S) \subseteq \pi^*(S)$. As $\pi^*(S) \in \mathcal{B}$, $|\pi^*(S)| = |S| = |\pi(S)|$. Thus $\pi(S) = \pi^*(S)$ for all $S \in \mathcal{B}$ and π is a symmetry of $(N, \mathcal{B})$.

To complete the proof let $\bar{i}, \bar{j} \in N$. Clearly, there is a bijection π^* of $\mathcal{B}$ such that $\pi^*(\mathcal{B}^{\bar{i}}) = \mathcal{B}^{\bar{j}}$. If π is defined by (6.2.4), then π is a symmetry of $(N, \mathcal{B})$ satisfying $\pi(\bar{i}) = \bar{j}$. Thus, $(N, \mathcal{B})$ is transitive. **q.e.d.**

Definition 6.2.6. *Let $\mathcal{H}_i = (N_i, (\mathcal{B}_{i,\ell})_{\ell \in L_i})$, $i = 1, 2$, be coalitional families. The* **product** *of $\mathcal{H}_1$ and $\mathcal{H}_2$ is the coalitional family*

$$\mathcal{H}^* = \left(N^*, (\mathcal{B}^*_{(i,\ell)})_{(i,\ell)\in L^*}\right)$$

defined by

$$\begin{aligned} N^* &= N_1 \times N_2, \\ L^* &= \{(1,\ell) \mid \ell \in L_1\} \cup \{(2,\ell) \mid \ell \in L_2\}, \\ \mathcal{B}^*_{(1,\ell)} &= \{S \subseteq N^* \mid S = T \times N_2 \textit{ for some } T \in \mathcal{B}_{1,\ell}\} \textit{ for all } \ell \in L_1, \\ \mathcal{B}^*_{(2,\ell)} &= \{S \subseteq N^* \mid S = N_1 \times T \textit{ for some } T \in \mathcal{B}_{2,\ell}\} \textit{ for all } \ell \in L_2. \end{aligned}$$

Lemma 6.2.7. *The product of two transitive coalitional families is itself a transitive coalitional family.*

Proof: Let $\mathcal{H}_i = (N_i, (\mathcal{B}_{i,\ell})_{\ell\in L_i})$, $i = 1, 2$, be two transitive coalitional families and let $\mathcal{H}^* = \left(N^*, (\mathcal{B}^*_{(i,\ell)})_{(i,\ell)\in L^*}\right)$ be the product of $\mathcal{H}_1$ and $\mathcal{H}_2$. If $(i_1^*, i_2^*), (j_1^*, j_2^*) \in N^* = N_1 \times N_2$, then there exist symmetries π_1 of $\mathcal{H}_1$ and π_2 of $\mathcal{H}_2$ such that $\pi_i(i_i^*) = j_i^*$, $i = 1, 2$. Now define a permutation $\pi^* : N^* \to N^*$ by $\pi^*(i_1, i_2) = (\pi_1(i_1), \pi_2(i_2))$. By Definition 6.2.6, π^* is a symmetry of $\mathcal{H}^*$. Clearly, $\pi^*(i_1^*, i_2^*) = (j_1^*, j_2^*)$. Thus, $\mathcal{H}^*$ is a transitive coalitional family. **q.e.d.**

Remark 6.2.8. Definition 6.2.6 and Lemma 6.2.7 can be generalized by induction to any finite number of coalitional families.

Lemma 6.2.9. *Let $\mathcal{H} = (N, (\mathcal{B}_\ell)_{\ell\in L})$ be a coalitional family which satisfies $\bigcup_{\ell\in L} \mathcal{B}_\ell = 2^N$ and $|L| > 1$. Let $\bar{\ell} \in L$. Then $\mathcal{H}$ is transitive if and only if the coalitional family $\mathcal{H}_* = \left(N, (\mathcal{B}_\ell)_{\ell\in L\setminus\{\bar{\ell}\}}\right)$ is transitive.*

Proof: The proof follows from the observation that a permutation of N is a symmetry of $\mathcal{H}$ if and only if it is a symmetry of $\mathcal{H}_*$. **q.e.d.**

Let $\mathcal{U}$ be a set of players and let, as usual, $\Gamma_\mathcal{U}$ be the set of games with player sets contained in $\mathcal{U}$.

Definition 6.2.10. *A solution σ on a set $\Gamma \subseteq \Gamma_\mathcal{U}$ of games is* **single-valued** *(*SIVA*) if $|\sigma(N, v)| = 1$ for every $(N, v) \in \Gamma$.*

Lemma 6.2.11. *Let σ be a solution on $\Gamma_\mathcal{U}$. If σ satisfies* SIVA, COV, *and* RGP, *then σ also satisfies* PO.

Proof: Let $(\{i\}, v) \in \Gamma_\mathcal{U}$ be a one-person game. If $v(\{i\}) = 0$, then, by COV,

$$\sigma(\{i\}, 0) = \sigma(\{i\}, 2 \cdot 0) = 2\sigma(\{i\}, 0).$$

Hence $\sigma(\{i\}, 0) = \{0\}$. Again by COV,

$$\sigma(\{i\}, v) = \sigma(\{i\}, 0 + v) = \sigma(\{i\}, 0) + v(\{i\}) = \{v(\{i\})\}$$

and PO is satisfied. Now let (N, v) be an n-person game and assume that $n \geq 2$. Let $x \in \sigma(N, v)$ and $i \in N$. The reduced game $(\{i\}, v_{\{i\},x})$ is a one-person game. By RGP, $x^i \in \sigma(\{i\}, v_{\{i\},x})$. Hence, by the definition of the reduced game, $x^i = v_{\{i\},x}(\{i\}) = v(N) - x(N \setminus \{i\})$. Thus $x(N) = v(N)$ and the proof is complete. **q.e.d.**

Exercises

Exercise 6.2.1. Let N be a finite set with at least two members. Show that $(N, \mathcal{B})$ satisfies (6.2.1) - (6.2.3) (1) if $\mathcal{B}$ is the set of all one-person coalitions and (2) if $\mathcal{B}$ is the set of all $n-1$-person coalitions.

Exercise 6.2.2. Let $|N| = 6$. Describe all coalitional families $(N, \mathcal{B})$ that satisfy (6.2.1) - (6.2.3).

6.3 An Axiomatization of the Prenucleolus

Let $\mathcal{U}$ be a set of players and $\Gamma_{\mathcal{U}}$ be the set of all games whose set of players is contained in $\mathcal{U}$.

Theorem 6.3.1 (Sobolev (1975)). *Let $\mathcal{U}$ be infinite. Then there is a unique solution on $\Gamma_{\mathcal{U}}$ that satisfies* SIVA, COV, AN, *and* RGP, *and it is the prenucleolus.*

Proof: Step 1. By Theorem 5.1.14 the prenucleolus satisfies SIVA. COV and AN follow from Exercise 5.1.2. Finally, by Corollary 5.2.8, the prenucleolus also satisfies RGP. Thus it remains to prove the uniqueness part of the theorem.

Step 2. Let σ be a solution on $\Gamma_{\mathcal{U}}$ that satisfies SIVA, COV, AN, and RGP, let $(N, v) \in \Gamma_{\mathcal{U}}$ and let $x = \nu(N, v)$. We have to prove that $\sigma(N, v) = \{x\}$. Define (N, w) by $w(S) = v(S) - x(S)$ for every $S \subseteq N$. By COV of the prenucleolus, $\nu(N, w) = 0 \in \mathbb{R}^N$. By COV of σ, it is sufficient to prove that $\sigma(N, w) = \{0\}$. Thus, we shall consider the game (N, w).

Step 3. Let $\{w(S) \mid S \subseteq N\} = \{\mu_1, \ldots, \mu_p\}$ where $\mu_1 > \cdots > \mu_p$. We denote

$$\mathcal{B}_k = \mathcal{D}(\mu_k, 0, w) \; (= \{S \subseteq N \mid w(S) \geq \mu_k\}) \text{ for all } k = 1, \ldots, p.$$

By Theorem 6.1.1, $\mathcal{B}_k$ is a balanced collection on N for $k = 1, \ldots, p$. Let $1 \leq k \leq p$. As $\mathcal{B}_k$ is balanced there exist balancing weights $\delta_S > 0$, $S \in \mathcal{B}_k$, such that

$$\sum_{S \in \mathcal{B}_k} \delta_S \chi_S = \chi_N. \tag{6.3.1}$$

These numbers δ_S, $S \in \mathcal{B}_k$ can be chosen to be rational. By (6.3.1) there exist natural numbers m and m_S, $S \in \mathcal{B}_k$, such that $\delta_S = m_S/m$. Hence we obtain

$$\sum_{S \in \mathcal{B}_k^i} m_S = m \text{ for every } i \in N \tag{6.3.2}$$

(see Notation 6.2.4). We denote $t = \sum_{S \in \mathcal{B}_k} m_S$.

We shall say that two players $i, j \in N$ are *equivalent with respect to* $\mathcal{B}_k$ if $\mathcal{B}_k^i = \mathcal{B}_k^j$. Let H_i be the equivalence class of player $i \in N$ and let $r = \max_{i \in N} |H_i|$.

Now we shall associate with $(N, \mathcal{B}_k)$ a new coalitional family $(N_k^*, \mathcal{B}_k^*)$. First, a set N_k^* is chosen such that $N \subseteq N_k^*$ and $|N_k^*| = r\binom{t}{m}$. Then $\mathcal{B}_k^*$ will be constructed in the following way. For each pair (S, q), where $S \in \mathcal{B}_k$ and $1 \leq q \leq m_S$, there will be a set $T_{S,q}^* \in \mathcal{B}_k^*$. That is,

$$\mathcal{B}_k^* = \{T_{S,q}^* \mid S \in \mathcal{B}_k, 1 \leq q \leq m_S\}.$$

Moreover, the following conditions will be satisfied:

$$T_{S,q}^* \cap N = S \text{ for all } S \in \mathcal{B}_k \text{ and } 1 \leq q \leq m_S; \tag{6.3.3}$$

$$|\mathcal{B}_k^{*i}| = m \text{ for every } i \in N_k^*; \tag{6.3.4}$$

$$|H_i^*| = r \text{ for every } i \in N_k^*. \tag{6.3.5}$$

Here $H_i^* = \{j \in N_k^* \mid \mathcal{B}_k^{*i} = \mathcal{B}_k^{*j}\}$. The actual construction of $\mathcal{B}_k^*$ is obtained by specifying the sets $\mathcal{B}_k^{*i}$ for $i \in N_k^*$. First, if $i \in N$ then

$$i \in T_{S,q}^* \text{ if and only if } i \in S. \tag{6.3.6}$$

By (6.3.6), (6.3.3) is satisfied. Also, by (6.3.2), (6.3.4) is satisfied for $i \in N$. Finally, by the definition of r, $|H_i| = |H_i^* \cap N| \leq r$ for every $i \in N$. Now we consider the members of $N_k^* \setminus N$. As $|N_k^*| = r\binom{t}{m}$ and $t = \sum_{S \in \mathcal{B}_k} m_S$, it is possible to add each $i \in N_k^* \setminus N$ to m coalitions such that (6.3.5) is satisfied.

Let $i \in N_k^*$. Then $|\mathcal{B}_k^{*i}| = m$ and $|H_i^*| = r$. Thus the number of different equivalence classes is $\binom{t}{m}$, the maximum possible. Therefore the sets $T_{S,q}^*$, $S \in \mathcal{B}_k$, $1 \leq q \leq m_S$, are distinct. Hence, $|\mathcal{B}_k^*| = t$.

Step 4. We shall prove that $(N_k^*, \mathcal{B}_k^*)$ is transitive. Let $\widetilde{N}_k = \{H_i^* \mid i \in N_k^*\}$ and let

$$\widetilde{\mathcal{B}}_k = \{\{H_i^* \mid i \in S\} \mid S \in \mathcal{B}_k^*\}.$$

Then, as the reader may easily verify, $(\widetilde{N}_k, \widetilde{\mathcal{B}}_k)$ satisfies all the conditions of Lemma 6.2.5. Thus $(\widetilde{N}_k, \widetilde{\mathcal{B}}_k)$ is transitive. As $|H_i^*| = r$ for every $i \in N_k^*$, $(N_k^*, \mathcal{B}_k^*)$ is transitive.

Step 5. Thus, for every $k = 1, \ldots, p$, we have a transitive coalitional family $(N_k^*, \mathcal{B}_k^*)$. We denote the product of the foregoing families by

$$\mathcal{H} = \left(\widehat{N}, (\widehat{B}_\ell)_{\ell \in \{1,\ldots,p\}}\right).$$

By Definition 6.2.6, $\widehat{N} = \prod_{k=1}^p N_k^*$ and, for every $k = 1, \ldots, p$,

$\widehat{B}_k = \{S \subseteq \widehat{N} \mid \widehat{S} = N_1^* \times \cdots \times N_{k-1}^* \times S \times N_{k+1}^* \times \cdots \times N_p^* \text{ for some } S \in \mathcal{B}_k^*\}.$

By Lemma 6.2.7, $\mathcal{H}$ is transitive. Define $\widehat{\mathcal{B}}_{p+1} = 2^{\widehat{N}} \setminus \bigcup_{k=1}^p \widehat{\mathcal{B}}_k$. By Lemma 6.2.9, $\widehat{\mathcal{H}} = \left(\widehat{N}, (\widehat{B}_\ell)_{\ell \in \{1,\ldots,p+1\}}\right)$ is transitive.

Using the coalitional family $\widehat{\mathcal{H}}$ we define $(\widehat{N}, \widehat{w})$ by the following rules:

$$\widehat{w}(S) = \begin{cases} 0 \text{ , if } S = \emptyset \text{ or } S = \widehat{N}, \\ \mu_k \text{ , if } S \in \widehat{\mathcal{B}}_k \setminus \{\emptyset, \widehat{N}\} \text{ for some } k = 1, \ldots, p, \\ \mu_p \text{ , if } S \in \widehat{\mathcal{B}}_{p+1} \setminus \{\emptyset, \widehat{N}\}. \end{cases}$$

As $\widehat{\mathcal{H}}$ is transitive, the symmetry group of $(\widehat{N}, \widehat{w})$ is transitive (see Definition 2.1.15). As $\mathcal{U}$ is an infinite set, there is an injective mapping $\pi : \widehat{N} \to \mathcal{U}$ such that $\pi(i, \ldots, i) = i$ for every $i \in N$. Let $M = \pi(\widehat{N})$ and $\widehat{u} = \pi\widehat{w}$. The game $(M, \widehat{u})$ is isomorphic to $(\widehat{N}, \widehat{w})$ (see Section 2.3). Then, clearly, the symmetry group of $(M, \widehat{u})$ is transitive.

Now we shall prove that $\sigma(M, \widehat{u}) = \{0\}$. Indeed, by SIVA there exists $z \in \mathbb{R}^M$ such that $\sigma(M, \widehat{u}) = z$. By AN and the transitivity of the symmetry group $z^i = z^j$ for all $i, j \in M$. Also, by Lemma 6.2.11, σ satisfies PO. Hence $z^i = \widehat{u}(M)/|M| = 0$ for every $i \in M$.

Step 6. Let $u = \widehat{u}_{N,0}$. By RGP, $\sigma(N, u) = \{0\}$. Hence it remains to prove that $u = w$. Let $\widetilde{S} = \pi^{-1}(S)$ for every $S \subseteq N$. Then

$$u(S) = \widehat{w}_{\widetilde{N},0}(\widetilde{S}) \text{ for every } S \subseteq N \tag{6.3.7}$$

(see Exercise 6.3.1). Hence $u(S) = w(S)$ if $S = \emptyset$ or $S = N$. Now, let $\emptyset \neq S \subsetneq N$. Let $k \in \{1, \ldots, p\}$ such that $w(S) = \mu_k$. Hence $S \in \mathcal{B}_k$. By (6.3.3) there exists $T^* \in \mathcal{B}_k^*$ such that $T^* \cap N = S$. Let

$$\widehat{S} = N_1^* \times \cdots \times N_{k-1}^* \times T^* \times N_{k+1}^* \times \cdots \times N_p^*.$$

Then $\widetilde{S} \subseteq \widehat{S}$ and $\widehat{Q} = \widehat{S} \setminus \widetilde{S} \subseteq \widehat{N} \setminus \widetilde{N}$. Also, by (6.3.7),

$$u(S) \geq \widehat{w}(\widetilde{S} \cup \widehat{Q}) = \mu_k = w(S).$$

In order to prove the opposite inequality, that is, $u(S) \leq w(S)$, let $Q \subseteq \widehat{N} \setminus \widetilde{N}$ and let $\bar{S} = \widetilde{S} \cup Q$. It remains to show that $\widehat{w}(\bar{S}) \leq w(S)$. Let $h \in \{1, \ldots, p+1\}$ such that $\bar{S} \in \widehat{\mathcal{B}}_h$. If $h = p+1$, then $\widehat{w}(\bar{S}) = \mu_p \leq w(S)$. If $h \leq p$, then there exists $T^* \in \mathcal{B}_h^*$ such that

$$\bar{S} = N_1^* \times \cdots \times N_{h-1}^* \times T^* \times N_{h+1}^* \times \cdots \times N_p^*.$$

Then $S = T^* \cap N$, so $S \in \mathcal{B}_h$. Hence, $w(S) \geq \mu_h = \widehat{w}(\bar{S})$. **q.e.d.**

The following examples show that the axioms SIVA and COV are each logically independent of the remaining axioms. The prekernel satisfies COV, AN, and RGP, but it violates SIVA, as four-person games show. The equal split solution (see Example 5.4.6) satisfies SIVA, AN, and RGP, but it violates COV. In Chapter 8 (Remark 8.3.3) we shall show that RGP is independent of the remaining axioms. The independence of AN is shown in Subsection 6.3.2.

Remark 6.3.2. Theorem 6.3.1 remains true, if $|\mathcal{U}| \leq 3$. Indeed, the proof of Lemma 6.2.11 shows that any solution that satisfies SIVA, COV, AN, and RGP coincides with the prekernel on games with at most three persons. By Exercise 5.4.2 the prekernel and the prenucleolus coincide for any three-person game.

Remark 6.3.3. Theorem 6.3.1 does not hold, if $4 \leq |\mathcal{U}| < \infty$. Indeed, in this case there exists a game $(\mathcal{U}, v)$ and $x \in \mathcal{PK}(\mathcal{U}, v)$, $x \neq \nu(\mathcal{U}, v)$, such that $x^S = \nu(S, v_{S,x})$ for every $\emptyset \neq S \subsetneqq \mathcal{U}$ (see Exercises 6.3.2 and 6.3.3). Now we define $\sigma(N, w)$ as follows: If $N = \mathcal{U}$ and if $(\mathcal{U}, w)$ is isomorphic to a game which is strategically equivalent to $(\mathcal{U}, v)$, that is, if $w = \pi(\alpha v + \beta)$ for some permutation π of $\mathcal{U}$, some $\alpha > 0$, and some $\beta \in \mathbb{R}^{\mathcal{U}}$, then let $\sigma(\mathcal{U}, w) = \{\pi(\alpha x + \beta)\}$. Otherwise, let $\sigma(N, w) = \mathcal{PN}(N, w)$. Then σ satisfies SIVA, COV, AN, and RGP.

Remark 6.3.4. Note that SIVA and AN imply ETP. Orshan (1993) shows that AN in Theorem 6.3.1 can be replaced by ETP. For $4 \leq |\mathcal{U}| < \infty$ let $(\mathcal{U}, v)$ and σ be defined as in Remark 6.3.3 and let $\widetilde{\sigma}$ on $\Gamma_{\mathcal{U}}$ be defined by $\widetilde{\sigma}(\mathcal{U}, w) = \sigma(\mathcal{U}, w)$ if $(\mathcal{U}, w)$ is strategically equivalent to $(\mathcal{U}, v)$, and by $\widetilde{\sigma}(N, w) = \mathcal{PN}(N, w)$ otherwise. Then $\widetilde{\sigma}$ satisfies SIVA, ETP, COV, and RGP, and violates AN. Hence, Orshan's axioms are weaker than Sobolev's axioms.

6.3.1 An Axiomatization of the Nucleolus

This subsection is devoted to an axiomatization of the nucleolus (see Definition 5.1.11). Let (N, v) be a game, let $\emptyset \neq S \subseteq N$ and let $x \in \mathbb{R}^N$. The *imputation saving reduced game* $(S, \widetilde{v}_{S,x})$ with respect to S and x is defined as follows: If $|S| = 1$, then $\widetilde{v}_{S,x} = v_{S,x}$. If $|S| \geq 2$, then

$$\widetilde{v}_{S,x}(T) = \begin{cases} v_{S,x}(T) & , \text{if } T \subseteq S \text{ and } |T| \neq 1, \\ \min\{x^j, v_{S,x}(\{j\})\} & , \text{if } T = \{j\} \text{ for some } j \in S. \end{cases}$$

Remark 6.3.5. If x is an imputation of (N, v) (that is, $x \in I(N, v)$), then $x^S \in I(S, \widetilde{v}_{S,x})$.

Definition 6.3.6. *Let $\Gamma \subseteq \Gamma_{\mathcal{U}}$ be a set of games. A solution σ on Γ satisfies the* **imputation saving reduced game property** *(*ISRGP*) if for every $(N, v) \in \Gamma$ and for every $\emptyset \neq S \subseteq N$ the following condition is satisfied: If $x \in \sigma(N, v)$, then $(S, \widetilde{v}_{S,x}) \in \Gamma$ and $x^S \in \sigma(S, \widetilde{v}_{S,x})$.*

Let $\Gamma^I_{\mathcal{U}} = \{(N,v) \in \Gamma_{\mathcal{U}} \mid I(N,v) \neq \emptyset\}$. We consider the nucleolus on $\Gamma^I_{\mathcal{U}}$.

Lemma 6.3.7. *The nucleolus satisfies* ISRGP *on* $\Gamma^I_{\mathcal{U}}$.

Proof: Let $(N,v) \in \Gamma^I_{\mathcal{U}}$, $\emptyset \neq S \subseteq N$, let x be the nucleolus of (N,v), and let $u = \widetilde{v}_{S,x}$. In order to show that x_S is the nucleolus of (S,u) we may assume that $|S| \geq 2$. For every $\mathcal{B} \subseteq 2^N \setminus \{\emptyset\}$ denote

$$\mathcal{B}^S = \{T \cap S \mid T \in \mathcal{B}\} \setminus \{\emptyset\}.$$

It is straightforward to prove that (a) if $\mathcal{B}$ is balanced (over N), then $\mathcal{B}^S$ is balanced (over S), and (b) if $\mathcal{B}^S \setminus \{S\} \neq \emptyset$, then $\mathcal{B}^S$ is balanced iff $\mathcal{B}^S \setminus \{S\}$ is balanced. Now, let $\alpha \in \mathbb{R}$ such that $\mathcal{D}(\alpha, x^S, u) \neq \emptyset$. Then, by Exercise 6.1.1, there exists $\mathcal{E} \subseteq \{\{j\} \mid j \in N, x^j = v(\{j\})\}$ such that $\mathcal{D} \cup \mathcal{E}$ is balanced, where $\mathcal{D} = \mathcal{D}(\alpha, x, v)$. If

$$\mathcal{E}' = \{\{j\} \in \mathcal{D}^S(\alpha_+, x, v) \mid j \in S\} \text{ and } \mathcal{E}(\alpha, x^S, u) = \mathcal{E}' \cup \mathcal{E}^S,$$

then $\mathcal{E}(\alpha, x^S, u) \subseteq \{\{j\} \mid j \in S, x^j = u(\{j\})\}$ and

$$\mathcal{D}^S \setminus \{S\} = \left(\mathcal{D}(\alpha, x^S, u) \setminus \{S\}\right) \cup \mathcal{E}'.$$

Hence, $\left(\mathcal{D}^S \cup \mathcal{E}^S\right) \setminus \{S\} = \left(\mathcal{D}(\alpha, x^S, u) \cup \mathcal{E}(\alpha, x^S, u)\right) \setminus \{S\}$ and (a), (b), and Exercise 6.1.1 complete the proof. **q.e.d.**

Remark 6.3.8. If σ is a solution on a set Γ of games that satisfies SIVA, COV, and ISRGP, then σ satisfies PO.

The easy proof of this remark is similar to the proof of Lemma 6.2.11 and, hence, it is left to the reader (Exercise 6.3.4). Now we are ready to prove the characterization of the nucleolus which is similar to Theorem 6.3.1.

Theorem 6.3.9. *Let $\mathcal{U}$ be infinite. Then there is a unique solution on $\Gamma^I_{\mathcal{U}}$ that satisfies* SIVA, COV, AN, *and* ISRGP, *and it is the nucleolus.*

Proof: The nucleolus satisfies SIVA and ISRGP by Remark 5.1.12 and Lemma 6.3.7. Clearly, it satisfies AN and COV. In order to prove the uniqueness part let σ be a solution on $\Gamma^I_{\mathcal{U}}$ that satisfies SIVA, COV, AN, and ISRGP. We continue as in the proof of Theorem 6.3.1 and we only indicate the necessary modifications to that proof. In Step 2, now x has to be the nucleolus of (N,v). In Step 3, we redefine the sets $\mathcal{B}_k$ as follows: By Exercise 6.1.1, for every $k = 1, \ldots, p$ there exists $\mathcal{E}_k \subseteq \{\{j\} \mid j \in N, x^j = w(\{j\})\}$ such that

$$\mathcal{B}_k = \mathcal{D}(\mu_k, 0, w) \cup \mathcal{E}_k$$

is balanced. In Step 5, σ satisfies PO now by Remark 6.3.8. Finally, a careful inspection of Step 6 shows that $u(S) = w(S)$ for every $S \subseteq N$ satisfying $|S| \neq 1$, because the imputation saving reduced game may differ from the reduced game only on one-person coalitions. Furthermore, the reader may convince himself that $u(\{j\}) = w(\{j\})$, $j \in N$, because $w(\{j\}) \leq 0$. **q.e.d.**

Theorem 6.3.9 is due to Snijders (1995).

Remark 6.3.10. The bargaining sets $\mathcal{M}, \mathcal{M}_r, \mathcal{M}_{sr}$ and the kernel satisfy ISRGP on $\Gamma_{\mathcal{U}}^I$, but none of these solutions (see Snijders (1995) and Sudhölter and Potters (2001)) satisfies the converse imputation saving reduced game property (which arises from CRGP by replacing reduced games by imputation saving reduced games).

6.3.2 The Positive Core

In this subsection a further solution is briefly discussed and used to show that AN is independent of the other axioms used in Theorem 6.3.1.

Definition 6.3.11. *Let (N, v) be a game and let $\nu = \nu(N, v)$. The* **positive core** *of (N, v), denoted by $\mathcal{C}_+(N, v)$, is defined by*

$$\mathcal{C}_+(N, v) = \{x \in X(N, v) \mid e(S, x, v) \leq (e(S, \nu, v))_+ \text{ for all } S \subseteq N\}.$$

Remark 6.3.12. Let (N, v) be a game. Then $\mathcal{C}_+(N, v)$ can be reached similarly as $\mathcal{PN}(N, v)$ by successively minimizing the excesses. The only difference is that this procedure is only applied as long as the excesses are positive. Hence, the positive core can be computed similarly to the prenucleolus by a sequence of linear programs following Remark 5.1.15 as long as the excesses are positive. Note that $\mathcal{C}_+(N, v)$ is a convex polytope.

Remark 6.3.13. $\mathcal{C}(N, v) = \mathcal{C}_+(N, v)$ for every balanced game (N, v). Also, $\nu(N, v) \in \mathcal{C}_+(N, v)$ for every game (N, v).

The easy proof of Remark 6.3.13 is omitted.

Theorem 6.3.14. *The positive core on $\Gamma_{\mathcal{U}}$ satisfies* NE, AN, COV, RGP, *and* RCP.

Proof: NE is implied by Remark 6.3.13, AN and COV are straightforward.

Let (N, v) be a game and $x \in X(N, v)$. We claim that $x \in \mathcal{C}_+(N, v)$ if and only if for all $\alpha > 0$ and $y \in \mathbb{R}^N$ satisfying $y(N) = 0$,

$$y(S) \geq 0 \text{ for all } S \in \mathcal{D}(\alpha, x, v) \Rightarrow y(S) = 0 \text{ for all } S \in \mathcal{D}(\alpha, x, v). \quad (6.3.8)$$

Indeed, the proof of the foregoing claim is similar to the proof of Theorem 5.2.6 and, hence, it is omitted.

The proof of RGP differs from the proof of Theorem 5.2.7 only inasmuch as $\alpha \in \mathbb{R}$ and Property I have to be replaced by $\alpha > 0$ and (6.3.8).

In order to show RCP, let $x \in \mathcal{C}_+(N,v)$, let $\emptyset \neq S \subseteq N$, let $y \in \mathcal{C}_+(S, v_{S,x})$, let $z = (y, x^{N\setminus S})$, and let $T \subseteq N$. It remains to show that $e(T,x,v)_+ = e(T,z,v)_+$. If $T \cap S = \emptyset$ or $S \subseteq T$, then $x(T \cap S) = y(T \cap S)$ and, hence, the proof is complete. Assume now that $\emptyset \neq T \cap S \neq S$. Let $Q \subseteq N \setminus S$ such that $v_{S,x}(T \cap S) = v((T \cap S) \cup Q) - x(Q)$. If $e(T \cap S, x^S, v_{S,x}) > 0$, then, by RGP, $y(T \cap S) = x(T \cap S)$ and, hence, $e(T,z,v) = e(T,x,v)$. If $e(T \cap S, x^S, v_{S,x}) \leq 0$, then, by RGP, $e(T \cap S, y, v_{S,x}) \leq 0$ and, thus, $e(T,z,v) \leq 0$ and $e(T,x,v) \leq 0$.
q.e.d.

Assume that $|\mathcal{U}| \geq 2$. We are now ready to construct a solution σ which satisfies SIVA, COV, and RGP, and which violates AN. Select any total order relation $\succeq$ of $\mathcal{U}$. For $(N,v) \in \Gamma_{\mathcal{U}}$ define

$$\sigma(N,v) = \{x \in \mathcal{C}_+(N,v) \mid x \succeq_{lex} y \text{ for all } y \in \mathcal{C}_+(N,v)\},$$

where $\succeq_{lex}$ is the lexicographic order induced by $\succeq$, that is, if $N \subseteq \mathcal{U}$ and $x, y \in \mathbb{R}^N$, then $x \succeq_{lex} y$ is defined by

$$i \in N,\ y^i > x^i \Rightarrow \text{there exists } j \in N \text{ such that } x^j > y^j \text{ and } j \succeq i.$$

Lemma 6.3.15. *The solution σ satisfies* SIVA, COV, *and* RGP.

Proof: By Remark 6.3.12 and Theorem 6.3.14, σ satisfies SIVA and COV. Let $(N,v) \in \Gamma_{\mathcal{U}}$, $y \in \mathbb{R}^N$, $\emptyset \neq S \subseteq N$, and $z \in \mathbb{R}^S$. If $z \succeq_{lex} y^S$, then $(z, y^{N\setminus S}) \succeq_{lex} y$. Let now $\{x\} = \sigma(N,v)$. Then $x^S \in \mathcal{C}_+(S, v_{S,x})$ by RGP of $\mathcal{C}_+$. Hence, by RCP of $\mathcal{C}_+$, $x^S \in \sigma(S, v_{S,x})$. Thus σ satisfies RGP. **q.e.d.**

Finally, we show by means of an example that σ does not satisfy AN. Indeed, let $(N,v) \in \Gamma_{\mathcal{U}}$ be a 0-1-normalized game, that is, $|N| = 2$, $v(S) = 0$ for $S \subsetneq N$ and $v(N) = 1$. Then (N,v) is symmetric. For simplicity assume that $N = \{1,2\}$ such that $2 \succ 1$. Then $\mathcal{C}_+(N,v) = \text{convh}\ \{(1,0),(0,1)\}$ and, hence, $\sigma(N,v) = \{(0,1)\}$ and σ satisfies neither AN nor ETP.

Theorem 6.3.14 and Lemma 6.3.15 are due to Sudhölter (1993). The expression "positive core" is due to Maschler (see Orshan (1994)).

Exercises

Exercise 6.3.1. Let (N,v) be a game, let $\pi : N \to N'$ be a bijection, let $\emptyset \neq S \subseteq N$, and let $x \in \mathbb{R}^N$. Prove that "reducing commutes with taking isomorphic games", that is, $(\pi v)_{\pi(S),\pi(x)} = \pi(v_{S,x})$.

Exercise 6.3.2. Let $N = \{1,\ldots,n\}$ for some $n \geq 4$. Let (N,v) be the coalitional game associated with the weighted majority game which has a representation $(\lambda; m)$ given by $\lambda = n-1$ and

$$m = \begin{cases} (\frac{n+1}{2}, \frac{n-1}{2}, \underbrace{1, \ldots, 1}_{n-2}) & \text{, if } n \text{ is odd,} \\ (\frac{n}{2}, \frac{n}{2}, \underbrace{1, \ldots, 1}_{n-2}) & \text{, if } n \text{ is even.} \end{cases}$$

Show that $\nu(N, v) = m/(2n-2)$.

Exercise 6.3.3. Let (N, v) be defined as in Exercise 6.3.2, let

$$x = (1/2, 1/2, 0, \ldots, 0) \in \mathbb{R}^N,$$

and let $\emptyset \neq S \subsetneqq N$. Show that $\nu(S, v_{S,x}) = x^S$.

Exercise 6.3.4. Prove Remark 6.3.8 (see Snijders (1995)).

6.4 The Prenucleolus of Games with Coalition Structures

Let $\mathcal{U}$ be a set of players and let $\Delta_{\mathcal{U}}$ be the set of all games with coalition structures whose players are members of $\mathcal{U}$.

Theorem 6.4.1. *Let $(N, v, \mathcal{R}) \in \Delta_{\mathcal{U}}$ and let $x \in X(N, v, \mathcal{R})$. Then $x = \nu(N, v, \mathcal{R})$ if and only if, for every $\alpha \in \mathbb{R}$, $\mathcal{D}(\alpha, x, v) \cup \mathcal{R}$ is a balanced collection over N.*

The proof of Theorem 6.4.1 is similar to the proof of Theorem 6.1.1. Hence the proof is omitted.

Definition 6.4.2. *A* **coalitional family with coalition structure** *is a triple $(N, (\mathcal{B}_\ell)_{\ell \in L}, \mathcal{R})$ where (i) N and L are finite nonempty sets, (ii) $\mathcal{B}_\ell \subseteq 2^N$ for every $\ell \in L$, (iii) $\mathcal{B}_\ell \cap \mathcal{B}_{\ell^*} = \emptyset$ if $\ell \neq \ell^*$, $\ell, \ell^* \in L$, and (iv) $\mathcal{R}$ is a coalition structure for N.*

Definition 6.4.3. *Let $\mathcal{H} = (N, (\mathcal{B}_\ell)_{\ell \in L}, \mathcal{R})$ be a coalitional family with coalition structure. A permutation π of N is a* **symmetry** *of $\mathcal{H}$ if (i) for every $\ell \in L$ and every $S \in \mathcal{B}_\ell$, $\pi(S) \in \mathcal{B}_\ell$, and (ii) $\pi(R) = R$ for every $R \in \mathcal{R}$. $\mathcal{H}$ is* **transitive** *if for every set of partners $\{i, j\} \in \mathcal{P}(\mathcal{R})$ there exists a symmetry π of $\mathcal{H}$ such that $\pi(i) = j$.*

Lemma 6.4.4. *Let $(N, \mathcal{B}, \mathcal{R})$ be a coalitional family with coalition structure, let $\mathcal{R} = \{R_1, \ldots, R_r\}$ where $r = |\mathcal{R}|$, and let $n_j = |R_j|$, $j = 1, \ldots, r$. If there exists $k_j \in \mathbb{N}$, $j = 1, \ldots, r$ such that*

$$|\mathcal{B}^i| = k_j \text{ for every } i \in R_j, \tag{6.4.1}$$

$$\mathcal{B}^i \neq \mathcal{B}^h \text{ if } i \neq h,\ i, h \in R_j,\ j = 1, \ldots, r, \text{ and} \tag{6.4.2}$$

$$n_j = \binom{t}{k_j} \text{ where } t = |\mathcal{B}|,\ j = 1, \ldots, r \tag{6.4.3}$$

then $(N, \mathcal{B}, \mathcal{R})$ is transitive.

The proof of Lemma 6.4.4 is similar to that of Lemma 6.2.5 and, hence, it is omitted.

Definition 6.4.5. *Let $\mathcal{H}_i = (N_i, (\mathcal{B}_{i,\ell})_{\ell \in L_i}, \mathcal{R}_i)$, $i = 1, 2$, be two coalitional families with coalition structures. The* **product** *of $\mathcal{H}_1$ and $\mathcal{H}_2$ is the coalitional family*

$$\mathcal{H}^* = \left(N^*, (\mathcal{B}^*_{(i,\ell)})_{(i,\ell) \in L^*}, \mathcal{R}^*\right),$$

where $\left(N^*, (\mathcal{B}^*_{(i,\ell)})_{(i,\ell) \in L^*}\right)$ is the product of $(N_i, (\mathcal{B}_{i,\ell})_{\ell \in L_i})$, $i = 1, 2$, and $\mathcal{R}^* = \mathcal{R}_1 \times \mathcal{R}_2$ is defined by

$$\mathcal{R}^* = \{R_1 \times R_2 \mid R_1 \in \mathcal{R}_1 \text{ and } R_2 \in \mathcal{R}_2\}.$$

Lemma 6.4.6. *The product of two transitive coalitional families with coalition structures is itself a transitive coalitional family with coalition structure.*

The proof of Lemma 6.4.6 is similar to that of Lemma 6.2.7

Lemma 6.4.7. *Let σ be a solution on $\Delta_{\mathcal{U}}$. If σ satisfies* SIVA, COV, *and* RGP, *then σ also satisfies* EFF.

The reader may prove Lemma 6.4.7 by following the steps of the proof of Lemma 6.2.11.

Using the lemmata of this section and the proof of Theorem 6.3.1 the following theorem can be proved.

Theorem 6.4.8. *Let $|\mathcal{U}| = \infty$. Then there is a unique solution on $\Delta_{\mathcal{U}}$ that satisfies* SIVA, COV, AN, *and* RGP, *and it is the prenucleolus.*

6.5 The Nucleolus of Strong Weighted Majority Games

Let $g = (N, \mathcal{W})$ be a simple game and let $G = (N, v)$ be the associated coalitional simple game. Let $y \in \mathbb{R}^N$. We denote

$$q(y, g) = q(y) = \min_{S \in \mathcal{W}^m} y(S). \tag{6.5.1}$$

We say that y is a *representation* of g, if $(q(y); y)$ is a representation of g (see Definition 2.2.11). Also, $x \in X(N, v)$ is a *normalized* representation of g if $(q(x); x)$ is a representation of g.

Remark 6.5.1. If g is a strong simple game, then its associated coalitional game G is a superadditive constant-sum game and $I(N, v) \neq \emptyset$ (see Subsection 2.2.3).

Lemma 6.5.2. *An imputation $x \in I(N, v)$ is a normalized representation of a strong simple game g if and only if $q(x) > 1/2$.*

Proof: Necessity. If x is a normalized representation of g, then $x(S) > 1/2$ for every $S \in \mathcal{W}$, because g is proper. Hence $q(x) > 1/2$.

Sufficiency. Let $x \in I(N,v)$ satisfy $q(x) > 1/2$. If $S \in \mathcal{W}$, then $x(S) \geq q(x)$. Conversely, if $x(S) \geq q(x)$, then $x(N \setminus S) \leq 1 - q(x) < q(x)$. Hence $N \setminus S \notin \mathcal{W}$. As g is strong, $S \in \mathcal{W}$. Thus, x is a normalized representation of g. **q.e.d.**

Let $(N, \mathcal{W})$ be strong. By Corollary 5.5.11, $\mathcal{N}(N,v) = \mathcal{PN}(N,v)$. We write $\mathcal{N}(N,\mathcal{W}) = \mathcal{N}(N,v) = \{\nu(N,\mathcal{W})\}$.

Lemma 6.5.3. *Let g be strong, let $\nu = \nu(N,\mathcal{W})$, and let $x \in I(N,v)$. Then $q(\nu) \geq q(x)$.*

Proof: For $y \in I(N,v)$ let $\mu(y) = \max_{S \subseteq N} e(S,y,v)$. Then $\mu(y) = 1 - q(y)$, so $1 - q(x) \geq 1 - q(\nu)$. Thus, $q(x) \leq q(\nu)$. **q.e.d.**

Theorem 6.5.4. *The nucleolus of a strong weighted majority game g is a normalized representation of g.*

Proof: Let $x \in I(N,v)$, where (N,v) is the associated coalitional game, be a normalized representation of g. By Lemma 6.5.2, $q(x) > 1/2$, and by Lemma 6.5.3, $q(\nu(N,v)) \geq q(x)$. Hence $\nu(N,v)$ is a normalized representation of g by Lemma 6.5.2. **q.e.d.**

Theorem 6.5.5. *Let $g = (N, \mathcal{W})$ be a strong homogeneous weighted majority game (see Definition 2.2.13) and let $G = (N,v)$ be the associated coalitional game. Then the nucleolus $\nu = \nu(g)$ of g is the unique normalized homogeneous representation of g which assigns a zero to each null player of G (see Definition 4.1.16).*

Proof: Denote by D the set of null players of G. Let y be a normalized homogeneous representation of g which satisfies $y^i = 0$ for all $i \in D$. As y is homogeneous,

$$y(S) = q(y) \text{ for all } S \in \mathcal{W}^m. \tag{6.5.2}$$

Let

$$Y = \{x \in I(G) \mid x(S) \geq q(y) \text{ for all } S \in \mathcal{W}^m \text{ and } x^i = 0 \text{ for all } i \in D\}.$$

By Lemma 6.5.3, $q(\nu) \geq q(y)$. Also, by RE of $\mathcal{PN}$, $\nu^i = 0$ for all $i \in D$. Hence, $\nu \in Y$. Assume that $\nu \neq y$. Then Y has an extreme point z such that $z \neq y$. By (6.5.2) there exists $j \in N \setminus D$ such that $z^j = 0$. As j is not a null player and z is a representation of G by Lemma 6.5.2, the desired contradiction has been obtained. **q.e.d.**

Remark 6.5.6. A vector $x \in \mathbb{R}^N$ satisfying $x \geq 0$ is a representation of the strong simple game $g = (N, \mathcal{W})$ if and only if $q(x,g) > x(N)/2$.

Lemma 6.5.2 implies Remark 6.5.6.

Definition 6.5.7. *A representation w of a weighted majority game g is an* **integral representation,** *if $w^i \in \mathbb{N} \cup \{0\}$ for all $i \in N$. Let w be an integral representation of g. Then w is a* **minimal** *integral representation if there does not exist any integral representation w_* of g such that $w_* \leq w$. If $w \leq w_*$ for every integral representation w_* of g then w is a (the)* **minimum** *integral representation of g.*

Let $g = (N, \mathcal{W})$ be a strong simple game and ν the nucleolus of g. By Exercise 6.1.4, ν has rational coordinates.

Notation 6.5.8. Let $\nu_*(g) = \nu_*$ be defined by $\nu = \nu_*/\nu_*(N)$, where the ν_*^i, $i \in N$, are integers whose greatest common divisor is 1.

If g is a strong weighted majority game, then ν_* is a representation of g. The following theorem provides a simple criterion for the minimality of ν_*.

Theorem 6.5.9. *Let $g = (N, \mathcal{W})$ be a strong simple game. Then $\nu_* = \nu_*(g)$ is a minimal integral representation of g if and only if $\nu_*(N) = 2q(\nu_*) - 1$.*

Proof: By Remark 6.5.6 we may assume that g is a weighted majority game. As ν_* is a representation of g, $\nu_*(N) \leq 2q(\nu_*) - 1$ by the same remark. If $\nu_*(N) < 2q(\nu_*) - 1$, then let $k \in N$ satisfy $\nu_*^k > 0$. Then $x \in \mathbb{R}^N$, defined by $x^k = \nu_*^k - 1$ and $x^i = \nu_*^i$ for all $i \in N \setminus \{k\}$, is an integral representation by Remark 6.5.6; hence ν_* is not a minimal integral representation. Thus, let $\nu_*(N) = 2q(\nu_*) - 1$ and assume, on the contrary, that ν_* is not a minimal integral representation. By Remark 6.5.6, there exists a minimal integral representation y_* satisfying $y_* \leq \nu_*$ and $y_*(N) < \nu_*(N)$. Clearly, $y_*(N) = 2q(y_*) - 1$ and, thus, $q(y_*) < q(\nu_*)$. Let $y = y_*/y_*(N)$. Then

$$q(y) = \frac{q(y_*)}{2q(y_*) - 1} > \frac{q(\nu_*)}{2q(\nu_*) - 1} = q(\nu_*),$$

which contradicts Lemma 6.5.3. **q.e.d.**

An example of a strong weighted majority game g for which $\nu_*(g)$ is **not** a minimal integral representation is given in Peleg (1968). However, if g is also homogeneous, then ν_* is a minimum integral representation. This claim will now be proved.

Lemma 6.5.10. *Let $g = (N, \mathcal{W})$ be a strong simple game. If $i \in N$ is not a null player, then there exist $S, T \in \mathcal{W}^m$ such that $\{i\} = S \cap T$.*

Proof: As i is not a null player, there exists $S \in \mathcal{W}^m$ such that $i \in S$. As g is strong,

$$N \setminus S \notin \mathcal{W} \ni (N \setminus S) \cup \{i\}. \tag{6.5.3}$$

Hence there exists $T \subseteq (N \setminus S) \cup \{i\}$ such that $T \in \mathcal{W}^m$. By (6.5.3), $i \in T$, so $i \in S \cap T$. **q.e.d.**

For $S \subseteq N$ we denote $S^c = N \setminus S$. For $y \in \mathbb{R}^N$ and $S, T \subseteq N$ the following equality holds:

$$y(S \cap T) = y(S^c \cap T^c) + \frac{y(S) - y(S^c) + y(T) - y(T^c)}{2}. \tag{6.5.4}$$

Lemma 6.5.11. *Let $g = (N, \mathcal{W})$ be a strong weighted majority game and let x be a representation of g which satisfies the following conditions:*
(1) If $i \in N$ is not a null player, then there exist $S, T \subseteq N$ such that

$$\{i\} = S \cap T \text{ and } x(S) = x(T) = x(S^c) + 1 = x(T^c) + 1. \tag{6.5.5}$$

(2) If i is a null player, then $x^i = 0$.
Then x is a minimum integral representation of g.

Proof: First we show that x is integral. Assume the contrary and let x^i be the smallest non-integer weight. By Condition 2, i is not a null player. Hence there exist $S, T \subseteq N$ which satisfy (6.5.5). By (6.5.4), $x(S^c \cap T^c) = x^i - 1$, which contradicts the choice of i. Thus, x is integral.

In order to prove that x is a minimum integral representation, assume, on the contrary, that there exists an integral representation y of g such that $y^j < x^j$ for some $j \in N$. Choose $i \in N$ such that

$$x^i = \min \left\{ x^\ell \mid y^\ell < x^\ell, \ell \in N \right\}.$$

Then $x^i > 0$ and, hence, i is not a null player. Let $S, T \subseteq N$ satisfy (6.5.5). By (6.5.4), $x(S^c \cap T^c) = x^i - 1$. By the choice of i,

$$y(S^c \cap T^c) \geq x(S^c \cap T^c) \text{ and } y^i \leq x^i - 1.$$

Hence $y^i \leq y(S^c \cap T^c)$. By (6.5.5), $S, T \in \mathcal{W}$. Hence $y(S) > y(S^c)$ and $y(T) > y(T^c)$. Thus, by (6.5.4),

$$y^i = y(S \cap T) = y(S^c \cap T^c) + \frac{y(S) - y(S^c) + y(T) - y(T^c)}{2} > y(S^c \cap T^c)$$

and the desired contradiction has been obtained. **q.e.d.**

Theorem 6.5.12. *Let g be a strong homogeneous weighted majority game and $\nu = \nu(g)$. Then ν_* is a minimum integral representation of g.*

Proof: Let $g = (N, \mathcal{W})$ and define $\widehat{\nu} = \nu / (2q(\nu) - 1)$. If $S \in \mathcal{W}^m$, then

$$\widehat{\nu}(S) = \frac{q(\nu)}{2q(\nu) - 1} \text{ and } \widehat{\nu}(S^c) = \frac{1 - q(\nu)}{2q(\nu) - 1},$$

so $\widehat{\nu}(S) = \widehat{\nu}(S^c) + 1$. By Lemma 6.5.10, $\widehat{\nu}$ satisfies Condition 1 of Lemma 6.5.11. Also, $\widehat{\nu}^i = 0$ for every null player i. Therefore, by Lemma 6.5.11, $\widehat{\nu}$ is a minimum integral representation of g. Hence, the greatest common divisor of the $\widehat{\nu}^i$, $i \in N$, is 1 and $\widehat{\nu} = \nu_*$. **q.e.d.**

Let g be a strong weighted majority game which has a minimum integral representation y_*. Isbell (1969) showed that $y_* \neq \nu_*(g)$ is possible. However, we can prove the following result.

Theorem 6.5.13. *Let $g = (N, \mathcal{W})$ be a strong weighted majority game that has a minimum integral representation y_*. Then $y = y_*/y_*(N)$ belongs to the kernel $\mathcal{K}(g) = \mathcal{K}(N, v)$ of the associated coalitional game (N, v).*

Proof: Suppose, on the contrary, that $y \notin \mathcal{K}(N, v)$. Then there exist $i, j \in N$ such that $s_{ij}(y, v) > s_{ji}(y, v)$ and $y^j > 0$. Clearly, $1 - q(y) \geq s_{ij}(y, v)$. Hence, if $S \in \mathcal{W}^m$ such that $i \notin S \ni j$, then $y(S) \geq q(y) + 1/y_*(N)$. Define $x \in \mathbb{R}^N$ by $x^i = y^i + 1/y_*(N)$, $x^j = y^j - 1/y_*(N)$, and $x^k = y^k$ for all $k \in N \setminus \{i, j\}$. By Remark 6.5.6, x is a representation of g. Also, $x_* = y_*(N)x$ is an integral representation of g and $x_*^j < y_*^j$. Thus, the desired contradiction has been obtained. **q.e.d.**

The results of this section, except Lemma 6.5.11, are due to Peleg (1968). Lemma 6.5.11 is proved in Isbell (1956).

6.6 The Modiclus

In this section the modiclus, which is a nucleolus as defined in Section 5.1, and some of its properties are discussed. We start with the definition of the modiclus.

Definition 6.6.1. *Let $(N, v, \mathcal{R})$ be a game and let*

$$H = (e(S, \cdot, v) - e(T, \cdot, v))_{(S,T) \in 2^N \times 2^N}.$$

Then $\mathcal{N}(H, X(N, v, \mathcal{R}))$ is the **modiclus** *of $(N, v, \mathcal{R})$ and it is denoted by $\Psi(N, v, \mathcal{R})$. The* **modiclus** *of (N, v) is $\Psi(N, v) = \Psi(N, v, \{N\})$.*

Recall that the prenucleolus of a game (see Definition 5.1.13) lexicographically minimizes the excesses of the coalitions within the set of feasible payoff vectors. The modiclus lexicographically minimizes the differences of excesses of the pairs of coalitions within the set of preimputations. Hence the vector of excesses $(e(S, x, v))_{S \in 2^N}$, $x \in X^*(N, v)$, is replaced by the *bi-excesses* $(e(S, x, v) - e(T, x, v))_{(S,T) \in 2^N \times 2^N}$, $x \in X(N, v)$ when comparing the definitions of the prenucleolus and the modiclus. The bi-excess $e(S, x, v) - e(T, x, v)$, $S, T \subseteq N$, may be regarded as the *envy* of S against T at x.

Theorem 6.6.2. *On every set of games with coalition structures the modiclus satisfies* SIVA.

Proof: Let $(N, v, \mathcal{R})$ be a game with coalition structure and let $y \in X(N, v, \mathcal{R})$. Let H be the collection of bi-excess mappings (see Definition 6.6.1). Define $\widehat{\mu} = (\max_{S \subseteq N} e(S, y, v)) - (\min_{T \subseteq N} e(T, y, v))$ and

$$X = \{x \in X(N, v, \mathcal{R}) \mid (e(S, x, v) - e(T, x, v)) \leq \widehat{\mu} \text{ for all } S, T \subseteq N\}.$$

Then $\Psi(N, v, \mathcal{R}) = \mathcal{N}(H, X)$. By Theorem 5.1.6, $\Psi(N, v, \mathcal{R}) \neq \emptyset$, because X is compact and the bi-excess mappings are continuous. As X is convex and the bi-excess mappings are convex, $\Psi(N, v, \mathcal{R})$ is convex by Theorem 5.1.8. Let $x, y \in \Psi(N, v, \mathcal{R})$. Again, by Theorem 5.1.8,

$$e(S, x, v) - e(\emptyset, x, v) = e(S, y, v) - e(\emptyset, y, v) \text{ for all } S \subseteq N;$$

thus $x = y$.

q.e.d.

Let $(N, v, \mathcal{R})$ be a game with coalition structure. The unique member of $\Psi(N, v, \mathcal{R})$ is again called the *modiclus* of $(N, v, \mathcal{R})$ and it is denoted by $\psi(N, v, \mathcal{R})$. Furthermore, denote $\psi(N, v) = \psi(N, v, \{N\})$.

Definition 6.6.3. *Let (N, v) be a game. The* **dual game** *of (N, v) is the game (N, v^*) defined by $v^*(S) = v(N) - v(N \setminus S)$ for every $S \subseteq N$.*

The value $v^*(S)$ describes the amount which can be given to $S \subseteq N$ if the complement $N \setminus S$ receives what it can reach by cooperation. Hence, the complement $N \setminus S$ cannot *prevent* S from the amount $v^*(S)$.

Let $\mathcal{U}$ be a set of players and let $\Gamma_{\mathcal{U}}$ denote the set of games.

Definition 6.6.4. *$\Gamma \subseteq \Gamma_{\mathcal{U}}$ is* **closed under duality**, *if $(N, v^*) \in \Gamma$ for every $(N, v) \in \Gamma$. A solution σ on a set $\Gamma \subseteq \Gamma_{\mathcal{U}}$ closed under duality is* **self dual** *(SD) if $\sigma(N, v) = \sigma(N, v^*)$ for every $(N, v) \in \Gamma_{\mathcal{U}}$.*

Lemma 6.6.5. *On every set $\Gamma \subseteq \Gamma_{\mathcal{U}}$ closed under duality the modiclus satisfies* SD.

Proof: Let $(N, v) \in \Gamma$, $x = \psi(N, v)$ and $S \subseteq N$. As $x(N) = v(N)$,

$$e(S, x, v^*) = v(N) - v(N \setminus S) - x(N) + x(N \setminus S) = -e(N \setminus S, x, v). \quad (6.6.1)$$

Hence, for all $(S, T) \in 2^N \times 2^N$,

$$e(S, x, v) - e(T, x, v) = e(N \setminus T, x, v^*) - e(N \setminus S, x, v^*). \quad (6.6.2)$$

By (6.6.2), H in Definition 6.6.1 can be replaced by

$$(e(S, \cdot, v^*) - e(T, \cdot, v^*))_{(S,T) \in 2^N \times 2^N}$$

without changing $\Psi(N, v)$.

q.e.d.

Theorem 6.6.6. *Let $(N, v, \mathcal{R})$ be a game with coalition structure and let $\psi = \psi(N, v, \mathcal{R})$. If $k, \ell \in R \in \mathcal{R}$ satisfy $k \succsim_v \ell$, then $\psi^k \geq \psi^\ell$.*

Proof: Let $(N, v, \mathcal{R})$ be a game with coalition structure, let $\{k, \ell\} \in \mathcal{P}(\mathcal{R})$ satisfy $k \succsim_v \ell$, and let $x \in X(N, v, \mathcal{R})$ satisfy $x^k < x^\ell$. It remains to show that $x \neq \psi$. Let $y \in \mathbb{R}^N$ be defined by $y^i = x^i$ for all $i \in N \setminus \{k, \ell\}$ and $y^k = y^\ell = \frac{x^k + x^\ell}{2}$. It remains to show that $\theta(x) >_{lex} \theta(y)$, where θ is the mapping defined by (5.1.1). It suffices to show that for every pair $(S, T) \in 2^N \times 2^N$ such that

$$e(S, x, v) - e(T, x, v) < e(S, y, v) - e(T, y, v) \tag{6.6.3}$$

there exists a pair $(S^*, T^*) \in 2^N \times 2^N$ such that

$$\begin{aligned} e(S^*, x, v) - e(T^*, x, v) &> e(S^*, y, v) - e(T^*, y, v) \\ &\geq e(S, y, v) - e(T, y, v). \end{aligned} \tag{6.6.4}$$

Let $(S, T) \in 2^N \times 2^N$ satisfy (6.6.3). Then $S \in \mathcal{T}_{\ell k}$ or $T \in \mathcal{T}_{k\ell}$. Let $\delta = x^\ell - y^\ell$. Hence $\delta > 0$, $y^\ell = x^\ell - \delta$ and $y^k = x^k + \delta$. Three cases may be distinguished:

(1) $S \in \mathcal{T}_{\ell k}$ and $T \in \mathcal{T}_{k\ell}$. Define $S^* = (S \setminus \{\ell\}) \cup \{k\}$ and $T^* = (T \setminus \{k\}) \cup \{\ell\}$. Then

$$e(S^*, x, v) = e(S^*, y, v) + \delta \geq e(S, y, v) + \delta \tag{6.6.5}$$

and

$$e(T^*, x, v) = e(T^*, y, v) - \delta \leq e(T, y, v) - \delta; \tag{6.6.6}$$

hence (6.6.4) holds.

(2) $S \in \mathcal{T}_{\ell k}$ and $T \notin \mathcal{T}_{k\ell}$. Define S^* as in (1) and $T^* = T$. Then (6.6.5) holds and $e(T^*, x, v) \leq e(T^*, y, v) = e(T, y, v)$; hence (6.6.4) holds.

(3) $S \notin \mathcal{T}_{\ell k}$ and $T \in \mathcal{T}_{k\ell}$. Define $S^* = S$ and T^* as in (1). Then (6.6.6) holds and $e(S^*, x, v) \geq e(S^*, y, v) = e(T, y, v)$; hence (6.6.4) holds.

q.e.d.

Corollary 6.6.7. *The modiclus satisfies* ETP *on every set of games and* RETP *on every set of games with coalition structures.*

The following relation of the modiclus and the prenucleolus is useful.

Theorem 6.6.8. *Let (N, v) be a game and let $^* : N \to N^*$, $i \mapsto i^*$, be a bijection such that $N^* \cap N = \emptyset$. If $(N \cup N^*, \widetilde{v})$ is defined by*

$$\widetilde{v}(S \cup T^*) = \max\{v(S) + v^*(T), v^*(S) + v(T)\} \textit{ for all } S, T \subseteq N,$$

then $\psi(N, v) = \nu(N \cup N^*, \widetilde{v})^N$.

Proof: Let $x = \nu(N \cup N^*, \widetilde{v})$. By AN and SIVA, $x^i = x^{i^*}$ for all $i \in N$. Let $H^1 = (e(S \cup T^*, \cdot, \widetilde{v}))_{(S,T) \in 2^N \times 2^N}$. Then $\mathcal{N}(N \cup N^*, \widetilde{v}) = \mathcal{N}(H^1, X)$, where

$$X = \{x \in X(N \cup N^*, \widetilde{v}) \mid x^i = x^{i^*} \text{ for all } i \in N\}.$$

Let $\mathcal{D} = \{\{S \cup T^*, T \cup S^*\} \mid S, T \subseteq N\}$ and define $Q : \mathcal{D} \to 2^{N \cup N^*}$ by

$$Q(\{S \cup T^*, T \cup S^*\}) = \begin{cases} S \cup T^* \text{ , if } v(S) + v^*(T) \geq v(T) + v^*(S), \\ T \cup S^* \text{ , otherwise.} \end{cases}$$

Also, let $\widetilde{\mathcal{D}} = \{Q(A) \mid A \in \mathcal{D}\}$ and $H^2 = (e(S \cup T^*, \cdot, \widetilde{v}))_{S \cup T^* \in \widetilde{\mathcal{D}}}$. Then $\mathcal{PN}(N \cup N^*, \widetilde{v}) = \mathcal{N}(H^2, X)$. Let

$$H^3 = (e(S, \cdot, v) + e(T, \cdot, v^*))_{(S,T) \in 2^N \times 2^N}.$$

By (6.6.1) (see Exercise 6.6.2), $\Psi(N, v) = \mathcal{N}(H^3, X(N, v))$. The proof is complete because $\mathcal{N}(H^3, X(N, v)) = \{x^N \mid x \in \mathcal{N}(H^2, X)\}$. **q.e.d.**

Lemma 6.6.9. *Let (N, v) be a game. Then*

$$b^i_{\max}(N, v) = b^i_{\max}(N, v^*) \text{ and } b^i_{\min}(N, v) = b^i_{\min}(N, v^*) \text{ for all } i \in N.$$

Proof: For every $i \in N$ and every $S \subseteq N \setminus \{i\}$,

$$v^*(S \cup \{i\}) - v(S) = v(N \setminus S) - v((N \setminus S) \setminus \{i\});$$

thus the proof is complete. **q.e.d.**

Theorem 6.6.10. *The modiclus is reasonable on every set of games.*

Proof: Let (N, v) be a game and let $(N \cup N^*, \widetilde{v})$ be defined as in Theorem 6.6.8. By Lemma 6.6.9,

$$b^i_{\max}(N \cup N^*, \widetilde{v}) \leq b^i_{\max}(N, v) \text{ and } b^i_{\min}(N \cup N^*, \widetilde{v}) \geq b^i_{\min}(N, v).$$

Theorems 5.1.17, 5.6.1, and 6.6.8 complete the proof. **q.e.d.**

Theorem 6.6.8 can be used to characterize the modiclus by means of a characterization of the prenucleolus by balanced collections (see Theorem 6.1.1). The following lemma is very useful.

Definition 6.6.11. *Let N be a finite nonempty set and $\widetilde{\mathcal{D}} \subseteq 2^N \times 2^N$. $\widetilde{\mathcal{D}}$ is* **balanced** *(over N), if a system (a system of balancing weights for $\widetilde{\mathcal{D}}$) $\left(\delta_{(S,T)}\right)_{(S,T) \in \mathcal{D}}$ exists such that $\sum_{(S,T) \in \widetilde{\mathcal{D}}} \delta_{(S,T)} (\chi_S + \chi_T) = \chi_N$.*

Lemma 6.6.12. *Let (N, v) be a game and $x \in X(N, v)$. Then $x = \psi(N, v)$ if and only if*

$$\widetilde{\mathcal{D}}(\alpha, x, v) = \left\{(S, T) \in 2^N \times 2^N \middle| e(S, x, v) + e(T, x, v^*) \geq \alpha\right\}$$

is balanced for every $\alpha \in \mathbb{R}$ such that $\widetilde{\mathcal{D}}(\alpha, x, v) \neq \emptyset$.

Proof: Let $(N \cup N^*, \widetilde{v})$ be defined as in Theorem 6.6.8 and let $\alpha \in \mathbb{R}$. Let $z = (x, x^*) \in \mathbb{R}^{N \cup N^*}$, that is, $z^i = z^{i^*}$ for all $i \in N$. For $\alpha \in \mathbb{R}$ let $\widetilde{\mathcal{D}}(\alpha) = \widetilde{\mathcal{D}}(\alpha, x, v)$ and $\mathcal{D}(\alpha) = \mathcal{D}(\alpha, z, \widetilde{v})$. Then the following assertions are valid:

$$(S,T) \in \widetilde{\mathcal{D}}(\alpha) \Rightarrow S \cup T^*, T \cup S^* \in \mathcal{D}(\alpha); \tag{6.6.7}$$

$$S \cup T^* \in \mathcal{D}(\alpha) \Rightarrow (S,T) \in \widetilde{\mathcal{D}}(\alpha) \text{ or } (T,S) \in \widetilde{\mathcal{D}}(\alpha); \tag{6.6.8}$$

$$S \cup T^* \in \mathcal{D}(\alpha) \Leftrightarrow T \cup S^* \in \mathcal{D}(\alpha). \tag{6.6.9}$$

Hence $\mathcal{D}(\alpha) \neq \emptyset$ if and only if $\widetilde{\mathcal{D}}(\alpha) \neq \emptyset$.

By Theorem 6.6.8, $\psi(N,v) = x$ if and only if $\nu(N \cup N^*, \widetilde{v}) = z$. In view of Theorem 6.1.1 it remains to show that $\mathcal{D}(\alpha)$ is balanced over $N \cup N^*$ if and only if $\widetilde{\mathcal{D}}(\alpha)$ is balanced over N.

First, assume that $\mathcal{D}(\alpha)$ is balanced. Let $(\delta_{S \cup T^*})_{S \cup T^* \in \mathcal{D}(\alpha)}$ be a system of balancing weights. For $(S,T) \in \widetilde{\mathcal{D}}(\alpha)$ define

$$\delta_{(S,T)} = \begin{cases} \frac{1}{2}\delta_{S \cup T^*} & , \text{if } (T,S) \in \widetilde{\mathcal{D}}(\alpha), \\ \frac{1}{2}\left(\delta_{S \cup T^*} + \delta_{T \cup S^*}\right) & , \text{if } (T,S) \notin \widetilde{\mathcal{D}}(\alpha). \end{cases}$$

Then

$$\begin{aligned} &\sum_{(S,T) \in \widetilde{\mathcal{D}}(\alpha)} \delta_{(S,T)} \left(\chi_S + \chi_T\right) \\ &= \sum_{\substack{(S,T) \in \widetilde{\mathcal{D}}(\alpha) \\ (T,S) \notin \widetilde{\mathcal{D}}(\alpha)}} \frac{1}{2}\left(\delta_{S \cup T^*} + \delta_{T \cup S^*}\right)\left(\chi_S + \chi_T\right) + \sum_{\substack{(S,T) \in \widetilde{\mathcal{D}}(\alpha) \\ (T,S) \in \widetilde{\mathcal{D}}(\alpha)}} \frac{1}{2}\delta_{S \cup T^*}\left(\chi_S + \chi_T\right) \\ &= \frac{1}{2} \sum_{S \cup T^* \in \mathcal{D}(\alpha)} \delta_{S \cup T^*}\left(\chi_S + \chi_T\right) \quad \text{(by (6.6.7) – (6.6.9))} \\ &= \chi_N \quad \text{(because the } \delta_{S \cup T^*} \text{ are balancing weights);} \end{aligned}$$

thus $\widetilde{\mathcal{D}}(\alpha)$ is balanced over N.

Conversely, if $\widetilde{\mathcal{D}}(\alpha)$ is balanced with balancing coefficients $\delta_{(S,T)}$ for $(S,T) \in \widetilde{\mathcal{D}}(\alpha)$, then define $(\delta_{S \cup T^*})_{S \cup T^* \in \mathcal{D}(\alpha)}$ by

$$\delta_{S \cup T^*} = \begin{cases} \delta_{(S,T)} + \delta_{(T,S)} & , \text{if } (S,T) \in \widetilde{\mathcal{D}}(\alpha) \ni (T,S), \\ \delta_{(S,T)} & , \text{if } (T,S) \notin \widetilde{\mathcal{D}}(\alpha) \ni (S,T), \\ \delta_{(T,S)} & , \text{if } (S,T) \notin \widetilde{\mathcal{D}}(\alpha) \ni (T,S). \end{cases}$$

Then, by (6.6.7) – (6.6.9), $(\delta_{S \cup T^*})_{S \cup T^* \in \mathcal{D}(\alpha)}$ is a system of balancing weights. **q.e.d.**

Example 6.6.13. Consider the family of market games (N,v) of Section 5.8, that is, let $N = P \cup Q$, $P = \{1,2\}$, $Q = \{3,4,5\}$, $a \geq 0$, and let v be given by (5.8.1). We claim that

$$\psi = \psi(N,v) = \frac{1}{6}(3v(N), 3v(N), 2v(N), 2v(N), 2v(N)); \tag{6.6.10}$$

thus P and Q receive the same payoffs when the modiclus is applied. Indeed, by RE and PO of the modiclus and by Corollary 6.6.7 there exists $c \in \mathbb{R}$ such that with $b = (v(N) - 3c)/2$,

$$0 \leq c \leq \frac{v(N)}{3} \text{ and } \psi = (b, b, c, c, c).$$

Let
$\mu = \max_{S \subseteq N} e(S, \psi, v)$, $\mu^* = \max_{S \subseteq N} e(S, \psi, v^*)$, $\mathcal{D} = \mathcal{D}(\mu, \psi, v)$, and $\mathcal{D}^* = \mathcal{D}(\mu^*, \psi, v^*)$. By Lemma 6.6.12, $\widetilde{\mathcal{D}} = \mathcal{D} \times \mathcal{D}^*$ is balanced. Let $(\delta_{(S,T)})_{(S,T) \in \widetilde{\mathcal{D}}}$ be a system of balancing weights and assume, on the contrary, that (6.6.10) does not hold, that is, $2b \neq 3c$. If $2b < 3c$, then $\mathcal{D}^* = \{P\}$. Hence

$$\sum_{(S,T) \in \widetilde{\mathcal{D}}} \delta_{(S,T)}(\chi_S + \chi_T) = \sum_{S \in \mathcal{D}} \delta_{(S,P)}(\chi_S + \chi_P) = \chi_N. \tag{6.6.11}$$

As $c > 0$ and $\mu \geq 0$, there exists $\widetilde{S} \in \mathcal{D}$ such that $\widetilde{S} \cap P \neq \emptyset$. Applied to any $i \in P \cap \widetilde{S}$ and any $j \in Q$, (6.6.11) yields

$$1 = \sum_{S \in \mathcal{D}} \delta_{(S,P)}(\chi_S^i + \chi_P^i) \geq \delta_{(\widetilde{S},P)} + \sum_{S \in \mathcal{D}} \delta_{(S,P)} > \sum_{S \in \mathcal{D}} \delta_{(S,P)}$$

and

$$1 = \sum_{S \in \mathcal{D}} \delta_{(S,P)}(\chi_S^j + \chi_P^j) \leq \sum_{S \in \mathcal{D}} \delta_{(S,P)},$$

which is impossible. If $2b > 3c$, then we obtain a contradiction in a similar way. Hence (6.6.10) is shown.

Remark 6.6.14. The modiclus assigns the same amount to P and Q in the games of Example 6.6.13. Hence it satisfies a kind of "equal treatment property for groups" on this one-parameter family of games. Sudhölter (2001) shows that the modiclus has the "equal treatment property for groups" on a wider class of games. Also, he shows that the groups of buyers and sellers in an assignment game are treated equally.

The results of this section concerning games are due to Sudhölter (1996a). The remaining results, concerning games with coalition structures, are deduced from the same paper.

6.6.1 Constant-Sum Games

Let (N, v) be a constant-sum game. We shall show that $\psi(N, v) = \nu(N, v)$.

Theorem 6.6.15. *If (N, v) is a constant-sum game, then $\psi(N, v) = \nu(N, v)$.*

Proof: As (N,v) is a constant-sum game, $v^*(S) = v(N) - v(N \setminus S) = v(S)$ for all $S \subseteq N$. Let $(N \cup N^*, \widetilde{v})$ be defined as in Theorem 6.6.8. Then, for all $S, T \subseteq N$,

$$\begin{aligned}\widetilde{v}^*(S \cup T^*) &= \widetilde{v}(N \cup N^*) - \widetilde{v}((N \setminus S) \cup (N \setminus T)^*)\\ &= 2v(N) - v(N \setminus S) - v(N \setminus T)\\ &= v(S) + v(T) = \widetilde{v}(S \cup T^*).\end{aligned}$$

Hence $(N \cup N^*, \widetilde{v})$ is a constant-sum game as well. Let $x = \nu(N \cup N^*, \widetilde{v})$ and let $w = \widetilde{v}_{N,x}$. Let $\mu = \max_{S \subseteq N} e(S, x, \widetilde{v})$ $(= \max_{S \subseteq N} e(S^*, x, \widetilde{v}))$. By Definition 2.3.11, $w(S) = v(S) + \mu$ for all $\emptyset \neq S \subsetneqq N$, $w(N) = v(N)$ and $w(\emptyset) = v(\emptyset)$. Hence $\nu(N, w) = \nu(N, v)$ (see Exercise 6.6.3). By RGP the proof is complete.
q.e.d.

Remark 6.6.16. Theorem 6.6.15 is due to Sudhölter (1996a) who shows that the modiclus is the prenucleolus of a game, if the prenucleoli of the game and of its dual game coincide.

6.6.2 Convex Games

In this subsection it is shown that the modiclus of a convex game is contained in the core of the game.

Theorem 6.6.17. *Let (N,v) be a convex game. Then $\psi(N,v) \in \mathcal{C}(N,v)$.*

Proof: Let (N,v) be a convex game. Put

$$\psi = \psi(N,v), \;\; \mu = \max_{S \subseteq N} e(S, \psi, v), \;\text{ and }\; \mu^* = \max_{S \subseteq N} e(S, \psi, v^*).$$

Also, let $\mathcal{D} = \mathcal{D}(\mu, \psi, v)$ and $\mathcal{D}^* = \mathcal{D}(\mu^*, \psi, v^*)$. We have to show that $\mu \leq 0$. Assume, on the contrary, that $\mu > 0$. By convexity of (N,v), $v^*(S) \geq v(S)$ for every $S \subseteq N$. Thus $\mu^* \geq \mu$. By Lemma 5.7.5, $\mathcal{D}$ is a near-ring. Hence, both $S_\cap = \bigcap_{S \in \mathcal{D}} S$ and $S_\cup = \bigcup_{S \in \mathcal{D}} S$ are members of $\mathcal{D}$. Also, $\emptyset \neq S_\cap \subseteq S_\cup \subsetneqq N$. Applying (5.7.8) to $T = S_\cap$ yields $e(S, \psi, v) < e(S \cup S_\cap, \psi, v)$ for every $S \subseteq N \setminus S_\cap$. By (6.6.1), $N \setminus (S \cup S_\cap) \notin \mathcal{D}^*$. Hence,

$$T \cap S_\cap \neq \emptyset \text{ for all } T \in \mathcal{D}^*. \tag{6.6.12}$$

By Lemma 6.6.12, $\widetilde{\mathcal{D}} = \mathcal{D} \times \mathcal{D}^*$ is balanced. Let $\delta_{(S,T)}$, $(S,T) \in \widetilde{\mathcal{D}}$, be balancing weights. Then

$$\sum_{(S,T) \in \widetilde{\mathcal{D}}} \delta_{(S,T)}(\chi_S + \chi_T) = \chi_N \tag{6.6.13}$$

applied to any $i \in N \setminus S_\cup$ implies that

$$\sum_{(S,T)\in\widetilde{\mathcal{D}}} \delta_{(S,T)} \geq 1. \tag{6.6.14}$$

Let $\widehat{T} \in \mathcal{D}^*$ and let $j \in \widehat{T} \cap S_\cap$. Applied to j, (6.6.13) yields

$$1 = \sum_{(S,T)\in\widetilde{\mathcal{D}}} \delta_{(S,T)}(\chi_S^j + \chi_T^j) \geq \sum_{(S,T)\in\widetilde{\mathcal{D}}} \delta_{(S,T)} + \sum_{S\in\mathcal{D}} \delta_{(S,\widehat{T})} > \sum_{(S,T)\in\widetilde{\mathcal{D}}} \delta_{(S,T)},$$

which contradicts (6.6.14). **q.e.d.**

Theorem 6.6.17 is due to Sudhölter (1997).

6.6.3 Weighted Majority Games

Let g be a simple game and G be the associated coalitional game. The *modiclus* (point) of g, denoted by $\psi(g)$, is defined by $\psi(g) = \psi(G)$.

Theorem 6.6.18. *If $g = (N, \mathcal{W})$ is a weighted majority game and $x = \psi(g)$, then x is a normalized representation of g.*

Proof: Let w be a representation of g and let (N, v) be the associated coalitional game. Then $w \geq 0$ and $w(N) > 0$, so $y = \frac{w}{w(N)}$ is a normalized representation of g. Let $z \in X(N, v)$ be such that $z \geq 0$. Then $\mu(z, v) = \max_{S\subseteq N} e(S, z, v)$ is attained by some $\widehat{S} \in \mathcal{W}$, because $z \geq 0$. Also $-\mu(z, v^*) = \min_{T\subseteq N} e(T, z, v)$ is attained by some $\widehat{T} \in 2^N \setminus \mathcal{W}$, because $N \in \mathcal{W}$, $\emptyset \in 2^N \setminus \mathcal{W}$, and $e(N, z, w) = e(\emptyset, z, w) = 0$. By monotonicity of (N, v) and the assumption that $z \geq 0$, $\mu(z, v) = 1 - q(z)$. Also, $\mu(z, v^*) = \max_{T\in 2^N\setminus\mathcal{W}} z(T)$. Hence, for all $S, T \in 2^N$,

$$e(S, y, v) - e(T, y, v) \leq 1 - q(y) + \max_{T\in 2^N\setminus\mathcal{W}} y(T) < 1. \tag{6.6.15}$$

Thus, by Definition 6.6.1, $e(S, x, v) - e(T, x, v) < 1$ for all $S, T \in 2^N$. In particular, for every $S \in \mathcal{W}$ and $T \in 2^N \setminus \mathcal{W}$,

$$e(S, x, v) - e(T, x, v) = 1 - x(S) + x(T) < 1;$$

hence $x(S) > x(T)$. By Theorem 6.6.10, $x \geq 0$. Thus, x is a normalized representation of g. **q.e.d.**

Thus, Theorem 6.5.4 is a consequence of Theorems 6.6.18 and 6.6.15.

Remark 6.6.19. Theorem 6.5.12, when the nucleolus is replaced by the modiclus, can be generalized to arbitrary homogeneous weighted majority games. Indeed, Ostmann (1987) and Rosenmüller (1987) showed that every homogeneous game has a minimum integral representation and that this representation is homogeneous. Moreover, after normalizing, it coincides with the modiclus of the game as shown in Sudhölter (1996a).

Exercises

Exercise 6.6.1. Prove that the modiclus satisfies COV and AN.

Exercise 6.6.2. Let (N, v) be a game. Show that

$$\Psi(N,v) = \mathcal{N}\left((e(S,x,v) + e(T,x,v^*))_{(S,T)\in 2^N \times 2^N}, X^*(N,v)\right).$$

(Equation (6.6.2) is helpful.)

Exercise 6.6.3. Let (N, v) be a game, $\alpha \in \mathbb{R}$ and (N, w) be defined by

$$w(S) = \begin{cases} v(S) & \text{, if } S = \emptyset, N, \\ v(S) + \alpha & \text{, if } \emptyset \neq S \subsetneqq N. \end{cases} \tag{6.6.16}$$

Show that $\nu(N, w) = \nu(N, v)$.

Exercise 6.6.4. Let (N, v) be a game. Show that there exists $\bar{\alpha} \in \mathbb{R}$ such that for every $\alpha \geq \bar{\alpha}$, $\nu(N, w) = \psi(N, w)$, where (N, w) is defined by (6.6.16). (See Sudhölter (1996a).)

Exercise 6.6.5. Use Remark 5.5.17 and Exercises 6.6.3 and 6.6.4 to prove that the modiclus does not satisfy aggregate monotonicity.

6.7 Notes and Comments

(1) Let $(N, v, \mathcal{R})$ be a game with coalition structure and $\nu = \nu(N, v, \mathcal{R})$. It is possible to prove the following generalization of Exercise 6.1.4: There exist rational numbers $r(i, S) \in \mathbb{Q}$, $i \in N$, $S \subseteq N$, such that

$$\nu^i = \sum_{S \subseteq N} r(i,S)v(S) \text{ for all } i \in N.$$

(2) Let N be a finite nonempty set, let $\mathcal{R}$ be a coalition structure for N, and let Γ_N be the set of coalition functions on 2^N. It is possible to prove that

$$\nu(N, \cdot, \mathcal{R}) : \Gamma_N \to \mathbb{R}^N$$

is piecewise linear.

(3) Clearly, (1) and (2) hold as well, when the prenucleolus is replaced by the modiclus.

(4) Sudhölter (1997) shows that the modiclus (on the set of all games with players in $\mathcal{U}$) is axiomatized by SIVA, COV, a weak variant of RGP, and a strong variant of self duality. In this characterization $|\mathcal{U}| = \infty$ is crucial.

7

Geometric Properties of the ε-Core, Kernel, and Prekernel

The ε-core is an immediate generalization of the core, which may be nonempty even when the core is empty. In Section 7.1 we define the ε-core of a coalitional game and study some of its geometric properties. The least-core of a game is the minimum nonempty ε-core. The individual rationality and reasonableness of the least-core are investigated in Section 7.2. In Section 7.3 we prove some basic results on the set of all reasonable payoff vectors of a game. Also, we find the minimum values of ε which guarantee that the ε-core contains the set of individually rational payoff vectors, or the set of reasonable payoffs, or the intersection of these two sets.

Section 7.4 is devoted to a geometric characterization of the intersection of the prekernel of a game with an ε-core. In particular, we prove that the foregoing intersection is completely determined by the shape of the ε-core. An algorithm for computing the prenucleolus of a game, which is based on the fact that the prenucleolus is the "lexicographic least-core", is provided in Section 7.5. We conclude with some notes on ε-cores of games with coalition structures.

7.1 Geometric Properties of the ε-Core

Let $G = (N, v)$ be a coalitional game. The following generalization of the core, due to Shapley and Shubik (1963) and (1966), may serve as a substitute for the core of G when $\mathcal{C}(G)$ is empty.

Definition 7.1.1. *Let ε be a real number. The* **ε-core** *of the game $G = (N, v)$, $\mathcal{C}_\varepsilon(G)$, is defined by*

$$\mathcal{C}_\varepsilon(G) = \{x \in X(G) \mid e(S, x, v) \leq \varepsilon \text{ for all } \emptyset \neq S \subsetneqq N\}.$$

Clearly, $\mathcal{C}_0(G) = \mathcal{C}(G)$. Now let $|N| \geq 2$. Then $\mathcal{C}_{\varepsilon'}(G) \subseteq \mathcal{C}_\varepsilon(G)$ if $\varepsilon' < \varepsilon$, with strict inclusion if $\mathcal{C}_\varepsilon(G) \neq \emptyset$. Also, $\mathcal{C}_\varepsilon(G) \neq \emptyset$ if ε is large enough, and $\mathcal{C}_\varepsilon(G) = \emptyset$ if ε is small enough.

The ε-core has the following interpretation. It is the set of efficient payoff vectors that cannot be improved upon by any coalition if coalition formation entails a "cost" of ε (or a "bonus" of $-\varepsilon$ if ε is negative). In order to be an "acceptable" substitute for the core, $\mathcal{C}_\varepsilon(G)$ has to be nonempty and ε should be small. This remark leads to the following definition.

Definition 7.1.2. *The* **least-core** *of the game $G = (N, v)$, denoted $\mathcal{LC}(G)$, is the intersection of all nonempty ε-cores of G.*

Remark 7.1.3. (1) The least-core can also be defined in the following way. Let $\varepsilon_0 = \varepsilon_0(G)$ be the smallest ε such that $\mathcal{C}_\varepsilon(G) \neq \emptyset$, that is,

$$\varepsilon_0 = \min_{x \in X(G)} \max_{\emptyset \neq S \subsetneqq N} e(S, x, v). \tag{7.1.1}$$

Then $\mathcal{LC}(G) = \mathcal{C}_{\varepsilon_0}(G)$.

(2) By (1), $\mathcal{LC}(G) \neq \emptyset$.

(3) If $\mathcal{C}(G) \neq \emptyset$, that is, if $\varepsilon_0 \leq 0$, then the least-core occupies a central position within the core. If $\mathcal{C}(G) = \emptyset$, that is, if $\varepsilon_0 > 0$, then the least-core may be regarded as revealing the "latent" position of the core.

Remark 7.1.4. In Shapley and Shubik (1966) the ε-core is called the *strong* ε-core, in order to distinguish it from another generalization of the core, called the *weak* ε-core. We shall not refer to the weak ε-core.

Let again $G = (N, v)$ be a coalitional game, let $n = |N|$, and let $\varepsilon \in \mathbb{R}$. We shall assume that $n \geq 2$. Then $\mathcal{C}_\varepsilon(G)$ is a convex compact polyhedron, bounded by not more than $2^n - 2$ hyperplanes of the form

$$H_S^\varepsilon = \{x \in X(G) \mid x(S) = v(S) - \varepsilon\},\ \emptyset \neq S \subsetneqq N.$$

We shall write H_S for H_S^0. Except for the least-core, all nonempty ε-cores have dimension $n-1$, that is, the dimension of $X(G)$. The dimension of $\mathcal{LC}(G)$ is always $n-2$ or less. We shall now prove that $\mathcal{C}_\varepsilon(G) \cap H_S^\varepsilon \neq \emptyset$ for every $\emptyset \neq S \subsetneqq N$ if ε is large enough.

Lemma 7.1.5. *Let $G = (N, v)$ be an n-person game and let $\emptyset \neq S \subsetneqq N$. Then there exists $\varepsilon^S \in \mathbb{R}$ such that $\mathcal{C}_\varepsilon(G) \cap H_S^\varepsilon \neq \emptyset$ if and only if $\varepsilon \geq \varepsilon^S$.*

Proof: First, we shall prove that there exists $\varepsilon' \in \mathbb{R}$ such that $\mathcal{C}_{\varepsilon'}(G) \cap H_S^{\varepsilon'} \neq \emptyset$. Let $y \in X(G)$ satisfy $y(S) = v(S)$. For $\varepsilon \in \mathbb{R}$ define $y_\varepsilon \in X(G)$ by

$$y_\varepsilon^i = \begin{cases} y^i - \frac{\varepsilon}{|S|} & , \text{ if } i \in S, \\ y^i + \frac{\varepsilon}{|N \setminus S|} & , \text{ if } i \in N \setminus S. \end{cases} \tag{7.1.2}$$

Clearly, $y_\varepsilon(S) = v(S) - \varepsilon$ and, thus, $y_\varepsilon \in H_S^\varepsilon$ for all $\varepsilon \in \mathbb{R}$. If $T \subseteq N,\ T \neq S$, and if $\varepsilon \geq 0$, then

$$y_\varepsilon(T) + \varepsilon \geq y(T) - (|S| - 1)\frac{\varepsilon}{|S|} + \varepsilon = y(T) + \frac{\varepsilon}{|S|}.$$

Thus, there exists $\varepsilon' \geq 0$ large enough such that $y_{\varepsilon'}(T) + \varepsilon' \geq v(T)$ for all $T \subseteq N$. Hence $y_{\varepsilon'} \in \mathcal{C}_{\varepsilon'}(G) \cap H_S^{\varepsilon'}$.

Now we shall prove that if $\mathcal{C}_{\varepsilon'}(G) \cap H_S^{\varepsilon'} \neq \emptyset$ and $\varepsilon'' \geq \varepsilon'$, then $\mathcal{C}_{\varepsilon''}(G) \cap H_S^{\varepsilon''} \neq \emptyset$. Indeed, let $y \in \mathcal{C}_{\varepsilon'}(G) \cap H_S^{\varepsilon'}$, let $\varepsilon = \varepsilon'' - \varepsilon'$, and let y_ε be given by (7.1.2). Then

$$y_\varepsilon(T) + \varepsilon = y_\varepsilon(T) + \varepsilon'' - \varepsilon' \geq y(T) \geq v(T) - \varepsilon'$$

for every $T \neq \emptyset, N$. Thus, $y_\varepsilon \in \mathcal{C}_{\varepsilon''}(G)$. Also, $y_\varepsilon(S) = v(S) - \varepsilon''$, that is, $y_\varepsilon \in H_S^{\varepsilon''}$. So, $\mathcal{C}_{\varepsilon''}(G) \cap H_S^{\varepsilon''} \neq \emptyset$.

Define now

$$\varepsilon^S = \min\{\varepsilon \in \mathbb{R} \mid \mathcal{C}_\varepsilon(G) \cap H_S^\varepsilon \neq \emptyset\}.$$

By the foregoing analysis, ε^S is well defined and $\mathcal{C}_\varepsilon(G) \cap H_S^\varepsilon \neq \emptyset$ if and only if $\varepsilon \geq \varepsilon^S$.

q.e.d.

The proof of the following remark is left to the reader (see Exercise 7.1.2).

Remark 7.1.6. If $\varepsilon > \varepsilon^S$, then $H_S^\varepsilon \cap \mathcal{C}_\varepsilon(G)$ has dimension $n - 2$.

Corollary 7.1.7. *If $\varepsilon > \max_{\emptyset \neq S \subsetneq N} \varepsilon^S$, then $\mathcal{C}_\varepsilon(G)$ has $2^n - 2$ facets, namely $H_S^\varepsilon \cap \mathcal{C}_\varepsilon(G),\ \emptyset \neq S \subsetneq N$.*

The proof of Lemma 7.1.5 yields the following interesting consequence.

Corollary 7.1.8. *Let $G_i = (N, v_i),\ i = 1, 2,$ be two games that, for some $\varepsilon', \varepsilon'' \in \mathbb{R}$, satisfy*

$$\mathcal{C}_{\varepsilon'}(G_1) = \mathcal{C}_{\varepsilon''}(G_2) \neq \emptyset.$$

Then, for every $\varepsilon > 0$,

$$\mathcal{C}_{\varepsilon' - \varepsilon}(G_1) = \mathcal{C}_{\varepsilon'' - \varepsilon}(G_2).$$

In particular, $\mathcal{LC}(G_1) = \mathcal{LC}(G_2)$.

Proof: Assume, on the contrary, that there exist $\varepsilon > 0$ such that $\mathcal{C}_{\varepsilon'-\varepsilon}(G_1) \neq \mathcal{C}_{\varepsilon''-\varepsilon}(G_2)$. Without loss of generality, let $y \in \mathcal{C}_{\varepsilon'-\varepsilon}(G_1) \setminus \mathcal{C}_{\varepsilon''-\varepsilon}(G_2)$. Then

$$y(T) \geq v_1(T) - \varepsilon' + \varepsilon \text{ for all } \emptyset \neq T \subsetneq N \text{ and}$$
$$y(S) < v_2(S) - \varepsilon'' + \varepsilon \text{ for at least one } \emptyset \neq S \subsetneq N.$$

Now define y_ε by (7.1.2). Then

$$y_\varepsilon(T) \geq v_1(T) - \varepsilon' \text{ for all } \emptyset \neq T \subsetneq N \text{ and}$$
$$y(S) < v_2(S) - \varepsilon''.$$

As $y(N) = v_1(N) = v_2(N) = y_\varepsilon(N)$, we find that $y_\varepsilon \in \mathcal{C}_{\varepsilon'}(G_1) \setminus \mathcal{C}_{\varepsilon''}(G_2)$, a contradiction.

q.e.d.

Exercises

Exercise 7.1.1. Let (N, v) be a game, $\varepsilon \in \mathbb{R}$, $x \in \mathbb{R}^N$, and $\alpha > 0$. Show that $\mathcal{C}_{\alpha\varepsilon}(N, \alpha v + x) = \alpha\mathcal{C}_\varepsilon(N, v) + x$.

Exercise 7.1.2. Prove Remark 7.1.6.

Exercise 7.1.3. Let $\Gamma_\mathcal{U}$ denote the set of coalitional games with players in $\mathcal{U}$. Show that $\mathcal{C}_\varepsilon$ satisfies RGP, RCP, and CRGP on $\Gamma_\mathcal{U}$.

7.2 Some Properties of the Least-Core

Let $\mathcal{U}$ be a set of players and let $\Gamma_\mathcal{U}$ be the set of coalitional games with players in $\mathcal{U}$. Let $\Gamma \subseteq \Gamma_\mathcal{U}$. Then the least-core satisfies COV and AN on Γ (see Exercise 7.2.1).

Example 7.2.1. Let (N, v) be defined by $N = \{1, 2, 3\}$, $v(\{1, 2\}) = 3$, $v(N) = v(\{1, 3\}) = v(\{2, 3\}) = 1$, and $v(S) = 0$ otherwise. Then $\mathcal{LC}(N, v) = \{(1, 1, -1)\}$. Thus $\mathcal{LC}$ does not satisfy IR.

Let $\Gamma_\mathcal{U}^w$ be the set of all weakly superadditive (0-monotonic) games. The following result is true.

Theorem 7.2.2. *The least-core satisfies* IR *on* $\Gamma_\mathcal{U}^w$.

Proof: Let $(N, v) \in \Gamma_\mathcal{U}^w$ and let $x \in \mathcal{LC}(N, v) = \mathcal{C}_{\varepsilon_0}(N, v)$ (see (7.1.1)). Assume, on the contrary, that there exists $i \in N$ such that $x^i < v(\{i\})$. Then $\varepsilon_0 \geq v(\{i\}) - x^i > 0$. If $\emptyset \neq S \subseteq N \setminus \{i\}$, then

$$\begin{aligned} x(S) = x(S \cup \{i\}) - x^i &\geq v(S \cup \{i\}) - \varepsilon_0 - x^i \\ &\geq v(S) + v(\{i\}) - \varepsilon_0 - x^i > v(S) - \varepsilon_0. \end{aligned}$$

Therefore, we may find $y \in X(N, v)$ such that (i) $y^k < x^k$ for all $k \in N \setminus \{i\}$ and (ii) $y(S) > v(S) - \varepsilon_0$ for all $\emptyset \neq S \subseteq N \setminus \{i\}$. Now, if $T \subsetneqq N$ and $i \in T$, then

$$y(T) > x(T) \geq v(T) - \varepsilon_0.$$

Thus, $y(T) > v(T) - \varepsilon_0$ for all $\emptyset \neq T \subsetneqq N$, contradicting the definition of ε_0. Hence $x^i \geq v(\{i\})$ and x is individually rational. **q.e.d.**

We now prove that the least-core is reasonable from above (see Definition 2.3.9).

Theorem 7.2.3. *The least-core satisfies* REAB *on every* $\Gamma \subseteq \Gamma_\mathcal{U}$.

Proof: Let $(N, v) \in \Gamma$ and assume that $|N| \geq 2$. Let ε_0 be defined by (7.1.1) and let $x \in \mathcal{C}_{\varepsilon_0}(N, v)$. Define

$$\mathcal{S} = \{S \mid e(S, x, v) = \varepsilon_0,\ \emptyset \neq S \subsetneqq N\}. \tag{7.2.1}$$

Then for every $i \in N$ there exists $S \in \mathcal{S}$ such that $i \in S$. Indeed, if $i \notin S$ for every $S \in \mathcal{S}$, then we could find $y \in X(N, v)$ such that (i) $y^i < x^i$, (ii) $y^k > x^k$ for every $k \in N \setminus \{i\}$, and (iii) $y(T) > v(T) - \varepsilon_0$ for every $\emptyset \neq T \subsetneqq N$, contradicting the definition of ε_0. Thus, for each $i \in N$, let $S_i \in \mathcal{S}$ contain i. If $S_i \neq \{i\}$ then, by (7.2.1),

$$e(S_i, x, v) \geq e(S_i \setminus \{i\}, x, v),$$

whence $x^i \leq v(S_i) - v(S_i \setminus \{i\})$ and the proof is complete. If $S_i = \{i\}$ and $\varepsilon_0 > 0$, then

$$x^i = v(\{i\} - \varepsilon_0 < v(\{i\}) = v(\{i\}) - v(\emptyset)$$

and the desired inequality is proved. Finally, if $\varepsilon_0 \leq 0$, then $x \in \mathcal{C}(N, v)$ and, therefore, by Lemma 2.3.10, the proof is complete. **q.e.d.**

The following theorem shows a relation between the least-core and the Mas-Colell prebargaining set (see Definition 4.4.11).

Theorem 7.2.4. *Let (N, v) be a game. Then $\mathcal{LC}(N, v) \subseteq \mathcal{PMB}(N, v)$.*

Proof: Let $x \in \mathcal{LC}(N, v)$ and let (P, y) be an objection at x (see Definition 4.4.11). Then $y > x^P$ and $y(P) = v(P)$. Thus, by Pareto optimality of the least-core, $\emptyset \neq P \subsetneqq N$. Let ε_0 be defined by (7.1.1) and let $\mathcal{S}$ be given by (7.2.1). Then, by Exercise 7.2.2, $\mathcal{S}$ contains a balanced subset $\mathcal{T}$. Let $i \in P$ such that $y^i > x^i$. If $P \in \mathcal{S}$, then, by balancedness of $\mathcal{T}$, there exists $Q \in \mathcal{T}$ such that $i \notin Q$. If $P \notin \mathcal{S}$ then let Q be an arbitrary member of $\mathcal{S}$. In both cases let $j \in Q$ and let $z \in \mathbb{R}^Q$ be defined by $z^{(Q \cap P)\setminus\{j\}} = y^{(Q \cap P)\setminus\{j\}}$, $z^{Q\setminus(P \cup \{j\})} = x^{Q\setminus(P \cup \{j\})}$, and $z^j = v(Q) - z(Q \setminus \{j\})$. Then $z^j > x^j$, if $j \in Q \setminus P$, and $z^j > y^j$, if $j \in Q \cap P$, because either (i) $e(P, x, v) = e(Q, x, v)$ and $i \notin Q$ or (ii) $e(P, x, v) < e(Q, x, v)$. Hence, (Q, z) is a counterobjection to (P, y) in both cases. **q.e.d.**

Theorem 7.2.4 is due to Einy, Holzman, and Monderer (1999).

Exercises

Exercise 7.2.1. Prove that the least-core satisfies COV and AN on every set $\Gamma \subseteq \Gamma_{\mathcal{U}}$ of games.

Exercise 7.2.2. Let $G = (N, v)$, $|N| \geq 2$, be a game, let $x \in X(N, v)$, and let $\varepsilon = \max_{S \in 2^N \setminus \{N, \emptyset\}} e(S, x, v)$. Show that $x \in \mathcal{LC}(N, v)$ if and only if $\mathcal{D}(\varepsilon, x, v) \setminus \{N, \emptyset\}$ (see Definition 5.2.5) contains a balanced subset.

Exercise 7.2.3. Let $N = \{1, \ldots, 4\}$ and let $v(S)$, $S \subseteq N$, be given by $v(\{1, 2, 3\}) = v(\{1, 2, 4\}) = v(\{1, 3, 4\}) = v(\{3, 4\}) = 1$, $v(\{2, 3, 4\}) = v(N) = 0$, and $v(S) = -|S|$ otherwise. Show that (N, v) is a superadditive game that satisfies $\mathcal{LC}(N, v) \not\subseteq \mathcal{PM}(N, v)$.

Exercise 7.2.4. Prove that the least-core satisfies RCP on every set $\Gamma \subseteq \Gamma_{\mathcal{U}}$.

Exercise 7.2.5. Show by means of an example that the least-core does not satisfy RGP or CRGP, respectively, on $\Gamma_{\mathcal{U}}$, provided that $|\mathcal{U}| \geq 3$ or $|\mathcal{U}| \geq 4$, respectively.

7.3 The Reasonable Set

Let $G = (N, v)$ be a game and let $i \in N$. We recall (see Section 2.3) that

$$b^i_{\max}(G) = \max_{S \subseteq N \setminus \{i\}} \Big(v(S \cup \{i\}) - v(S) \Big). \tag{7.3.1}$$

In the current section we define $b \in \mathbb{R}^N$ by $b^i = b^i_{\max}(G),\ i \in N$.

Definition 7.3.1. *The* **reasonable set** *of* $G = (N, v)$, $\varrho(G)$, *is the set of all members of* $X(G)$ *that are reasonable from above, that is,*

$$\varrho(G) = \{x \in X(N, v) \mid x^i \leq b^i \textit{ for all } i \in N\}.$$

Remark 7.3.2. The reasonable set satisfies AN and COV on every set of games.

Lemma 7.3.3. *Let* (N, v) *be an* n*-person game. Then* $\varrho(N, v)$ *is a simplex of dimension* $n - 1$ *unless* (N, v) *is inessential, in which case it consists of the unique imputation, that is,* $\varrho(N, v) = I(N, v)$.

Proof: Let $(i_1, \ldots, i_n)$ be any ordering of N, let $S^k = \{i_1, \ldots, i_k\}$ for $k = 1, \ldots, n$, and let $S^0 = \emptyset$. By (7.3.1),

$$b^{i_k} \geq v(S^k) - v(S^{k-1}) \text{ for all } k = 1, \ldots, n. \tag{7.3.2}$$

Summing (7.3.2) yields

$$b(N) \geq v(N). \tag{7.3.3}$$

By (7.3.3), $\varrho(N, v) \neq \emptyset$. Furthermore, the dimension of $\varrho(N, v)$ is $n - 1$ unless $b(N) = v(N)$. In this case the inequalities in (7.3.2) have to be equalities for all orderings of N, which makes the coalition function v additive and yields $I(N, v) = \varrho(N, v) = \{b\}$. **q.e.d.**

We shall now assume that $I(N, v) \neq \emptyset$, that is,

$$v(N) \geq \sum_{i \in N} v(\{i\}). \tag{7.3.4}$$

Under this condition we shall analyze the relationship between $I(N, v)$ and $\varrho(N, v)$.

Theorem 7.3.4. *Let $G = (N, v)$ satisfy (7.3.4). Then*

(1) $I(G) \cap \varrho(G) \neq \emptyset$; furthermore, the intersection has dimension $n-1$ unless $I(G)$ is a single point or $b^i = v(\{i\})$ for some $i \in N$;

(2) no extreme point of $\varrho(G)$ is in the interior[1] of $I(G)$;

(3) if G is 0-monotonic, then no extreme point of $I(G)$ is in the interior of $\varrho(G)$.

Proof: (1) By (7.3.4) and (7.3.3), $b(N) \geq v(N) \geq \sum_{i\in N} v(\{i\})$. So there exist $\alpha \in \mathbb{R}$ such that $0 \leq \alpha \leq 1$ and

$$\alpha b(N) + (1-\alpha)\sum_{i\in N} v(\{i\}) = v(N).$$

Define $y \in \mathbb{R}^N$ by $y^i = \alpha b^i + (1-\alpha)v(\{i\})$, $i \in N$. Then $y \in X(G)$ and, by (7.3.2),

$$b^i \geq y^i \geq v(\{i\}),\ i \in N.$$

Thus y belongs to both $\varrho(G)$ and $I(G)$. Moreover, if $b^i > v(\{i\})$ for all $i \in N$, then $b^i > y^i$ for all $i \in N$, because $b(N) > v(N)$ implies that $\alpha < 1$. If, in addition, $v(N) > \sum_{i\in N} v(\{i\})$, then $\alpha > 0$ and, thus, $y^i > v(\{i\})$ for all $i \in N$. Hence, if both of the foregoing conditions are satisfied, then the dimension of $I(G) \cap \varrho(G)$ is $n-1$.

(2) For $j \in N$ let $x_j \in \mathbb{R}^N$ be defined by

$$x_j^i = \begin{cases} b^i & \text{, if } i \in N \setminus \{j\}, \\ v(N) - b(N \setminus \{j\}) & \text{, if } i = j. \end{cases} \tag{7.3.5}$$

Then the vectors x_j, $j \in N$, are the extreme points of $\varrho(G)$. Let $j \in N$. Summing the $n-1$ inequalities of (7.3.2) with $i_1 = j$ corresponding to $k = 2, \ldots, n$ yields

$$b(N \setminus \{j\}) \geq v(N) - v(\{j\}).$$

Hence $x_j^j \leq v(\{j\})$. Therefore none of the x_j, $j \in N$, is an interior point of $I(G)$.

(3) Analogously to (7.3.5) the vectors y_j, $j \in N$, defined by

$$y_j^i = \begin{cases} v(\{i\}) & \text{, if } i \in N \setminus \{j\}, \\ v(N) - \sum_{i\in N\setminus\{j\}} v(\{i\}) & \text{, if } i = j, \end{cases} \tag{7.3.6}$$

are the extreme points of $I(G)$. Let $S \subseteq N$. The 0-monotonicity of G yields

$$v(S) \geq \sum_{i\in S} v(\{i\}) \text{ and } v(N) \geq v(S) + \sum_{i\in N\setminus S} v(\{i\}). \tag{7.3.7}$$

[1] In the relative topology of $X(G)$.

Let $j \in N$. Then there exists $T \subseteq N$ containing j such that $b^j = v(T) - v(T \setminus \{j\})$. Applying (7.3.7) we obtain

$$b^j \le v(T) - \sum_{i \in T \setminus \{j\}} v(\{i\}) \le v(N) - \sum_{i \in N \setminus T} v(\{i\}) - \sum_{i \in T \setminus \{j\}} v(\{i\}) = y^j_j.$$

So, $y^j_j \ge b^j$ and y^j does not belong to the interior of $\varrho(G)$. **q.e.d.**

Throughout the rest of this section we shall assume that every coalitional game considered has at least two players. Let $G = (N, v)$ satisfy (7.3.4). We denote

$$\varepsilon_1 = \varepsilon_1(G) = \max_{\emptyset \ne S \subsetneqq N} \left(v(S) - \sum_{i \in S} v(\{i\}) \right). \tag{7.3.8}$$

Lemma 7.3.5. *$I(G) \subseteq \mathcal{C}_\varepsilon(G)$ if and only if $\varepsilon \ge \varepsilon_1(G)$.*

Proof: Let $I(G) \subseteq \mathcal{C}_\varepsilon(G)$, let $S \subseteq N$, $S \ne \emptyset, N$, and let $j \in N \setminus S$. Then the extreme point y_j of $I(G)$, defined by (7.3.6), belongs to $\mathcal{C}_\varepsilon(G)$ as well. Therefore, $v(S) - \sum_{i \in S} v(\{i\}) \le \varepsilon$. Thus, $\varepsilon \ge \varepsilon_1(G)$.

Conversely, let $y \in I(G)$ and $\emptyset \ne S \subsetneqq N$. Then

$$v(S) - y(S) \le v(S) - \textstyle\sum_{i \in S} v(\{i\}) \le \varepsilon_1(G).$$ **q.e.d.**

Let $G = (N, v)$ be a coalitional game. We shall now determine the minimum ε that still implies that $\varrho(G) \subseteq \mathcal{C}_\varepsilon(G)$. Let

$$\varepsilon_2 = \varepsilon_2(G) = \max_{\emptyset \ne S \subsetneqq N} \left(v(S) - v(N) + b(N \setminus S) \right). \tag{7.3.9}$$

Lemma 7.3.6. *$\varrho(G) \subseteq \mathcal{C}_\varepsilon(G)$ if and only if $\varepsilon \ge \varepsilon_2(G)$.*

Proof: Let x_j, $j \in N$, be defined by (7.3.5) and let $\varepsilon \in \mathbb{R}$. Then $\varrho(G) \subseteq \mathcal{C}_\varepsilon(G)$ if and only if $x_j \in \mathcal{C}_\varepsilon(G)$ for all $j \in N$. For every $j \in N$ and every $S \subseteq N$ containing j we have

$$\begin{aligned} x_j(S) &= x^j_j + b(S \setminus \{j\}) \\ &= v(N) - b(N \setminus \{j\}) + b(S \setminus \{j\}) = v(N) - b(N \setminus S). \end{aligned} \tag{7.3.10}$$

Suppose first that $x_j \in \mathcal{C}_\varepsilon(G)$ for all $j \in N$ and let $\emptyset \ne S \subsetneqq N$. Choose $j \in S$. By (7.3.10),

$$x_j(S) = v(N) - b(N \setminus S) \ge v(S) - \varepsilon.$$

Thus, $\varepsilon \ge v(S) - v(N) + b(N \setminus S)$ and the inequality $\varepsilon \ge \varepsilon_2(G)$ is proved.

Conversely, assume now that $\varepsilon \ge \varepsilon_2(G)$. Let $j \in N$ and let $\emptyset \ne S \subsetneqq N$. Two cases may occur.

(1) $j \in S$. Then, by (7.3.10),

$$x_j(S) = v(N) - b(N \setminus S) \geq v(S) - \varepsilon,$$

because $\varepsilon \geq \varepsilon_2(G) \geq v(S) - v(N) + b(N \setminus S)$.

(2) $j \in N \setminus S$. Then

$$x_j(S) = b(S) \geq b(S) + v(N) - b(N) = v(N) - b(N \setminus S) \geq v(S) - \varepsilon.$$

Thus, $x_j \in \mathcal{C}_\varepsilon(G)$ for all $j \in N$ and the proof is complete. **q.e.d.**

Let $G = (N, v)$ satisfy (7.3.4). We are interested in finding the minimum ε-core that still contains the intersection of $I(G)$ and $\varrho(G)$. Let

$$\varepsilon_*(G) = \max_{\emptyset \neq S \subsetneqq N} \min\left\{v(S) - \sum_{i \in S} v(\{i\}), v(S) - v(N) + b(N \setminus S)\right\}. \tag{7.3.11}$$

The following result will be useful in the next section.

Lemma 7.3.7. *$\varrho(G) \cap I(G) \subseteq \mathcal{C}_\varepsilon(G)$ if and only if $\varepsilon \geq \varepsilon_*(G)$.*

Proof: Without loss of generality (see Exercise 7.1.1 and Remark 7.3.2) we may assume that G is 0-normalized. Thus, $x \in \varrho(G) \cap I(G)$ if and only if $x \in X(G)$ and $0 \leq x^i \leq b^i$ for all $i \in N$.

Assume first that $x \in \varrho(G) \cap I(G)$. Let $S \subseteq N$. Then we have both $x(S) \geq 0$ and

$$x(S) = x(N) - x(N \setminus S) \geq v(N) - b(N \setminus S).$$

Therefore,

$$\begin{aligned} v(S) - x(S) &\leq v(S) - \max\{0, v(N) - b(N \setminus S)\} \\ &= \min\{v(S), v(S) + b(N \setminus S) - v(N)\} \leq \varepsilon_*(G) \end{aligned}$$

for all $\emptyset \neq S \subsetneqq N$. Thus, $x \in \mathcal{C}_\varepsilon(G)$ for all $\varepsilon \geq \varepsilon_*(G)$, proving the sufficiency part of the lemma.

To prove necessity let $\varrho(G) \cap I(G) \subseteq \mathcal{C}_\varepsilon(G)$ and let the maximum in (7.3.11) be attained at S. Two cases may occur.

(1) $b(N \setminus S) \geq v(N)$. If $v(N) > 0$ define $x \in \mathbb{R}^N$ by

$$x^i = \begin{cases} 0 & \text{, if } i \in S, \\ b^i \frac{v(N)}{b(N \setminus S)} & \text{, if } i \in N \setminus S. \end{cases}$$

If $v(N) = 0$ let $x = 0 \in \mathbb{R}^N$. Then $x \in \varrho(G) \cap I(G)$ and

$$x(S) = 0 = v(S) - \varepsilon_*(G) \geq v(S) - \varepsilon.$$

Hence $\varepsilon \geq \varepsilon_*(G)$.

(2) $v(N) > b(N \setminus S)$. This time define $x \in \mathbb{R}^N$ by

$$x^i = \begin{cases} b^i \frac{v(N)-b(N\setminus S)}{b(S)} & \text{, if } i \in S, \\ b^i & \text{, if } i \in N \setminus S. \end{cases}$$

Again, as the reader may easily verify, $x \in \varrho(G) \cap I(G)$. Also, $\varepsilon_*(G) = v(S) + b(N \setminus S) - v(N)$. Hence,

$$x(S) = v(N) - b(N \setminus S) = v(S) - \varepsilon_*(G) \geq v(S) - \varepsilon$$

and, thus, $\varepsilon \geq \varepsilon_*(G)$. **q.e.d.**

Exercises

Exercise 7.3.1. Provide a three-person game G such that there exists an extreme point of $I(G)$ which is in the interior of $\varrho(G)$.

Exercise 7.3.2. Let $G = (N, v)$ be an n-person game and let d denote the number of dummies (see Remark 4.1.18) of G. Assume that $d < n$. Show that $n - 1 - d$ is the dimension of (see Definition 2.3.9)

$$\varrho_1(G) = \{x \in \varrho(G) \mid x^i \geq b^i_{\min}(N, v) \text{ for all } i \in N\}.$$

7.4 Geometric Characterizations of the Prekernel and Kernel

Let $G = (N, v)$ be a coalitional game. Throughout this section we shall assume that $n = |N| \geq 2$. Let $\varepsilon \geq \varepsilon_0(G)$ (see (7.1.1)), let $x \in \mathcal{C}_\varepsilon(G)$, and let $i, j \in N$, $i \neq j$. Consider the half-line emanating from x obtained by letting the j-th coordinate increase and the i-the coordinate decrease by the same amount. Let $\delta^\varepsilon_{ij}(x) = \delta^\varepsilon_{ij}(x, v)$ denote the maximum amount which can be transferred from i to j in this way while remaining in $\mathcal{C}_\varepsilon(G)$. Thus,

$$\delta^\varepsilon_{ij}(x) = \max\{\delta \in \mathbb{R} \mid x - \delta\chi_{\{i\}} + \delta\chi_{\{j\}} \in \mathcal{C}_\varepsilon(G)\}. \tag{7.4.1}$$

Note that $\delta^\varepsilon_{ij}(x)$ is well defined for all $x \in \mathcal{C}_\varepsilon(G)$.

Lemma 7.4.1. *If $x \in \mathcal{C}_\varepsilon(G)$, then $\delta^\varepsilon_{ij}(x) = \varepsilon - s_{ij}(x, v)$ for all $i, j \in N$, $i \neq j$.*

Proof: If $x_\delta = x - \delta\chi_{\{i\}} + \delta\chi_{\{j\}}$ and $S \subseteq N$, then

$$e(S, x_\delta, v) = \begin{cases} e(S, x, v) + \delta & \text{, if } S \in \mathcal{T}_{ij}; \\ e(S, x, v) - \delta & \text{, if } S \in \mathcal{T}_{ji}; \\ e(S, x, v) & \text{, otherwise.} \end{cases}$$

Thus

$$\begin{aligned}\delta^{\varepsilon}_{ij}(x) &= \max\{\delta \mid x_\delta \in \mathcal{C}_\varepsilon(G)\}\\ &= \max\{\delta \mid e(S,x,v)+\delta \le \varepsilon \text{ for all } S \in \mathcal{T}_{ij}\}\\ &= \max\{\delta \mid \delta \le \varepsilon - e(S,x,v) \text{ for all } S \in \mathcal{T}_{ij}\} = \varepsilon - s_{ij}(x,v).\end{aligned}$$

q.e.d.

The proof of Lemma 7.4.1 yields the following corollary on the structure of the ε-core.

Corollary 7.4.2. *Let $x \in \mathcal{C}_\varepsilon(G)$, let $i \in N$, $j \in N \setminus \{i\}$, and let $S \in \mathcal{T}_{ij}$ satisfy $e(S,x,v) = s_{ij}(x,v)$. Then $H^{\varepsilon}_{S} \cap \mathcal{C}_\varepsilon(G) \neq \emptyset$.*

Let $x \in \mathcal{C}_\varepsilon(G)$. Let $R_{ij}(x,\varepsilon)$ denote the following line segment:

$$R_{ij}(x,\varepsilon) = \left[x + \delta^{\varepsilon}_{ij}(x)\left(\chi_{\{j\}} - \chi_{\{i\}}\right), x + \delta^{\varepsilon}_{ji}(x)\left(\chi_{\{i\}} - \chi_{\{j\}}\right)\right]. \tag{7.4.2}$$

$R_{ij}(x,\varepsilon)$ will be called the *segment through x in the $i-j$ direction.*

Now we are ready for the geometric characterizations of the prekernel and the kernel.

Theorem 7.4.3. *Let $G = (N,v)$ be a coalitional game, let $\varepsilon \in \mathbb{R}$, and let $x \in \mathcal{C}_\varepsilon(G)$. Then $x \in \mathcal{PK}(G)$ if and only if for each pair of players $i,j \in N$, $i \neq j$, x bisects the segment $R_{ij}(x,\varepsilon)$.*

Proof: Lemma 7.4.1. **q.e.d.**

Theorem 7.4.3 has the following interesting corollary.

Corollary 7.4.4. *Let $G = (N,v)$ and $G' = (N,v')$ be games and let $\varepsilon, \varepsilon' \in \mathbb{R}$. If $\mathcal{C}_\varepsilon(G) = \mathcal{C}_{\varepsilon'}(G')$, then*

$$\mathcal{PK}(G) \cap \mathcal{C}_\varepsilon(G) = \mathcal{PK}(G') \cap \mathcal{C}_{\varepsilon'}(G').$$

The foregoing results provide a complete geometric description of the intersection of the prekernel with an ε-core. Thus, we would like to know when an ε-core contains the prekernel. The following lemma answers the foregoing question.

Lemma 7.4.5. *Let $G = (N,v)$ be a game. Then*

(1) if $\varepsilon \ge \varepsilon_2(G)$, then $\mathcal{PK}(G) \subseteq \mathcal{C}_\varepsilon(G)$;

(2) if G is 0-monotonic and $\varepsilon \ge \varepsilon_*(G)$, then $\mathcal{PK}(G) \subseteq \mathcal{C}_\varepsilon(G)$.

Proof: (1) Theorem 5.6.1 and Lemma 7.3.6.

(2) Theorems 5.5.6 and 5.6.1 and Lemma 7.3.7. **q.e.d.**

Another connection between the prekernel and the ε-core is seen in the following result.

Lemma 7.4.6. *Let $G = (N, v)$ be a game. If $\mathcal{C}_\varepsilon \neq \emptyset$, then $\mathcal{PK}(G) \cap \mathcal{C}_\varepsilon(G) \neq \emptyset$.*

Proof: If $\mathcal{C}_\varepsilon(G) \neq \emptyset$, then, by its definition, $\nu(G) \in \mathcal{C}_\varepsilon(G)$. By Theorem 5.1.17, $\nu(G) \in \mathcal{PK}(G)$. **q.e.d.**

Remark 7.4.7. Let $G = (N, v)$ be a game and let $\mathcal{C}_\varepsilon(G) \neq \emptyset$. Consider the *bisecting hyperfaces* B_{ij} obtained by taking the midpoints of all segments in the $i - j$ direction through the points in $\mathcal{C}_\varepsilon(G)$. As $B_{ij} = B_{ji}$, there are $\binom{n}{2}$ such hyperfaces, and according to Theorem 7.4.3, $\mathcal{C}_\varepsilon(G) \cap \mathcal{PK}(G)$ is their intersection. By Lemma 7.4.6 this set is nonempty. Thus, we have discovered an interesting geometric property of $\mathcal{C}_\varepsilon(G)$, namely, its bisecting hyperfaces B_{ij} must have a nonempty intersection.

The geometric characterization of the kernel is somewhat more complicated than that of the prekernel.

Theorem 7.4.8. *Let $G = (N, v)$ be a game and let $x \in \mathcal{C}_\varepsilon(G)$. Then $x \in \mathcal{K}(G)$ if and only if $x \in I(G)$ and, for each ordered pair of players $(i, j) \in N \times N$, $i \neq j$, either x bisects the segment $R_{ij}(x, \varepsilon)$, or $x^j = v(\{j\})$ and $\delta^\varepsilon_{ji}(x) > \delta^\varepsilon_{ij}(x)$.*

Proof: Definition 5.5.7 and Lemma 7.4.1. **q.e.d.**

Theorem 7.4.8 yields the following corollary.

Corollary 7.4.9. *Let $G = (N, v)$ and $G' = (N, v')$ be two games. If $I(G) = I(G')$ and $\mathcal{C}_\varepsilon(G) = \mathcal{C}_{\varepsilon'}(G')$, then*

$$\mathcal{K}(G) \cap \mathcal{C}_\varepsilon(G) = \mathcal{K}(G') \cap \mathcal{C}_{\varepsilon'}(G').$$

Let $G = (N, v)$ be a game and let $\varepsilon \geq \varepsilon_0(G)$. The description of $\mathcal{PK}(G) \cap \mathcal{C}_\varepsilon(G)$ is simpler than that of $\mathcal{K}(G) \cap \mathcal{C}_\varepsilon(G)$ (see Theorems 7.4.3 and 7.4.8). Hence it is interesting to know when the foregoing sets coincide. The following lemmata answer this question.

Lemma 7.4.10. *Let $G = (N, v)$ be a 0-monotonic game. Then, for every $\varepsilon \in \mathbb{R}$,*

$$\mathcal{PK}(G) \cap \mathcal{C}_\varepsilon(G) = \mathcal{K}(G) \cap \mathcal{C}_\varepsilon(G). \tag{7.4.3}$$

Proof: Theorem 5.5.10. **q.e.d.**

Lemma 7.4.11. *Let $G = (N, v)$ be a game and $\varepsilon \leq 0$. Then (7.4.3) is true.*

Proof: Let $x \in \mathcal{PK}(G) \cap \mathcal{C}_\varepsilon(G)$. Then

$$x^i \geq v(\{i\}) - \varepsilon \geq v(\{i\}) \text{ for all } i \in N,$$

because $\varepsilon \leq 0$. Thus, $x \in I(G) \cap \mathcal{PK}(G)$. By Remark 5.5.8, $x \in \mathcal{K}(G)$.

Conversely, let $x \in \mathcal{K}(G) \cap \mathcal{C}_\varepsilon(G)$ and let $i, j \in N$, $i \neq j$. We claim that $s_{ij}(x,v) = s_{ji}(x,v)$. Indeed, if, say, $s_{ij}(x,v) > s_{ji}(x,v)$, then $x^j = v(\{j\})$. Therefore,

$$\varepsilon \geq s_{ij}(x,v) > s_{ji}(x,v) \geq e(\{j\},x,v) = 0,$$

which is a contradiction. Thus, $s_{ij}(x,v) = s_{ji}(x,v)$ for all $i, j \in N$, $i \neq j$, and $x \in \mathcal{PK}(G)$. **q.e.d.**

The following results are analogues to Lemmata 7.4.5 and 7.4.6. Their proofs are left to the reader (Exercises 7.4.1 and 7.4.2).

Remark 7.4.12. Let $G = (N,v)$ satisfy (7.3.4). If $\varepsilon \geq \varepsilon_*(G)$, then $\mathcal{K}(G) \subseteq \mathcal{C}_\varepsilon(G)$.

Remark 7.4.13. Let $G = (N,v)$ satisfy (7.3.4). If $\mathcal{C}_\varepsilon(G) \cap I(G) \neq \emptyset$, then $\mathcal{C}_\varepsilon(G) \cap \mathcal{K}(G) \neq \emptyset$.

The following example shows that two games may have the same core and yet have prekernels that differ outside the core.

Example 7.4.14. Let $N = \{1,\ldots,5\}$ and let

$$\mathcal{T} = \{\{1,2,4\},\{1,3,4\},\{2,3,4\},\{1,2,5\},\{1,3,5\},\{2,3,5\},\{4,5\}\}.$$

Define v_1 on 2^N by the following rules:

(1) $v_1(N) = 7$;

(2) $v_1(S) = 4$ if $S \subsetneqq N$ and there exists $T \in \mathcal{T}$ such that $T \subseteq S$; and

(3) $v_1(S) = 0$ otherwise.

The game (N,v_1) is superadditive and, thus, $\mathcal{PK}(N,v_1) = \mathcal{K}(N,v_1)$. Players 1, 2 and 3 are substitutes (see Definition 3.8.5) as well as players 5 and 6. Using ETP (see Definition 5.3.9) the reader may verify that

$$\mathcal{PK}(N,v_1) = \left\{ \left(t,t,t,\frac{7-3t}{2},\frac{7-3t}{2}\right) \middle| \, 0 \leq t \leq 1 \right\}.$$

Now consider the coalition function v_2 on 2^N which differs from v_1 only inasmuch as $v_2(\{2,3,5\}) = 0$. As the reader may prove by direct computations,

$$\mathcal{C}(N,v_1) = \mathcal{C}(N,v_2) = \{(1,1,1,2,2)\}.$$

We claim that $\mathcal{PK}(N,v_2) = \mathcal{C}(N,v_2)$. Indeed, let $x \in \mathcal{PK}(N,v_2)$. Then $x \in \mathcal{K}(N,v_2)$, because the game is superadditive. Thus, $x^i \geq 0$ for all $i \in N$. This implies that $e(\{2,3,5\},x,v_2) \leq 0$ and, thus, x is a member of the positive prekernel of (N,v_1), that is, $x \in \mathcal{PK}_+(N,v_1)$. With the help of Exercise 5.1.5 it is easy to show (see Exercise 7.4.3) that

$$\mathcal{PK}_+(N, v_1) = \mathcal{PK}(N, v_1). \tag{7.4.4}$$

Every $y \in \mathcal{PK}(N, v_1) \setminus \{(1,1,1,2,2)\}$ satisfies

$$s_{1,5}(y, v_2) = e(\{1,2,4\}, y, v_2) > s_{5,1}(y, v_2);$$

thus $\mathcal{PK}(N, v_2) = \{(1,1,1,2,2)\}$.

Exercises

Exercise 7.4.1. Prove Remark 7.4.12.

Exercise 7.4.2. Prove Remark 7.4.13.

Exercise 7.4.3. Apply Exercise 5.1.5 to verify (7.4.4).

7.5 A Method for Computing the Prenucleolus

Let (N, v) be an n-person game. We shall show that the prenucleolus point $\nu(N, v)$ may be obtained as a solution of a sequence of at most $n - 1$ linear programs. The first linear program yields the least-core and is described as follows. Let $n \geq 2$, let $\mathcal{S}^0 = \{S \mid \emptyset \neq S \subsetneqq N\}$, and let $X^0 = X(N, v)$. Then we consider the problem

$$(\mathrm{P}^0) \quad \begin{cases} \min t \\ \text{subject to } e(S, x, v) \leq t \text{ for } S \in \mathcal{S}^0 \text{ and } x \in X^0. \end{cases}$$

The value of (P^0) is $\varepsilon^0 := \varepsilon_0(N, v)$ (see (7.1.1)) and its set of optimal vectors is $X^1 = \mathcal{LC}(N, v)$ (see Definition 7.1.2). We remark that X^1 is a nonempty convex polytope. Let

$$\mathcal{Q}^0 = \{S \in \mathcal{S}^0 \mid e(S, x, v) = \varepsilon^0 \text{ for all } x \in X^1\}.$$

The following lemmata will be used.

Lemma 7.5.1. $\mathcal{Q}^0 \neq \emptyset$.

Proof: Assume on the contrary that for each $S \in \mathcal{S}^0$ there exists $y_S \in X^1$ such that $e(S, y_S, v) < \varepsilon^0$. Let

$$y = \sum_{S \in \mathcal{S}^0} \frac{y_S}{|\mathcal{S}^0|}.$$

Then $y \in X^1$, because X^1 is convex. As the reader may easily verify, $e(S, y, v) < \varepsilon^0$ for all $S \in \mathcal{S}^0$. So, the value of (P^0) is smaller than ε^0 and, thus, the desired contradiction has been obtained. **q.e.d**

We denote

$$\mathcal{R}^0 = \{S \in \mathcal{S}^0 \mid e(S,x,v) \text{ is constant on } X^1\}.$$

Clearly, $\mathcal{Q}^0 \subseteq \mathcal{R}^0$.

Lemma 7.5.2. *Let $x \in X^1$ and $\nu = \nu(N,v)$. Then $e(S,x,v) = e(S,\nu,v)$ for all $S \in \mathcal{R}^0$.*

Proof: The prenucleolus belongs to X^1. **q.e.d**

Now we distinguish the following possibilities. If $\mathcal{R}^0 = \mathcal{S}^0$, then, by Lemma 7.5.2, $X^1 = \{\nu\}$ and the computation of ν has been accomplished. However, if $\mathcal{S}^1 = \mathcal{S}^0 \setminus \mathcal{R}^0 \neq \emptyset$, then we consider the following linear program:

$$(\mathrm{P}^1) \quad \begin{cases} \min t \\ \text{subject to } e(S,x,v) \leq t \text{ for } S \in \mathcal{S}^1 \text{ and } x \in X^1. \end{cases}$$

We continue by induction. Let $1 \leq k < n-1$ and assume that

$$X^0, \ldots, X^k, \mathcal{S}^0, \ldots, \mathcal{S}^k, \mathcal{R}^0, \ldots, \mathcal{R}^{k-1}, \text{ and } \varepsilon^0, \ldots, \varepsilon^{k-1}$$

have been defined such that, for all $i = 1, \ldots, k$,

$$\varepsilon^{i-1} = \min_{x \in X^{i-1}} \max_{S \in \mathcal{S}^{i-1}} e(S,x,v); \tag{7.5.1}$$

$$X^i = \{x \in X^{i-1} \mid e(S,x,v) \leq \varepsilon^{i-1} \text{ for all } S \in \mathcal{S}^{i-1}\}; \tag{7.5.2}$$

$$\mathcal{R}^{i-1} = \{S \in \mathcal{S}^{i-1} \mid e(S,x,v) \text{ is constant on } X^i\}; \tag{7.5.3}$$

$$\mathcal{S}^i = \mathcal{S}^{i-1} \setminus \mathcal{R}^{i-1}; \tag{7.5.4}$$

$$\nu \in X^k. \tag{7.5.5}$$

Again, two cases may be distinguished. If $\mathcal{S}^k = \emptyset$, then we claim that $X^k = \{\nu\}$. Indeed, if $x \in X^k$, then, by (7.5.2)–(7.5.4), $e(S,x,v) = e(S,\nu,v)$ for all $S \subseteq N$. Hence, $x = \nu$. When $\mathcal{S}^k \neq \emptyset$ we consider the linear program which computes ε^k, namely

$$\left(\mathrm{P}^k\right) \quad \begin{cases} \min t \\ \text{subject to } e(S,x,v) \leq t \text{ for } S \in \mathcal{S}^k \text{ and } x \in X^k. \end{cases}$$

By (7.5.2), $X^1, \ldots, X^k$ are convex and nonempty polytopes. Therefore $\left(\mathrm{P}^k\right)$ is, indeed, a linear program. Let ε^k be the value of $\left(\mathrm{P}^k\right)$ and let X^{k+1} be its set of optimal vectors. We define

$$\mathcal{Q}^k = \{S \in \mathcal{S}^k \mid e(S,x,v) = \varepsilon^k \text{ for all } x \in X^{k+1}\}.$$

By repeating the proof of Lemma 7.5.1 we may show that $\mathcal{Q}^k \neq \emptyset$. Thus

$$\mathcal{R}^k = \{S \in \mathcal{S}^k \mid e(S,x,v) \text{ is constant on } X^{k+1}\} \neq \emptyset.$$

Also, by (7.5.4), $\mathcal{R}^{k-1} \cap \mathcal{R}^k = \emptyset$. Therefore, $\varepsilon^k < \varepsilon^{k-1}$ and none of the vectors χ_S, where $S \in \mathcal{R}^k$, is a linear combination of the vectors in $\left\{\chi_T \,\middle|\, T \in \bigcup_{i=0}^{k-1} \mathcal{R}^i\right\}$. Hence, by (7.5.5) and the definition of $\left(\mathrm{P}^k\right)$, $\nu \in X^{k+1}$. Furthermore, the dimension of X^{k+1} is smaller than the dimension of X^k. Finally, let $\mathcal{S}^{k+1} = \mathcal{S}^k \setminus \mathcal{R}^k$. Then the description of the k-th step of the algorithm is complete.

For $i = 1, \ldots, k$ let $\dim(X^i)$ denote the dimension of the nonempty convex polytope X^i. As $\dim(X^0) = n-1$ and $\dim(X^i) < \dim(X^{i-1})$ for $i = 1, \ldots, k$, there exists $1 \leq m \leq n-1$ such that $X^m = \{\nu\}$.

Remark 7.5.3. If $I(N,v) \neq \emptyset$, then a similar procedure computes the nucleolus of (N,v). Indeed, if we replace the initial condition $X^0 = X(N,v)$ by $X^0 = I(N,v)$, then we obtain the desired algorithm.

Remark 7.5.4. The foregoing procedure was applied by Kopelowitz (1967). Kohlberg (1972) and Owen (1974) also provide algorithms for computing the nucleolus. Kohlberg's method can be used to characterize the prenucleolus as the unique maximizer of a function that respects the Lorenz order applied to the vectors of excesses at preimputations (see Sudhölter and Peleg (1998)). However, both papers do not contain reports on the applicability of their results.

Remark 7.5.5. Potters, Reijnierse, and Ansing (1996) describe a modification of the foregoing procedure and transform it into an implementable sequence of simplex algorithms. Each new initial tableau arises from the final tableau of the preceding step in a simple way.

Remark 7.5.6. In view of the foregoing algorithm the prenucleolus may be called the "lexicographic least-core". Indeed, the linear program $\left(\mathrm{P}^k\right)$, $k \geq 1$, is solved by the least-core of the constrained game, namely a game whose set of feasible vectors is X^k and whose set of permissible coalitions is $\mathcal{S}^k$. Nevertheless, the prenucleolus is **not** determined solely by the least-core (or the core) of a game. More precisely, two games may have the same least-core (or the same nonempty core) and yet their prenucleoli may differ. This is shown by the following example.

Example 7.5.7. Let $N = \{1,2,3,4\}$ and let v be defined on 2^N by $v(N) = 2$, $v(\{1,3\}) = 1/2$, $v(S) = 1$ if $2 \leq |S| \leq 3$ and $S \notin \{\{1,3\},\{2,4\}\}$, and $v(S) = 0$ otherwise. In addition, let v' be the coalition function which differs from v only inasmuch as $v'(\{1,2,3\}) = 5/4$. Any element x of the core of (N,v) or (N,v'), respectively, must be of the form $x = (t, 1-t, t, 1-t)$ (see the example (M,u) defined by (3.7.1)), and yields $e(S,x,v) = 0$ or $e(S,x,v') = 0$, respectively, for all $S \in \{\{1,2\},\{2,3\},\{3,4\},\{4,1\}\}$. A careful inspection of the excesses of the other coalitions shows that

$$\mathcal{C}(N,v) = \mathcal{C}(N,v') = \left[\left(\frac{1}{4},\frac{3}{4},\frac{1}{4},\frac{3}{4}\right),(1,0,1,0)\right] = \mathcal{LC}(N,v) = \mathcal{LC}(N,v').$$

Also, both games (N, v) and (N, v') are superadditive. Nevertheless, using, e.g., Corollary 6.1.3, it can easily be verified that

$$\mathcal{PN}(N, v) = \mathcal{N}(N, v) = \left\{ \left(\frac{1}{2}, \frac{1}{2}, \frac{1}{2}, \frac{1}{2} \right) \right\} \text{ and}$$
$$\mathcal{PN}(N, v') = \mathcal{N}(N, v') = \left\{ \left(\frac{5}{8}, \frac{3}{8}, \frac{5}{8}, \frac{3}{8} \right) \right\}.$$

7.6 Notes and Comments

We remark that Gérard-Varet and Zamir (1987) define, in contrast to Definition 7.3.1, the set of *reasonable outcomes* of a game (N, v), $\mathcal{R}(N, v)$, by

$$\mathcal{R}(N, v) = \prod_{i \in N} \left[b^i_{\min}(N, v), b^i_{\max}(N, v) \right].$$

Also, they characterize $\mathcal{R}$ as the maximal correspondence that satisfies AN, COV, and a certain monotonicity property, and they investigate whether various solutions of games (with coalition structures) are contained in $\mathcal{R}$.

Some of the results of this chapter may be generalized to games with coalition structures. First, we give the definition of the ε-core of a game with coalition structure.

Definition 7.6.1. *Let $(N, v, \mathcal{R})$ be a game with coalition structure and $\varepsilon \in \mathbb{R}$. The* ε**-core** *of $(N, v, \mathcal{R})$ is the set*

$$\mathcal{C}_\varepsilon(N, v, \mathcal{R}) = \{x \in X(N, v, \mathcal{R}) \mid e(S, x, v) \leq \varepsilon \text{ for all } S \in 2^N \setminus \mathcal{R}, S \neq \emptyset\}.$$

All the results of Section 7.4, except Lemmata 7.4.5 and 7.4.10 and Remark 7.4.13, hold for games with coalition structures. For example, the following generalization of Theorem 7.4.3 is true.

Theorem 7.6.2. *Let $(N, v, \mathcal{R})$ be a game with coalition structure and let $x \in \mathcal{C}_\varepsilon(N, v, \mathcal{R})$. Then $x \in \mathcal{PK}(N, v, \mathcal{R})$ if and only if, for each pair of partners $(i, j) \in \mathcal{P}(\mathcal{R})$ (see (3.8.1)), x bisects the segment $R_{ij}(x, \varepsilon)$.*

Finally, we remark that this chapter is based on Maschler, Peleg, and Shapley (1979).

8

The Shapley Value

Shapley (1953) writes: "At the foundation of the theory of games is the assumption that the players of a game can evaluate, in their utility scales, every "prospect" that might arise as a result of a play. In attempting to apply the theory to any field, one would normally expect to be permitted to include, in the class of "prospects", the prospect of having to play a game. The possibility of evaluating the game is therefore of critical importance." In this chapter we study the Shapley value which provides an a priori evaluation of every coalitional game.

In Section 8.1 we prove that the Shapley value is the unique single-valued solution which satisfies the following four axioms: the equal treatment property, the null player property, Pareto optimality, and additivity. These are the axioms of Shapley (1953). Also, uniqueness is proved for some important families of games. Monotonicity properties of solutions are investigated in Section 8.2. The example of Megiddo (1974) which proves that the nucleolus is not monotonic is presented. Then we discuss two variants of monotonicity due to Young (1985b) and provide an axiomatization of the Shapley value by means of strong monotonicity. Section 8.3 is devoted to a consistency property of single-valued solutions introduced by Hart and Mas-Colell (1989). It enables us to prove a characterization theorem for the Shapley value which is similar to Sobolev's result for the prenucleolus (see Theorem 6.3.1). Only the reduced game property has to be replaced by the consistency property.

The potential function for the Shapley value was defined and investigated by Hart and Mas-Colell (1989). In Section 8.4 we prove some of their results on the potential. An axiomatization of the Shapley value by means of a reduced game property based on an "ordinary" reduced game due to Sobolev (1975) is given in Section 8.5. The axiomatization of the Shapley value of (monotonic) simple games due to Dubey (1975) is presented in Section 8.6.

The Aumann and Drèze (1974) value for games with coalition structures is defined and axiomatized in Section 8.7. Games with a priori unions are analyzed in the next section. The main result is an axiomatization of the Owen value (Owen (1977)). Section 8.9 is devoted to multilinear extensions of coalitional games (Owen (1972)). The formula that relates the Shapley value of a game to its multilinear extension is proved.

8.1 Existence and Uniqueness of the Value

In this section we shall be interested in a notion of an *a priori* evaluation of a coalitional game by each of its players. Clearly, such an evaluation should be a single-valued solution, that is, it has to satisfy SIVA (see Definition 6.2.10). Shapley (1953) contains three axioms, besides SIVA, that uniquely determine the value to every player of each possible superadditive game. We shall now present the definition of a value of a game. Also, after proving the main results, we shall briefly discuss the minor differences between Shapley's and our approach (see Remark 8.1.11).

Let N be a coalition and $\mathcal{V}$ be the set of all coalition functions on 2^N. Clearly, $\mathcal{V}$ may be identified with the set of all games whose player set is N. As in Section 3.5 we shall assume throughout that $N = \{1, \ldots, n\}$. Let $\mathcal{K} \subseteq \mathcal{V}$.

Definition 8.1.1. *A solution σ on $\mathcal{K}$ is* **additive** (ADD) *if*

$$\sigma(v_1 + v_2) = \sigma(v_1) + \sigma(v_2) \text{ when } v_1, v_2, v_1 + v_2 \in \mathcal{K}. \tag{8.1.1}$$

Additivity was already discussed in Chapter 2 after the definition of superadditivity (see Definition 2.3.8). If σ satisfies SIVA and $\sigma^i(v)$, $i \in N$, is interpreted as i's expected payoff when he plays the game v, then (8.1.1) is a desirable property.

We now introduce the solution studied in this chapter. Let $\mathcal{K} \subseteq \mathcal{V}$.

Definition 8.1.2. *A* **value** *on $\mathcal{K}$ is a solution σ on $\mathcal{K}$ that satisfies* SIVA, ETP, NP, PO, *and* ADD *(see Definitions 6.2.10, 5.3.9, 4.1.17, 2.3.6, and 8.1.1, respectively).*

Most of the results of this section may be deduced from the following theorem.

Theorem 8.1.3. *There is a unique solution σ on $\mathcal{V}$ that satisfies* NE, ETP, NP, *and* ADD.

We postpone the proof of Theorem 8.1.3 and shall prove a useful lemma. For $T \subseteq N$, $T \neq \emptyset$, define the *unanimity game on T*, u_T, by

$$u_T(S) = \begin{cases} 1 \text{ , if } S \supseteq T, \\ 0 \text{ , otherwise.} \end{cases}$$

Lemma 8.1.4. *The set* $\{u_T \mid \emptyset \neq T \subseteq N\}$ *is a linear basis of* $\mathcal{V}$.

Proof: There are $2^n - 1$ unanimity games and the dimension of $\mathcal{V}$ is also $2^n - 1$. Thus, we have only to prove that the unanimity games are linearly independent. Assume, on the contrary, that $\sum_{\emptyset \neq T \subseteq N} \alpha_T u_T = 0$, where $\alpha_T \in \mathbb{R}$ are not all zero. Let T_0 be a minimal set in

$$\{T \subseteq N \mid T \neq \emptyset, \alpha_T \neq 0\}.$$

Then $\left(\sum_{\emptyset \neq T \subseteq N} \alpha_T u_T\right)(T_0) = \alpha_{T_0} \neq 0$, which is the desired contradiction.

q.e.d.

Proof of Theorem 8.1.3: Uniqueness. Let ϕ be a solution on $\mathcal{V}$ that satisfies NE, ETP, NP, and ADD. We first show that ϕ is a value, that is, SIVA and PO. Indeed, if $v = 0 \in \mathcal{V}$ then, by NE and NP, $0 \in \mathbb{R}^N$ is the unique member of $\phi(v)$. If $v \in \mathcal{V}$ is an arbitrary game, then also $-v \in \mathcal{V}$. Let $x \in \phi(v)$ and $y \in \phi(-v)$. Then, by ADD, $x + y \in \phi(v - v)$; thus $x = -y$ and SIVA follows from NE. Moreover, $x(N) = -y(N) \leq v(N)$ and $y(N) \leq -v(N)$, because ϕ is a solution. Hence $x(N) = v(N)$ and PO is shown.

Now we may assume that ϕ is a value. Let $\emptyset \neq T \subseteq N$. If $\alpha \in \mathbb{R}$, then the game αu_T has the following properties: (i) If $i \in N \setminus T$, then i is a null player, and (ii) if $i, j \in T$, then i and j are substitutes. Hence, by SIVA, NP, ETP, and PO, $\phi(\alpha u_T) = \alpha \chi_T / |T|$. (For simplicity we write $\sigma(v) = x$ instead of $\sigma(v) = \{x\}$ when σ is a single-valued solution.) Thus, ϕ is uniquely determined on multiples of unanimity games.

Now let $v \in \mathcal{V}$. By Lemma 8.1.4 there exist $\alpha_T \in \mathbb{R}$, $\emptyset \neq T \subseteq N$, such that $v = \sum_{\emptyset \neq T \subseteq N} \alpha_T u_T$. Therefore, by ADD, $\phi(v) = \sum_{\emptyset \neq T \subseteq N} \phi(\alpha_T u_T)$, and ϕ is uniquely determined on $\mathcal{V}$.

Existence. Let $\pi \in \mathcal{SYM}_N$ (the group of all permutations of N) and let $i \in N$. We recall that

$$P^i_\pi = \{j \in N \mid \pi(j) < \pi(i)\} \text{ and} \tag{8.1.2}$$
$$a^i_\pi(v) = v(P^i_\pi \cup \{i\}) - v(P^i_\pi) \tag{8.1.3}$$

(see Section 3.5). Now we define a function $\phi : \mathcal{V} \to \mathbb{R}^N$ by

$$\phi^i(v) = \sum_{\pi \in \mathcal{SYM}_N} \frac{a^i_\pi(v)}{n!} \text{ for every } i \in N. \tag{8.1.4}$$

For every permutation π the solution $a_\pi = (a^1_\pi, \ldots, a^n_\pi)$ satisfies ADD and NP. Thus, ϕ satisfies the same axioms. Moreover, it is straightforward to show that ϕ satisfies ETP.

q.e.d.

The value ϕ defined by (8.1.4) is called the *Shapley value*.

Corollary 8.1.5. *The Shapley value ϕ on $\mathcal{V}$ is given by*

$$\phi^i(v) = \sum_{S \subseteq N \setminus \{i\}} \frac{|S|!(n-|S|-1)!}{n!}(v(S \cup \{i\}) - v(S)) \tag{8.1.5}$$

for every $v \in \mathcal{V}$ and for every $i \in N$.

Shapley (1953) gives the following interpretation of (8.1.5): "The players in N agree to play the game v in a grand coalition, formed in the following way: 1. Starting with a single member, the coalition adds one player at a time until everybody has been admitted. 2. The order in which the players are to join is determined by chance, with all arrangements equally probable. 3. Each player, on his admission, demands and is promised the amount which his adherence contributes to the value of the coalition (as determined by the function v). The grand coalition then plays the game "efficiently" so as to obtain $v(N)$ – exactly enough to meet all the promises."

Remark 8.1.6. The four axioms used in Theorem 8.1.3 are each logically independent of the remaining axioms and of PO, provided that $|N| \geq 2$.

The easy proof of this remark is left to the reader (Exercise 8.1.3).

We now recall that a subset $\mathcal{K}$ of $\mathcal{V}$ is a *convex cone* if for all $x, y \in \mathcal{K}$ and $\alpha, \beta \in \mathbb{R},\ \alpha, \beta \geq 0,\ \alpha x + \beta y \in \mathcal{K}$.

Theorem 8.1.7. *Let $\mathcal{K} \subseteq \mathcal{V}$ be a convex cone which contains all the unanimity games. Then the only value on $\mathcal{K}$ is the Shapley value.*

Proof: Let ϕ' be a value on $\mathcal{K}$. By the proof of Theorem 8.1.3, $\phi'(\alpha u_T) = \phi(\alpha u_T)$ for all $\emptyset \neq T \subseteq N$ and $\alpha \geq 0$. Now let $v \in \mathcal{K}$. By Lemma 8.1.4 v can be written as $v = \sum_{\emptyset \neq T \subseteq N} \alpha_T u_T$, where $\alpha_T \in \mathbb{R}$ are suitable constants. Denote $\mathcal{P} = \{T \mid \alpha_T > 0\}$ and $\mathcal{Q} = \{T \mid \alpha_T < 0\}$. Then

$$v + \sum_{T \in \mathcal{Q}} (-\alpha_T) u_T = \sum_{T \in \mathcal{P}} \alpha_T u_T.$$

As $\mathcal{K}$ is a convex cone, $\sum_{T \in \mathcal{Q}} (-\alpha_T) u_T \in \mathcal{K}$ and $\sum_{T \in \mathcal{P}} \alpha_T u_T \in \mathcal{K}$. Applying the additivity of ϕ' yields

$$\phi'\left(v + \sum_{T \in \mathcal{Q}} (-\alpha_T) u_T\right) = \phi'(v) + \sum_{T \in \mathcal{Q}} \phi'(-\alpha_T u_T) = \sum_{T \in \mathcal{P}} \phi'(\alpha_T u_T).$$

Hence

$$\begin{aligned} \phi'(v) &= \textstyle\sum_{T \in \mathcal{P}} \phi'(\alpha_T u_T) - \sum_{T \in \mathcal{Q}} \phi'(-\alpha_T u_T) \\ &= \textstyle\sum_{T \in \mathcal{P}} \phi(\alpha_T u_T) - \sum_{T \in \mathcal{Q}} \phi(-\alpha_T u_T) \\ &= \phi\left(\textstyle\sum_{\emptyset \neq T \subseteq N} \alpha_T u_T\right) = \phi(v). \end{aligned}$$

q.e.d.

Corollary 8.1.8. *Let $\mathcal{V}^s$ be the set of all superadditive games in $\mathcal{V}$. Then ϕ is the only value on $\mathcal{V}^s$.*

Corollary 8.1.9. *Let $\mathcal{V}^m$ be the set of all monotonic games in $\mathcal{V}$. Then ϕ is the only value on $\mathcal{V}^m$.*

Corollary 8.1.10. *The Shapley value is the unique value on the set of all convex games in $\mathcal{V}$.*

Remark 8.1.11. Originally Shapley (1953) proved Corollary 8.1.8 using a slightly different notation. He considered a player set $\mathcal{U}$ and defined a superadditive game as a function v which assigns to each $S \subseteq \mathcal{U}$ a real number $v(S)$ such that $v(S) \geq v(S \cap T) + v(S \setminus T)$ for all $T \subseteq \mathcal{U}$ (superadditivity) and $v(\emptyset) = 0$. A *carrier* of v is a set $N \subseteq \mathcal{U}$ such that $v(S \cap N) = v(S)$ for all $S \subseteq \mathcal{U}$. Now, according to Shapley, a mapping ϕ which assigns to each superadditive game and each $i \in \mathcal{U}$ a real number $\phi^i(v)$ is a *value*, if it satisfies (1) AN (for each permutation π of $\mathcal{U}$ and each superadditive game v, $\pi\phi(v) = \phi(\pi v)$), (2) ADD (as defined by (8.1.1)), and (3) for each carrier N of a superadditive game v, $\sum_{i \in N} \phi^i(v) = v(N)$. He then proved that there is a unique value on the class of superadditive games with finite carriers. In our notation (3) implies NP and PO. Also, SIVA and AN imply ETP.

The following strong result on the uniqueness is contained in Neyman (1989). Let $v \in \mathcal{V}$. For $S \subseteq N$ let $v_S \in \mathcal{V}$ be defined by $v_S(T) = v(S \cap T)$ for all $T \subseteq N$. Then the members of $N \setminus S$ are null players of v_S. If $S = \emptyset$, then $v_S = 0 \in \mathcal{V}$. If $S \neq \emptyset$, then v_S is obtained from the subgame (S, v) of (N, v) by adding the null players $i \in N \setminus S$. Finally, let $\mathcal{G}(v)$ be the additive group generated by the games v_S, $S \subseteq N$, that is,

$$\mathcal{G}(v) = \left\{ w \in \mathcal{V} \,\middle|\, w = \sum_{S \subseteq N} k_S v_S,\ k_S \text{ is an integer for } S \subseteq N \right\}.$$

Notice that $v_N = v$. Then the Shapley value is the only value on $\mathcal{G}(v)$. (For a proof see Neyman (1989).)

Exercises

Exercise 8.1.1. Show that the Shapley value is self dual (see Definition 6.6.4).

Exercise 8.1.2. Let $v \in \mathcal{V}$ and let

$$\alpha_T = \sum_{S \subseteq T} (-1)^{|T|-|S|} v(S) \text{ for all } \emptyset \neq T \subseteq N. \tag{8.1.6}$$

Prove that $v = \sum_{\emptyset \neq T \subseteq N} \alpha_T u_T$ (see Lemma 8.1.4).

Let $\omega \in \mathbb{R}^N$, $\omega^i > 0$ for all $i \in N$. We define the *ω-weighted Shapley value* (see Kalai and Samet (1988)) ϕ_ω on $\mathcal{V}$ by the following formulae. If $\emptyset \neq T \subseteq N$, then

$$\phi^i_\omega(u_T) = \begin{cases} \frac{\omega^i}{\omega(T)} & \text{, if } i \in T, \\ 0 & \text{, if } i \in N \setminus T \end{cases} \tag{8.1.7}$$

and if $v \in \mathcal{V}$, then let $\alpha_T, \emptyset \neq T \subseteq N$, be given by (8.1.6), and let

$$\phi_\omega(v) = \sum_{\emptyset \neq T \subseteq N} \alpha_T \phi_\omega(u_T). \tag{8.1.8}$$

Exercise 8.1.3. Prove Remark 8.1.6 by means of examples. (Note that any weighted Shapley value satisfies all of the axioms of a value except ETP.)

Exercise 8.1.4. Prove that if $v \in \mathcal{V}$ is convex then $\phi(v) \in \mathcal{C}(v)$ (see Theorem 3.5.1).

Exercise 8.1.5. Show by means of an example that PO is logically independent of the remaining axioms in Corollary 8.1.8.

8.2 Monotonicity Properties of Solutions and the Value

Let N be a finite nonempty set, let $n = |N|$, and let $\mathcal{V}$ be the set of all coalition functions on 2^N. Intuitively, a single-valued solution σ on a subset $\mathcal{K}$ of $\mathcal{V}$ is monotonic if it has the following property. Let $u, v \in \mathcal{K}$ and let $i \in N$. If u is obtained from v by "improving the position" of i, then $\sigma^i(u) \geq \sigma^i(v)$. If σ is set-valued, then a possible generalization is as follows: For every $x \in \sigma(v)$ there exists $y \in \sigma(u)$ such that $y^i \geq x^i$. However, the improvement of a player's position in a game may be defined in several different ways which lead to distinct definitions of monotonicity.

One version of monotonicity, namely *(strong) aggregate monotonicity*, (S)AM, has been defined in Chapters 2 and 4 (see Definitions 2.3.23 and 4.5.1). None of the variants of the bargaining set satisfies AM even on $\mathcal{V}^s$ (the set of superadditive games), provided that $n \geq 6$ (see Corollary 4.5.6). Also, the (pre)kernel, the (pre)nucleolus, and the modiclus do not satisfy this property (see Remark 5.5.17 and Exercise 6.6.5). Clearly, the Shapley value satisfies SAM and hence shares this property with the core.

The next variant of monotonicity is due to Young (1985b).

Definition 8.2.1. *Let σ be a single-valued solution on a set $\mathcal{K} \subseteq \mathcal{V}$. σ is* **coalitionally monotonic** *if the following condition is satisfied: If $u, v \in \mathcal{K}$, $u(T) \geq v(T)$ for some $T \subseteq N$ and $u(S) = v(S)$ for all $S \in 2^N \setminus \{T\}$, then $\sigma^i(u) \geq \sigma^i(v)$ for all $i \in T$.*

Definition 8.2.1 may be generalized to a set-valued solution σ by replacing the statement "$\sigma^i(u) \geq \sigma^i(v)$" by the statement "for every $x \in \sigma(v)$ there exists $y \in \sigma(u)$ such that $y^i \geq x^i$". Then coalition monotonicity implies AM.

Remark 8.2.2. Coalitional monotonicity is equivalent to the following condition: If $u, v \in \mathcal{K}$, $i \in N$, $u(S \cup \{i\}) \geq v(S \cup \{i\})$ and $u(S) = v(S)$ for all $S \subseteq N \setminus \{i\}$, then $\sigma^i(u) \geq \sigma^i(v)$. Thus, coalitional monotonicity is a highly intuitive property.

A single-valued solution σ on a set $\mathcal{K} \subseteq \mathcal{V}$ is a *core allocation* if $\sigma(v) \in \mathcal{C}(v)$ when $\mathcal{C}(v) \neq \emptyset$ and $v \in \mathcal{K}$.

Theorem 8.2.3. *Let $n \geq 4$ and σ be a core allocation on a set $\mathcal{K} \subseteq \mathcal{V}$ which contains the set of balanced games in $\mathcal{V}$. Then σ is not coalitionally monotonic.*

Proof: Let $M = \{1, \ldots, 4\}$ and (M, u) be the game given by (3.7.1). By Lemma 3.7.3,

$$\mathcal{C}(M, u) = \{(\gamma, -\gamma, \gamma, -\gamma) \mid -1 \leq \gamma \leq 1\}.$$

Let (M, v) and (M, w), respectively, be the games which differ from (M, u) only inasmuch as $v(\{1, 2, 3\}) = 1$ and $w(\{2, 3, 4\}) = 1$, respectively. Then $\mathcal{C}(M, v) = \{(1, -1, 1, -1)\}$ and $\mathcal{C}(M, w) = \{(-1, 1, -1, 1)\}$. If σ were coalitionally monotonic, then $\sigma^2(M, u) \leq -1 = \sigma^2(M, v)$ and $\sigma^3(M, u) \leq -1 = \sigma^3(M, w)$, which is impossible. The extension of the result to $n > 4$ is obvious.

q.e.d.

For $n \geq 5$ Theorem 8.2.3 is due to Young (1985b). Housman and Clark (1998) proved the result for $n = 4$ and showed the following remark, the proof of which is left to the reader (Exercise 8.2.1).

Remark 8.2.4. If $n = 3$, then the prenucleolus is coalitionally monotonic.

The strongest version of monotonicity is the following. Let $\mathcal{K} \subseteq \mathcal{V}$.

Definition 8.2.5. *A single-valued solution σ on $\mathcal{K}$ is* **strongly monotonic** *if the following condition is satisfied: If $u, v \in \mathcal{K}$, $i \in N$, and*

$$u(S \cup \{i\}) - u(S) \geq v(S \cup \{i\}) - v(S) \text{ for all } S \subseteq N,$$

then $\sigma^i(u) \geq \sigma^i(v)$.

Definition 8.2.5 may be generalized to a set-valued solution σ in the same way as Definition 8.2.1 was generalized. The Shapley value is strongly monotonic. Furthermore, strong monotonicity may be used in the following axiomatization of the Shapley value.

Theorem 8.2.6 (Young (1985b)). *Let σ be a single-valued solution on $\mathcal{V}$. If σ is strongly monotonic and satisfies* ETP *and* PO, *then σ is the Shapley value.*

Proof: Let $v \in \mathcal{V}$. We denote

$$\mathcal{D}(v) = \{S \subseteq N \mid \text{ there exists } T \subseteq S \text{ with } v(T) \neq 0\}$$

and prove that $\sigma(v) = \phi(v)$ by induction on $|\mathcal{D}(v)|$. If $|\mathcal{D}(v)| = 0$, then $v = 0$ and $\sigma(v) = 0 = \phi(v)$ by ETP and PO. Assume now that $\sigma(w) = \phi(w)$ if $w \in \mathcal{V}$ and $|\mathcal{D}(w)| \leq k$. Assume that $v \in \mathcal{V}$ satisfies $|\mathcal{D}(v)| = k+1$. Denote by $\mathcal{D}^m(v)$ the set of minimal coalitions in $\mathcal{D}(v)$. Let $S \in \mathcal{D}^m(v)$, let $v_S \in \mathcal{V}$ be defined by $v_S(T) = v(S \cap T)$ for all $T \subseteq N$, and let $w = v - v_S$. Then $|\mathcal{D}(w)| \leq k$. Also, if $i \in N \setminus S$, then

$$w(T \cup \{i\}) - w(T) = v(T \cup \{i\}) - v(T) \text{ for all } T \subseteq N.$$

Therefore, by strong monotonicity of σ, $\sigma^i(w) = \sigma^i(v)$, and by the induction hypothesis, $\sigma^i(w) = \phi^i(w)$ for all $i \in N \setminus S$. Applying strong monotonicity of ϕ yields $\sigma^i(v) = \sigma^i(w) = \phi^i(w) = \phi^i(v)$. Let $S_0 = \bigcap\{S \mid S \in \mathcal{D}^m(v)\}$. We have shown that

$$\sigma^i(v) = \phi^i(v) \text{ for all } i \in N \setminus S_0. \tag{8.2.1}$$

We now observe that $v(T) = 0$ if $S_0 \setminus T \neq \emptyset$. Hence, if $i, j \in S_0$, then i and j are substitutes (see Definition 3.8.5). By ETP of σ and ϕ, $\sigma^i(v) = \sigma^j(v)$ and $\phi^i(v) = \phi^j(v)$ for all $i, j \in S_0$. Hence, by (8.2.1) and Pareto optimality (of σ and ϕ), $\sigma^i(v) = \phi^i(v)$ for all $i \in S_0$, and the proof is complete. **q.e.d.**

Remark 8.2.7. The proof of Theorem 8.2.6 is similar to the proof of Theorem B in Neyman (1989). A close examination of the proof reveals that strong monotonicity may be replaced by the following weaker condition: If $u, v \in \mathcal{K}$, $i \in N$, and

$$u(S \cup \{i\}) - u(S) = v(S \cup \{i\}) - v(S) \text{ for all } S \subseteq N,$$

then $\sigma^i(u) = \sigma^i(v)$.

Exercises

Exercise 8.2.1. Prove Remark 8.2.4. (Let u, v be three-person games which satisfy the assumptions of Definition 8.2.1. Let $x \in \mathcal{PK}(N, v)$. Show that, if $y \in X(N, u)$ such that $y^k < x^k$ for some $k \in T$, then $y \notin \mathcal{PK}(N, u)$.)

Exercise 8.2.2. Let $\mathcal{V}^s$ be the set of all superadditive games in $\mathcal{V}$. Prove that the Shapley value is the unique single-valued solution on $\mathcal{V}^S$ that satisfies strong monotonicity, ETP, and PO (see Young (1985b)).

Exercise 8.2.3. Prove that any weighted Shapley value is strongly monotonic. (It is helpful to consider the probabilistic representation of a weighted Shapley value as presented in Kalai and Samet (1988).)

Exercise 8.2.4. Show that each axiom in Theorem 8.2.6 is logically independent of the remaining axioms, provided that $n \geq 2$.

8.3 Consistency

Let $\mathcal{U}$ be a set of players, let Γ be a subset of the set of all games $\Gamma_{\mathcal{U}}$, and let σ be a single-valued solution on Γ. If $(N,v) \in \Gamma$ and $\emptyset \neq S \subseteq N$, then we associate with S and v a new type of reduced game $(S, v_{S,\sigma})$ which depends on σ. If $T \subseteq S$ then the worth of T, $v_{S,\sigma}(T)$, is computed in the following way. The members of $T \cup (N \setminus S) = Q$ consider the subgame (Q,v) (see Definition 3.2.2). We assume that $(Q,v) \in \Gamma$. Then the players in $N \setminus S$ are paid according to $\sigma(Q,v)$. The remainder, that is, $v(Q) - \sum_{i \in N\setminus S} \sigma^i(Q,v)$, is the worth of T in the reduced game. Thus, if all members of $\mathcal{U}$ adopt σ as a solution for coalitional games, then the foregoing definition is intuitively acceptable. The precise definition of the new reduced game is the following.

Definition 8.3.1. *Let Γ be a set of games and let σ be a single-valued solution on Γ. If $(N,v) \in \Gamma$ and $\emptyset \neq S \subseteq N$, then the* **$\sigma$-reduced game** *$(S, v_{S,\sigma})$ is defined by the following rule. If $\emptyset \neq T \subseteq S$, then*

$$v_{S,\sigma}(T) = v(T \cup (N \setminus S)) - \sum_{i \in N\setminus S} \sigma^i(T \cup (N \setminus S), v), \tag{8.3.1}$$

and $v_{S,\sigma}(\emptyset) = 0$.

Remark 8.3.2. Notice that the new reduced game is defined only relative to a single-valued solution, unlike Definition 2.3.11 which associates a reduced game with every feasible payoff vector. Furthermore, let $(N,v) \in \Gamma$, let $x = \sigma(N,v)$, and let $\emptyset \neq S \subsetneqq N$. Then the coalition functions $v_{S,\sigma}$ and $v_{S,x}$ (see Definition 2.3.11) may not be equal, because they are defined by different formulae. Clearly, $v_{N,\sigma} = v = v_{N,x}$.

Remark 8.3.3. Definition 2.3.11 is very useful in investigations of the core and its relatives, the prebargaining sets, the prekernel, and the prenucleolus (see Chapters 2 – 6). Now, as the reader may easily verify by means of an example, the Shapley value does not satisfy RGP (see Definition 2.3.14). Hence, it shows that RGP is logically independent of the remaining axioms of Theorem 6.3.1. However, Definition 8.3.1 will enable us to find a substitute to RGP for the Shapley value.

Definition 8.3.4. *A single-valued solution σ on a set Γ of games is* **consistent** *(CON) if the following condition is satisfied: If $(N,v) \in \Gamma$ and $\emptyset \neq S \subseteq N$, then $(S, v_{S,\sigma}) \in \Gamma$ and*

$$\sigma^i(S, v_{S,\sigma}) = \sigma^i(N,v) \text{ for all } i \in S. \tag{8.3.2}$$

The interpretation of CON is similar to that of RGP (see Remark 2.3.15).

Theorem 8.3.5. *The Shapley value on $\Gamma_{\mathcal{U}}$ is consistent.*

Proof: Let $N \subseteq \mathcal{U}$ be a finite and nonempty set of players and $\emptyset \neq S \subseteq N$. Denote by Γ_* the set of all games (N,v) which satisfy (8.3.2) with $\sigma = \phi$. Let $(N,v), (N,w) \in \Gamma_*$ and $\alpha \in \mathbb{R}$. By the linearity of ϕ and Definition 8.3.1, $(N, v+w)$ and $(N, \alpha v)$ are in Γ_*. By Lemma 8.1.4, using again the linearity of ϕ, it is sufficient to prove that every unanimity game (on a subset of N) is in Γ_*. Let $u = u_Q$ be the coalition function of the unanimity game on Q where $\emptyset \neq Q \subseteq N$. We distinguish the following possibilities.

(1) $Q \cap S = \emptyset$. In this case $Q \subseteq N \setminus S$ and therefore, by (8.3.1), $u_{S,\sigma} = 0 \in \mathbb{R}^N$. Hence $\phi^j(S, u_{S,\phi}) = 0 = \phi^j(N, u)$ for all $j \in S$.

(2) $Q \cap S \neq \emptyset$. In this case (8.3.1) yields the following formula for $u_{S,\sigma}$:

$$u_{S,\sigma}(T) = \begin{cases} 1 - \frac{|Q \setminus S|}{|Q|} & \text{, if } T \supseteq (S \cap Q), \\ 0 & \text{, otherwise.} \end{cases}$$

Thus, $\phi^j(S, u_{S,\phi}) = 1/|Q| = \phi^j(N,u)$ if $j \in S \cap Q$, and $\phi^j(S, u_{S,\phi}) = 0 = \phi^j(N,u)$ if $j \in S \setminus Q$, and the proof is complete. **q.e.d.**

In the following theorem consistency will be used to characterize the Shapley value.

Theorem 8.3.6. *There is a unique single-valued solution on $\Gamma_{\mathcal{U}}$ that satisfies* COV, PO, ETP, *and* CON, *and it is the Shapley value.*

Proof: The Shapley value ϕ is single-valued and satisfies PO and ETP by Definition 8.1.2. By Theorem 8.3.5, it is consistent, and by linearity, it satisfies COV. Thus, we only have to prove the uniqueness part.

Let σ be a single-valued solution on $\Gamma_{\mathcal{U}}$ that satisfies COV, PO, ETP, and CON. Let (N,v) be an n-person game. By induction on n we shall prove that $\sigma(N,v) = \phi(N,v)$. If $n=1$ then $\sigma(N,v) = \phi(N,v)$ by PO. For $n = 2$ the desired equality follows from Lemma 5.4.3. Thus, let $k \geq 3$ and assume that $\sigma(N,v) = \phi(N,v)$ whenever $n < k$. Now, if $n = k$ let $S \subseteq N$ satisfy $|S| = 2$. Denote $S = \{i, j\}$. Then $v_{S,\sigma}(\{i\})$ and $v_{S,\phi}(\{i\})$ are determined by the subgame $(N \setminus \{j\}, v)$. Therefore, by the induction hypothesis, $v_{S,\sigma}(\{i\}) = v_{S,\phi}(\{i\})$. Analogously, $v_{S,\sigma}(\{j\}) = v_{S,\phi}(\{j\})$. Now, by (5.4.2),

$$\begin{aligned} \sigma^i(S, v_{S,\sigma}) - \sigma^j(S, v_{S,\sigma}) &= v_{S,\sigma}(\{i\}) - v_{S,\sigma}(\{j\}) \\ = v_{S,\phi}(\{i\}) - v_{S,\phi}(\{j\}) \quad &= \phi^i(S, v_{S,\phi}) - \phi^j(S, v_{S,\phi}). \end{aligned} \tag{8.3.3}$$

By CON, $\sigma^\ell(S, v_{S,\sigma}) = \sigma^\ell(N,v)$ and $\phi^\ell(S, v_{S,\phi}) = \phi^\ell(N,v)$ for $\ell = i,j$. Therefore, by (8.3.3),

$$\phi^i(N,v) - \sigma^i(N,v) = \phi^j(N,v) - \sigma^j(N,v) \text{ for all } i,j \in N. \tag{8.3.4}$$

By PO and (8.3.4), $\sigma(N,v) = \phi(N,v)$, and the proof is complete. **q.e.d.**

So far all the material of this section is based on Hart and Mas-Colell (1989). However, the proof of Theorem 8.3.6 that we presented is due to M. Maschler.

Remark 8.3.7. Note that each of the axioms used in Theorem 8.3.6 except PO is logically independent of the remaining axioms (Exercise 8.3.2). The following lemma shows that Theorem 8.3.6 is true without the assumption of PO.

Lemma 8.3.8. *Let σ be a single-valued solution on $\Gamma_{\mathcal{U}}$ that satisfies* COV *and* CON. *Then σ satisfies* PO.

Proof: By COV and SIVA, σ is Pareto optimal when applied to any one-person game (see the first part of the proof of Lemma 6.2.11). Now, let $(N, v) \in \Gamma_{\mathcal{U}}$, let $x = \sigma(N, v)$, and let $i \in N$. Then, by (8.3.1), $v_{\{i\},\sigma}(\{i\}) = v(N) - x(N \setminus \{i\})$. By CON, $\sigma(\{i\}, v_{\{i\},\sigma}) = \sigma^i(N, v)$ and thus $\sigma^i(N, v) = x^i = v(N) - x(N \setminus \{i\})$.

q.e.d.

Exercises

In order to define *weighted Shapley values* on $\Gamma_{\mathcal{U}}$ let $\omega : \mathcal{U} \to \mathbb{R}$ satisfy $\omega^i > 0$ for all $i \in \mathcal{U}$. Then the ω-weighted Shapley value may be defined by (8.1.7) and (8.1.8) for every nonempty finite $N \subseteq \mathcal{U}$.

Exercise 8.3.1. Prove that any weighted Shapley value satisfies CON.

Exercise 8.3.2. Show by means of three examples that each of the axioms COV, ETP, and CONS are logically independent of **all** the remaining axioms in Theorem 8.3.6, provided that $|\mathcal{U}| \geq 3$.

8.4 The Potential of the Shapley Value

Let $\mathcal{U}$ be a set of players and let $\Gamma_{\mathcal{U}}$ be the set of games with players in $\mathcal{U}$.

Definition 8.4.1. *Let $P : \Gamma_{\mathcal{U}} \to \mathbb{R}$ and let $(N, v) \in \Gamma_{\mathcal{U}}$. The* **marginal contribution** *of a player $i \in N$ in the game (N, v) (according to P) is*

$$D^i P(N, v) = \begin{cases} P(N, v) & , \text{ if } |N| = 1, \\ P(N, v) - P(N \setminus \{i\}, v) & , \text{ if } |N| \geq 2. \end{cases}$$

Here $(N \setminus \{i\}, v)$ is a subgame of (N, v) (see Definition 3.2.2) when $|N| \geq 2$.

Definition 8.4.2. *A function $P : \Gamma_{\mathcal{U}} \to \mathbb{R}$ is called a* **potential** *if for every game $(N, v) \in \Gamma_{\mathcal{U}}$,*

$$\sum_{i \in N} D^i P(N, v) = v(N). \tag{8.4.1}$$

Thus, P is a potential if the allocation of marginal contributions (according to P) is always Pareto optimal.

Lemma 8.4.3. *There exists a unique potential P. Moreover, the potential of a game (N,v) is uniquely determined by applying (8.4.1) only to the subgames (S,v), $\emptyset \neq S \subseteq N$, of (N,v).*

Proof: Let (N,v) be a game. If $|N|=1$, then $P(N,v)=D^i(N,v)=v(N)$ by definition. If $|N|\geq 2$, then, by (8.4.1),

$$P(N,v)=\frac{v(N)+\sum_{i\in N}P(N\setminus\{i\},v)}{|N|}. \tag{8.4.2}$$

Hence, the lemma follows from (8.4.2) by induction on $|N|$. **q.e.d.**

Henceforth we denote by P the unique potential function (on $\Gamma_{\mathcal{U}}$).

Theorem 8.4.4. *For every game $(N,v)\in\Gamma_{\mathcal{U}}$ and for every $i\in N$, $D^iP(N,v)=\phi^i(N,v)$ where $\phi(N,v)$ is the Shapley value of (N,v).*

Proof: We define a function $P^*:\Gamma_{\mathcal{U}}\to\mathbb{R}$ by the following rule. Let $(N,v)\in\Gamma_{\mathcal{U}}$. Then, by Lemma 8.1.4, (N,v) is a linear combination of unanimity games (N,u_T), that is, $v=\sum_{\emptyset\neq T\subseteq N}\alpha_T u_T$. Let

$$P^*(N,v)=\sum_{\emptyset\neq T\subseteq N}\frac{\alpha_T}{|T|}. \tag{8.4.3}$$

Let $i\in N$. If $|N|=1$, then $P^*(N,v)=D^iP^*(N,v)=v(N)$. If $|N|\geq 2$, then $P^*(N\setminus\{i\},v)=\sum_{\emptyset\neq T\subseteq N\setminus\{i\}}\alpha_T/|T|$; thus

$$\sum_{i\in N}D^iP^*(N,v)=\sum_{i\in N}\sum_{T\ni i}\frac{\alpha_T}{|T|}=\sum_{\emptyset\neq T\subseteq N}\sum_{i\in T}\frac{\alpha_T}{|T|}=\sum_{\emptyset\neq T\subseteq N}\alpha_T=v(N).$$

Thus, P^* satisfies (8.4.1) and therefore, by Lemma 8.4.3, $P^*=P$. Now if $i\in N$, then

$$D^iP(N,v)=D^iP^*(N,v)=\sum_{T\ni i}\frac{\alpha_T}{|T|}=\phi^i(N,v)$$

and the proof is complete. **q.e.d.**

Remark 8.4.5. By Theorem 8.4.4 the Shapley value is the (discrete) gradient of P. Thus, P is a potential function for the Shapley value. This explains the choice of the name for it. Ortmann (1998) describes parallels between the Shapley value in the theory of cooperative games and the potential in physics.

Exercises

Exercise 8.4.1. Prove that P is given by

$$P(N,v) = \sum_{\emptyset \neq S \subseteq N} \frac{(|S|-1)!(|N|-|S|)!}{|N|!} v(S) \tag{8.4.4}$$

for every $(N,v) \in \Gamma_{\mathcal{U}}$. Provide a probabilistic interpretation for (8.4.4) (see Hart and Mas-Colell (1988)).

Exercise 8.4.2. Provide alternative proofs of Theorems 8.3.5 and 8.3.6 by means of Theorem 8.4.4 (see Hart and Mas-Colell (1989)).

Exercise 8.4.3. Let N be a finite nonempty subset of $\mathcal{U}$ and let Γ_N denote the set of all coalition functions on 2^N. We define an operator $P_* : \Gamma_N \to \Gamma_N$ by $(P_*v)(S) = P(S,v)$ for all $\emptyset \neq S \subseteq N$ and $(P_*v)(\emptyset) = 0$ for all $v \in \Gamma_N$. Prove the following assertion for all $v, w \in \Gamma_N$:

(1) P_* is *linear*: $P_*(\alpha v + \beta w) = \alpha P_* v + \beta P_* w$ for all $\alpha, \beta \in \mathbb{R}$.

(2) P_* is *symmetric*: $P_*(\pi v) = \pi(P_* v)$ for every permutation π of N (see Definition 2.3.4).

(3) P_* is *positive*: If $v \geq 0$ (that is, $v(S) \geq 0$ for all $S \subseteq N$), then $P_* v \geq 0$.

(4) P_* is a bijection.

(5) $P_* v = v$ if and only if v is inessential (see Definition 2.1.9).

(See Hart and Mas-Colell (1988).)

Exercise 8.4.4. Let $(N,v) \in \Gamma_{\mathcal{U}}$. Prove that if $\mathcal{C}(N,v) \neq \emptyset$, then $P(N,v) \leq v(N)$. (Hint: See Exercise 8.4.3.)

8.5 A Reduced Game for the Shapley Value

In Section 8.3 a reduced game depending on the single-valued solution is used to characterize the Shapley value. It is the aim of the present section to show that the Shapley value may be characterized by a reduced game property based on a reduced game which does not depend on the solution of subgames. Let $\mathcal{U}$ be a set of players and $\Gamma_{\mathcal{U}}$ the set of all games whose players are members of $\mathcal{U}$. Let $(N,v) \in \Gamma_{\mathcal{U}}$, let $x \in \mathbb{R}^N$, let $i \in N$, and let $|N| \geq 2$. Define $v^{SH}_{N\setminus\{i\},x}(T)$, $T \subseteq N \setminus \{i\}$, by

$$v^{SH}_{N\setminus\{i\},x}(T) = \frac{|T|}{|N|-1}(v(T \cup \{i\}) - x^i) + \frac{|N|-|T|-1}{|N|-1} v(T). \tag{8.5.1}$$

It should be noted that $(N \setminus \{i\}, v^{SH}_{N\setminus\{i\},x})$ is a game, because $v^{SH}_{N\setminus\{i\},x}(\emptyset) = 0$. Throughout this section it is called the *reduced game* of (N,v) (with respect to $N \setminus \{i\}$ and x).

Remark 8.5.1. By (8.5.1), $v^{SH}_{N\setminus\{i\},x}(N\setminus\{i\}) = v(N) - x^i$. Hence, $x^{N\setminus\{i\}}$ is Pareto optimal according to the reduced game if and only if $x \in X(N,v)$.

The game $(N\setminus\{i\}, v^{SH}_{N\setminus\{i\},x})$ has the following simple interpretation. An arbitrator selects one player j of $N\setminus\{i\}$, each player equally probable. If $j \in T$, then i joins T and T has to pay x^i to its new member i. Hence the worth of T according to the reduced game is $v(T\cup\{i\}) - x^i$. If $j \notin T$, then i joins $(N\setminus\{i\})\setminus T$ and the worth of T remains $v(T)$. If the members of $N\setminus\{i\}$ agree on this rule, then the expected worth of T is given by (8.5.1).

Remark 8.5.2. Let $(N,v) \in \Gamma_{\mathcal{U}}$, $x \in \mathbb{R}^N$, and $i,j \in N$, $i \neq j$. If $|N| \geq 3$, then

$$\left(v^{SH}_{N\setminus\{i\},x}\right)^{SH}_{N\setminus\{i,j\},x^{N\setminus\{i\}}} = \left(v^{SH}_{N\setminus\{j\},x}\right)^{SH}_{N\setminus\{i,j\},x^{N\setminus\{j\}}}, \tag{8.5.2}$$

that is, reducing does not depend on the order of the players which are deleted.

The simple proof of the foregoing remark is left to the reader (Exercise 8.5.1). With the help of Remark 8.5.2 we may define the reduced game $(S, v^{SH}_{S,x})$ with respect to an arbitrary coalition $\emptyset \neq S \subseteq N$ recursively on $|N| - |S|$. Let $v^{SH}_{N,x} = v$ and assume that $v^{SH}_{S,x}$ is defined when $|S| \geq k$ for some $1 < k \leq |N| - 1$. If $|S| = k-1$, let $i \in N\setminus S$. Then $u = v^{SH}_{S\cup\{i\},x}$ is already defined. Let $(S, v^{SH}_{S,x})$ be defined by (8.5.1), that is, $v^{SH}_{S,x} = u^{SH}_{S,x^{S\cup\{i\}}}$. By Remark 8.5.1, $(S, v^{SH}_{S,x})$ is well defined.

Definition 8.5.3. *A solution σ on a set $\Gamma \subseteq \Gamma_{\mathcal{U}}$ satisfies* RGP^{SH} *if $(S, v^{SH}_{S,x}) \in \Gamma$ and $x^S \in \sigma(S, v^{SH}_{S,x})$ for every $(N,v) \in \Gamma$, for every $x \in \sigma(N,v)$, and for every $\emptyset \neq S \subseteq N$.*

Lemma 8.5.4. *The Shapley value satisfies* RGP^{SH} *on $\Gamma_{\mathcal{U}}$.*

Proof: Let $(N,v) \in \Gamma_{\mathcal{U}}$ satisfy $|N| \geq 2$ and let $i,j \in N$, $i \neq j$. Let $x = \phi(N,v)$ and $u = v^{SH}_{N\setminus\{i\},x}$. It suffices to show that $\phi^j(N\setminus\{i\}, u) = x^j$. Let $t = |T|$ for every $T \subseteq N$. If $t = 0, \ldots, n-2$, denote

$$a_t = \frac{(t+1)!(n-t-2)!}{n!} \text{ and } b_t = \frac{t!(n-t-1)!}{n!}.$$

By (8.1.5),

$$x^i = \sum_{T\subseteq N\setminus\{i,j\}} a_t v(T\cup\{i,j\}) + b_t v(T\cup\{i\}) - a_t v(T\cup\{j\}) - b_t v(T) \tag{8.5.3}$$

and

$$x^j = \sum_{T\subseteq N\setminus\{i,j\}} a_t v(T\cup\{i,j\}) - a_t v(T\cup\{i\}) + b_t v(T\cup\{j\}) - b_t v(T). \tag{8.5.4}$$

Denote

$$a_t^1 = \frac{t!(n-t-2)!(n-t-2)}{(n-1)!(n-1)} \text{ and } b_t^1 = \frac{t!(n-t-2)!t}{(n-1)!(n-1)}.$$

By (8.5.1),

$$\begin{aligned}
&\phi^j(N \setminus \{i\}, u) \\
&= \textstyle\sum_{T \subseteq N\setminus\{i,j\}} \frac{n}{n-1} a_t(v(T \cup \{i,j\}) - x^i) - b_t^1(v(T \cup \{i\}) - x^i) \\
&\qquad + a_t^1 v(T \cup \{j\}) - \frac{n}{n-1} b_t v(T) \\
&= \Big(\textstyle\sum_{T \subseteq N\setminus\{i,j\}} \frac{n}{n-1} a_t v(T \cup \{i,j\}) - b_t^1 v(T \cup \{i\}) \\
&\qquad + a_t^1 v(T \cup \{j\}) - \frac{n}{n-1} b_t v(T) \Big) - \frac{x^i}{n-1}.
\end{aligned}$$

As the reader may easily verify,

$$\frac{na_t}{n-1} - \frac{a_t}{n-1} = a_t, b_t^1 + \frac{b_t}{n-1} = a_t, a_t^1 + \frac{a_t}{n-1} = b_t, \frac{nb_t}{n-1} - \frac{b_t}{n-1} = b_t,$$

and, by (8.5.3) and (8.5.4), the proof is complete. **q.e.d.**

Theorem 8.5.5. *The Shapley value is the unique solution on $\Gamma_{\mathcal{U}}$ that satisfies* SIVA, COV, ETP, *and* RGP^{SH}.

Proof: By Theorem 8.3.6 and Lemma 8.5.4, ϕ satisfies SIVA, COV, ETP, and RGP^{SH}. Let σ be a solution on $\Gamma_{\mathcal{U}}$ which satisfies the same axioms. Let $(N, v) \in \Gamma_{\mathcal{U}}$. If $|N| = 1$, then, by SIVA and COV, $\sigma(N, v) = \phi(N, v)$. Hence, by Remark 8.5.1, σ satisfies PO. If $|N| = 2$, then $\sigma(N, v) = \phi(N, v)$ by Lemma 5.4.3. Let $|N| \geq 3$ and let $S \subseteq N$ satisfy $|S| = 2$. Let $x, y \in \mathbb{R}^N$ and let $S = \{i, j\}$. Then

$$v_{S,x}^{SH}(\{i\}) - v_{S,x}^{SH}(\{j\}) = v_{S,y}^{SH}(\{i\}) - v_{S,y}^{SH}(\{j\}) \tag{8.5.5}$$

(see Exercise 8.5.2). Now, let $x = \sigma(N, v)$, $y = \phi(N, v)$, $u = v_{S,x}^{SH}$, and $w = v_{S,y}^{SH}$. As σ and ϕ coincide with the standard solution,

$$\begin{aligned}
x^i &= \frac{1}{2}(u(\{i\}) - u(\{j\}) + u(S)) \,,\; y^i = \frac{1}{2}(w(\{i\}) - w(\{j\}) + w(S)), \\
x^j &= \frac{1}{2}(u(\{j\}) - u(\{i\}) + u(S)) \,,\; y^j = \frac{1}{2}(w(\{j\}) - w(\{i\}) + w(S)).
\end{aligned}$$

By (8.5.5), $x^i - x^j = y^i - y^j$. Hence $x^k - x^\ell = y^k - y^\ell$ for all $k, \ell \in N$. Hence by PO, for every $\ell \in N$,

$$v(N) - |N|x^\ell = \sum_{k\in N}(x^k - x^\ell) = \sum_{k\in N}(y^k - y^\ell) = v(N) - |N|y^\ell,$$

that is, $y = x$. **q.e.d.**

Theorem 8.5.5 is due to Sobolev (1975).

The empty solution (see Example 5.4.4), the equal split solution (see Example 5.4.6), and the prenucleolus are examples which show that each of the axioms SIVA, COV, and RGP^{SH} is logically independent of the remaining axioms employed in the foregoing theorem, provided that $|\mathcal{U}| \geq 2$. For $|\mathcal{U}| = 2$, any weighted Shapley value satisfies SIVA, COV, and RGP^{SH}, but it may not satisfy ETP. If $|\mathcal{U}| \geq 3$, then the following theorem shows that ETP can be deduced from SIVA, COV, and RGP^{SH}.

Theorem 8.5.6. *Let $|\mathcal{U}| \geq 3$ and let σ be a solution on $\Gamma_{\mathcal{U}}$. Then $\sigma = \phi$ if and only if σ satisfies* SIVA, COV, *and* RGP^{SH}.

Proof: The Shapley value satisfies the desired axioms. In order to prove the opposite direction, let σ satisfy SIVA, COV, and RGP^{SH}. By SIVA and COV, σ is Pareto optimal on one-person games and hence, by Remark 8.5.1, σ satisfies PO. A careful inspection of the proof of Theorem 8.5.5 shows that it suffices to show that σ coincides with the standard solution on two-person games. Let $(N, v) \in \Gamma_{\mathcal{U}}$ satisfy $|N| = 2$, let us say $N = \{1, 2\}$. In order to show that $\sigma(N, v) = \phi(N, v)$ we may assume, by COV, that (N, v) is 0-normalized. If $v(N) > 0$ or $v(N) < 0$, respectively, then, again by COV, we may assume that $v(N) = 1$ (that is, (N, v) is 0-1-normalized) or $v(N) = -1$ (that is, (N, v) is 0-(-1)-normalized), respectively. We distinguish three cases.

(1) $v(N) = 0$. Then $\alpha v = v$ for all $\alpha > 0$, so $\sigma(N, v) = 0$ by SIVA and COV.

(2) $v(N) = -1$. Let $\ell \in \mathcal{U} \setminus N$, let us say $\ell = 3$. Let $M = N \cup \{3\}$ and let (M, w) be the game which arises from (N, v) by adding the null player 3, that is, $w(S) = v(S \cap N)$ for all $S \subseteq M$. Let $\widehat{x} = (-1/2, -1/2, 0)$. Then $\widehat{x}^N = \phi(N, v)$ and, by (8.5.1), $w^{SH}_{N,\widehat{x}} = v$. Therefore, by RGP^{SH}, it suffices to show that $x := \sigma(M, w) = \widehat{x}$. Let $P = \{1, 3\}$ and $Q = \{2, 3\}$. We claim that the reduced games $(N, w^{SH}_{N,x})$, $(P, w^{SH}_{P,x})$, and $(Q, w^{SH}_{Q,x})$, respectively, are strategically equivalent to the 0-(-1)-normalized two-person games (N, v), (P, v'), and (Q, v''), respectively. Indeed, by (8.5.1),

$$w^{SH}_{N,x}(\{1\}) = -\frac{x^3}{2} = w^{SH}_{N,x}(\{2\}) \text{ and } w^{SH}_{N,x}(N) = v(N) - x^3 \tag{8.5.6}$$

and, thus, $v = \left(w^{SH}_{N,x} + 1/2(x^3, x^3)\right)$. By COV and RGP^{SH},

$$x^N + \frac{1}{2}(x^3, x^3) = \left(x^1 + \frac{x^3}{2}, x^2 + \frac{x^3}{2}\right) = \sigma(N, v). \tag{8.5.7}$$

Similarly, we receive

$$v' = 2\left(w^{SH}_{P,x} + \frac{1}{2}(1 + x^2, x^2)\right) \text{ and } v'' = 2\left(w^{SH}_{Q,x} + \frac{1}{2}(1 + x^1, x^1)\right)$$

and thus

$$2(x^P + \frac{1}{2}(1 + x^2, x^2)) = (2x^1 + x^2 + 1, 2x^3 + x^2) = \sigma(P, v') \tag{8.5.8}$$

and

$$2(x^Q + \frac{1}{2}(1 + x^1, x^1)) = (x^1 + 2x^2 + 1, x^1 + 2x^3) = \sigma(Q, v''). \tag{8.5.9}$$

Now, let (M, w') and (M, w''), respectively, be the games which arise from (P, v') and from (Q, v''), respectively, by adding the null player 2 and 1, respectively, that is, $w'(S) = v'(S \cap P)$ and $w''(S) = v''(S \cap Q)$ for all $S \subseteq M$. Let $y = \sigma(M, w')$ and $z = \sigma(M, w'')$. Analogously as before, by showing that the two-person reduced games of (M, w') with respect to y as well as the two-person reduced games of (M, w'') with respect to z are strategically equivalent to (N, v), (P, v'), or (Q, v''), we receive

$$\begin{array}{rcl} (2z^1 + z^3, 2z^2 + z^3 + 1) = (2y^1 + y^3 + 1, 2y^2 + y^3) = \sigma(N, v), \\ (2z^1 + z^2, z^2 + 2z^3 + 1) = (y^1 + y^2/2, y^2/2 + y^3) = \sigma(P, v'), \\ (z^1/2 + z^2, z^1/2 + z^3) = (y^1 + 2y^2, y^1 + 2y^3 + 1) = \sigma(Q, v''). \end{array} \tag{8.5.10}$$

By PO we may substitute $x^3 = -1 - x(N), y^3 = -1 - y(N)$, and $z^3 = -1 - z(N)$. By (8.5.8) – (8.5.10),

$$\begin{array}{rcl} \frac{x^1}{2} - \frac{x^2}{2} - \frac{1}{2} = y^1 - y^2 = z^1 - z^2 - 1, \\ 2x^1 + x^2 + 1 = y^1 + \frac{y^2}{2} = 2z^1 + z^2, \text{ and} \\ x^1 + 2x^2 + 1 = y^1 + 2y^2 = \frac{z^1}{2} + z^2. \end{array} \tag{8.5.11}$$

Each line of 8.5.11 yields 3 equations. We may choose 6 equations, two from each line, which are linearly independent:

$$\begin{array}{rrrrrrl} x^1 & -x^2 & -2y^1 & +2y^2 & & & = 1, \\ x^1 & -x^2 & & & -2z^1 & +2z^2 & = -1, \\ 4x^1 & +2x^2 & -2y^1 & -y^2 & & & = -2, \\ 2x^1 & +x^2 & & & -2z^1 & -z^2 & = -1, \\ x^1 & +2x^2 & -y^1 & -2y^2 & & & = -1, \text{ and} \\ 2x^1 & +4x^2 & & & -z^1 & -2z^2 & = -2. \end{array} \tag{8.5.12}$$

As the reader may easily verify, the unique solution of the system (8.5.12) is $x^1 = x^2 = y^1 = z^2 = -\frac{1}{2}, y^2 = z^1 = 0$, that is,

$$x = (-\frac{1}{2}, -\frac{1}{2}, 0), y = (-\frac{1}{2}, 0, -\frac{1}{2}), z = (0, -\frac{1}{2}, -\frac{1}{2}).$$

Hence $x = \widehat{x}$.

(3) $v(N) = 1$. This case can be treated in the same manner as Case (2). Only "0-(−1)-normalized" has to be replaced by "0-1-normalized" wherever it occurs.

q.e.d

Exercises

Exercise 8.5.1. Prove Remark 8.5.2.

Exercise 8.5.2. Let $(N,v) \in \Gamma_{\mathcal{U}}$, $\emptyset \neq S \subseteq N$ and $x, y \in \mathbb{R}^N$. Use (8.5.1) and induction on $|N| - |S|$ to prove that

$$v_{S,x}^{SH}(R) - v_{S,x}^{SH}(T) = v_{S,y}^{SH}(R) - v_{S,y}^{SH}(T) \text{ for all } R, T \subseteq S \text{ with } |R| = |T|.$$

8.6 The Shapley Value for Simple Games

The Shapley value of a committee, that is a simple game (see Definition 2.2.6), has the following probabilistic interpretation. Let $g = (N, \mathcal{W})$ be a simple game and let π be a permutation of N. By (2.2.1) – (2.2.3) there exists exactly one player $j \in N$ such that

$$P_\pi^j \notin \mathcal{W} \text{ and } P_\pi^j \cup \{j\} \in \mathcal{W} \tag{8.6.1}$$

(see (8.1.2)). Player j is called the *pivot* of π. Now let $i \in N$. The Shapley value of i in the coalitional game (N,v) associated with g (see Definition 2.2.9) is given by

$$\phi^i(N,v) = \frac{|\{\pi \in \mathcal{SYM}_N \mid i \text{ is a pivot of } \pi\}|}{n!} \tag{8.6.2}$$

where $n = |N|$. By (8.6.2), $\phi^i(G) = \phi^i(N,v)$ is the probability that i is a pivot under the assumption that all permutations of N have the same probability.

Let again $g = (N, \mathcal{W})$ be a simple game and let I be an issue which has to be approved or rejected by the members of N. Then I determines an ordering (i.e., a permutation) π of the players of g according to their degree of support of I. If the strongest supporter of I is the first player according to π, the next-strongest the second, and so forth, then the pivot of π, $j = j(\pi)$, is decisive for the approval of I. Indeed, if j is persuaded to support I, then all the members of P_π^j will also support I; together they will form the winning coalition $P_\pi^j \cup \{j\}$ of supporters. Similarly, if j votes against I, then $N \setminus P_\pi^j$ will block I. Thus, if $i \in N$ then $\phi^i(G)$ is the probability that player i determines the outcome of a vote on some issue. Here it is assumed that all possible "orders of support" of the players, induced by all relevant issues, are equiprobable.

Let $N = \{1, \ldots, n\}$ be a set of players and let Σ^N be the set of all simple games with player set N. We first notice that there may be more than one value (see Definition 8.1.2) on Σ^N (that is, on the set of coalitional games associated with the simple games of Σ^N).

Example 8.6.1. Let $N = \{1,2,3\}$ and $g_0 \widehat{=} (3; 2, 1, 1)$ (see Definition 2.2.11) and let $\phi_* : \Sigma^N \to \mathbb{R}^N$ satisfy $\phi_*(g_0) = (1,1,1)/3$ and $\phi_*(g) = \phi(g)$ if $g \neq g_0$

(we recall that $\phi(g)$ is the Shapley value of the associated coalitional game). Then ϕ_* is a value on Σ^N. Nevertheless, $\phi_* \neq \phi$ on Σ^N.

We shall now replace ADD (see Definition 8.1.1) by a similar axiom that will enable us to characterize ϕ on Σ^N. Let $\mathcal{V}$ be the set of all coalition functions on 2^N. If $u, v \in \mathcal{V}$ then we define $u \vee w$ and $u \wedge v$ in $\mathcal{V}$ by

$$(u \vee v)(S) = \max\{u(S), v(S)\} \text{ and } (u \wedge v)(S) = \min\{u(S), v(S)\}$$

for all $S \subseteq N$. A subset $\mathcal{L}$ of $\mathcal{V}$ is a *lattice* if it satisfies the following conditions. If $u, v \in \mathcal{L}$, then $u \vee v, u \wedge v \in \mathcal{L}$. The set Σ^N is a lattice. (Here again we identify simple games with their associated coalitional games.) We now introduce the following modified form of additivity.

Definition 8.6.2. *Let $\mathcal{K} \subseteq \mathcal{V}$. A single-valued solution σ on $\mathcal{K}$ is a* **valuation** *if the following equation is satisfied for all $u, v \in \mathcal{K}$ such that $u \vee v, u \wedge v \in \mathcal{K}$:*

$$\sigma(u \vee v) + \sigma(u \wedge v) = \sigma(u) + \sigma(v). \tag{8.6.3}$$

Remark 8.6.3. Let $\mathcal{K} \subseteq \mathcal{V}$, let $\mathcal{K}^* \subseteq \mathcal{V}$ be a convex cone which contains $\mathcal{K}$, and let σ be a single-valued additive solution on $\mathcal{K}^*$. Then σ is a valuation on $\mathcal{K}$. Indeed, if $u, v, u \vee v, u \wedge v \in \mathcal{K}$, then $u + v \in \mathcal{K}^*$ and

$$\sigma(u \vee v) + \sigma(u \wedge v) = \sigma((u \vee v) + (u \wedge v)) = \sigma(u + v) = \sigma(u) + \sigma(v).$$

The following theorem is due to Dubey (1975).

Theorem 8.6.4. *There is a unique valuation on Σ^N that satisfies* ETP, NP, *and* PO, *and it is the Shapley value ϕ.*

Proof: By Remark 8.6.3 and Theorem 8.1.3, the Shapley value on Σ^N is a valuation and satisfies ETP, NP, and PO. Thus, only the uniqueness part of the theorem has to be shown. Let σ be a valuation on Σ^N that satisfies ETP, NP, and PO, and let $g = (N, \mathcal{W}) \in \Sigma^N$. We prove that $\sigma(g) = \phi(g)$ by induction on $|\mathcal{W}^m| = t$ (see Notation 2.2.12). If $t = 1$, then g is the unanimity game u_S where S is the unique minimal winning coalition of g. Hence, by ETP, NP, and PO, $\sigma(g) = \phi(g)$ (see the proof of Theorem 8.1.3). Now assume that $t \geq 2$. Let v be the associated coalition function of g, let $S \in \mathcal{W}^m$, let $v_1 = \bigvee_{T \in \mathcal{W}^m \setminus \{S\}} u_T$ and $v_2 = u_S$. Then

$$v = v_1 \vee v_2 \text{ and } v_1 \wedge v_2 = \bigvee_{T \in \mathcal{W}^m \setminus \{S\}} u_{T \cup S}.$$

Hence, by the induction hypothesis, $\sigma(v_1) = \phi(v_1)$ and $\sigma(v_1 \wedge v_2) = \phi(v_1 \wedge v_2)$. As σ is a valuation we obtain the equalities

$$\begin{aligned}\sigma(v) = \sigma(v_1 \vee v_2) &= \sigma(v_1) + \sigma(v_2) - \sigma(v_1 \wedge v_2)\\ &= \phi(v_1) + \phi(v_2) - \phi(v_1 \wedge v_2)\\ &= \phi(v_1 + v_2 - (v_1 \wedge v_2)) = \phi(v_1 \vee v_2) = \phi(v)\end{aligned}$$

and the proof is complete. **q.e.d.**

There are simple examples of valuations which show that each of the axioms ETP and NP are independent of the remaining axioms in Theorem 8.6.4, provided that $|N| \geq 2$. Also, PO is logically independent, because the solution which assigns $0 \in \mathbb{R}^N$ to every $g \in \Sigma^N$ is a valuation that satisfies ETP and NP.

Exercises

Exercise 8.6.1. Compute the Shapley value for the UN Security Council (see Example 2.2.15 and Monjardet (1972)).

Exercise 8.6.2. Compute the Shapley value of the games represented by $(2k+1; 2k, \underbrace{1, \ldots, 1}_{2k+1})$ and $(2k+1; k, k, \underbrace{1, \ldots, 1}_{2k+1})$ for any $k \in \mathbb{N}$.

Exercise 8.6.3. Let N be a coalition and let A be a set of nonnegative numbers that contains 0. Denote

$$\mathcal{V}^m(A) = \{v \in \mathcal{V}^m \mid v(S) \in A \text{ for all } S \subseteq N\}$$

(see Corollary 8.1.9). Prove that Theorem 8.6.4 remains true, if Σ^N is replaced by $\mathcal{V}^m(A)$. This result is due to Einy (1988).

Exercise 8.6.4. Let $g = (N, \mathcal{W})$ be a simple game with $|\mathcal{W}^m| = k$ and let $\mathcal{W}^m = \{T_1, \ldots, T_k\}$. Prove that

$$\phi(g) = \sum_{i=1}^{k} \frac{1}{|T_i|}\chi_{T_i} - \sum_{1 \leq i < j \leq k} \frac{1}{|T_i \cup T_j|}\chi_{T_i \cup T_j} + \cdots + \frac{(-1)^{k-1}}{|\bigcup_{i=1}^{k} T_i|}\chi_{\bigcup_{i=1}^{k} T_i}$$

(see Notation 3.1.1 and Einy (1988)).

Exercise 8.6.5. Let Σ_s^N be the set of all proper simple games with player set N. Prove that the Shapley value is the unique valuation on Σ_s^N that satisfies ETP, NP, and PO.

8.7 Games with Coalition Structures

Let $\mathcal{U}$ be a set of players and $\Delta_{\mathcal{U}}$ be the set of all games with coalition structures and with players in $\mathcal{U}$. Aumann and Drèze (1974) define a value on $\Delta_{\mathcal{U}}$ in the following way. Let $\Delta \subseteq \Delta_{\mathcal{U}}$.

Definition 8.7.1. *A solution σ on Δ is* **additive** *(ADD) if*

$$\sigma(N, u+v, \mathcal{R}) = \sigma(N, u, \mathcal{R}) + \sigma(N, v, \mathcal{R}) \tag{8.7.1}$$

when $(N, u, \mathcal{R}), (N, v, \mathcal{R}), (N, u+v, \mathcal{R}) \in \Delta$.

Definition 8.7.2. *A* **value** *on Δ is a single-valued solution on Δ that satisfies* EFF, RETP, NP, *and* ADD *(see Definitions 5.4.10, 5.3.11, 4.1.17, and 8.7.1, respectively).*

The following theorem is due to Aumann and Drèze (1974).

Theorem 8.7.3. *There is a unique value ϕ_* on $\Delta_{\mathcal{U}}$ and, for every $(N, v, \mathcal{R}) \in \Delta_{\mathcal{U}}$ and for all $i \in N$, it is given by*

$$\phi_*^i(N, v, \mathcal{R}) = \phi^i(R, v) \tag{8.7.2}$$

where $R \in \mathcal{R}$ is determined by $i \in R$ and where $\phi(R, v)$ is the Shapley value of the subgame (R, v) of (N, v).

Proof: Formula (8.7.2) defines a single-valued solution ϕ_* on $\Delta_{\mathcal{U}}$. Also, as the reader may easily verify, ϕ_* satisfies EFF, RETP, NP, and ADD, because ϕ is a value. Thus, it remains to prove the uniqueness part of the theorem.

Let σ be a value on $\Delta_{\mathcal{U}}$, let $N \subseteq \mathcal{U}$ be finite and nonempty, and let $\mathcal{R}$ be a coalition structure for N. Denote by $\mathcal{V}$ the set of all coalition functions on 2^N. By (8.7.1), σ is additive on $\mathcal{V}$. Hence, if we prove that $\sigma(N, \alpha u_T, \mathcal{R}) = \phi_*(N, \alpha u_T, \mathcal{R})$ for every unanimity game u_T, $\emptyset \neq T \subseteq N$, and every $\alpha \in \mathbb{R}$, then the proof is completed by Lemma 8.1.4. So, let $\emptyset \neq T \subseteq N$ and $\alpha \in \mathbb{R}$. If $i \in N$, then, by EFF, RETP, and NP,

$$\sigma^i(N, \alpha u_T, \mathcal{R}) = \begin{cases} \alpha/|T| & \text{, if } i \in T \text{ and } T \subseteq R \text{ for some } R \in \mathcal{R}, \\ 0 & \text{, otherwise.} \end{cases}$$

Thus, by (8.7.2), $\sigma(N, v, \mathcal{R}) = \phi_*(N, v, \mathcal{R})$ for every $v \in \mathcal{V}$. **q.e.d.**

Remark 8.7.4. For every game $(N, v) \in \Gamma_{\mathcal{U}}$, $\phi(N, v) = \phi_*(N, v, \{N\})$.

Remark 8.7.5. Let $N \subseteq \mathcal{U}$ be a coalition, let $\mathcal{R}$ be a coalition structure for N, and let $\Delta_{\mathcal{R}}^N$ be the set of games on 2^N with the coalition structure $\mathcal{R}$. The following stronger result may be proved similarly to Theorem 8.1.3: There is a unique solution σ on $\Delta_{\mathcal{R}}^N$ that satisfies NE, RETP, NP, and ADD, and it is ϕ_*. Also, each of the axioms is logically independent of the remaining axioms and of PO, provided that there exists $R \in \mathcal{R}$ such that $|R| \geq 2$.

The Aumann-Drèze value ϕ_* may be characterized by consistency in the following way. Let $\Delta \subseteq \Delta_{\mathcal{U}}$ and let σ be a single-valued solution on Δ. If $(N, v, \mathcal{R}) \in \Delta$ and $\emptyset \neq S \subseteq R \in \mathcal{R}$, then we define the *first σ-reduced game* $(S, v_{S,\sigma}^{\mathcal{R}})$ by

$$v_{S,\sigma}^{\mathcal{R}}(T) = v((R \setminus S) \cup T) - \sum_{i \in R \setminus S} \sigma^i((R \setminus S) \cup T, v) \text{ for every } \emptyset \neq T \subseteq S \tag{8.7.3}$$

and $v_{S,\sigma}^{\mathcal{R}}(\emptyset) = 0$. (Here $((R \setminus S) \cup T, v)$ is a subgame of (N, v).) Now we are ready for the following definition.

Definition 8.7.6. *A solution σ on Δ is* **consistent** *(CON1) with respect to (8.7.3), if the following condition is satisfied. If $(N, v, \mathcal{R}) \in \Delta$ and $\emptyset \neq S \subseteq R \in \mathcal{R}$, then*

$$(S, v^{\mathcal{R}}_{S,\sigma}) \in \Delta \text{ and } \sigma^j(N, v, \mathcal{R}) = \sigma^j(S, v^{\mathcal{R}}_{S,\sigma}) \text{ for all } j \in S. \tag{8.7.4}$$

The following theorem is proved in Winter (1988).

Theorem 8.7.7. *There is a unique single-valued solution on $\Delta_{\mathcal{U}}$ that satisfies* EFF, RETP, COV, *and* CON1, *and it is the Aumann-Drèze value.*

Remark 8.7.8. EFF can be deduced from COV and CON1, so it is not needed as an assumption of the preceding theorem.

Exercises

Exercise 8.7.1. Prove Remark 8.7.5.

Exercise 8.7.2. Prove Theorem 8.7.7.

8.8 Games with A Priori Unions

Owen (1977) and Hart and Kurz (1983) have investigated another value for games with coalition structures. Unlike the Aumann-Drèze value, their value is Pareto optimal. Hart and Kurz (1983) write: "Our view is that the reason coalitions form is not in order to get their worth, but to be in a better position when bargaining with the others on how to divide the maximum amount available (i.e., the worth of the grand coalition)." The following example illustrates the foregoing remark.

Example 8.8.1. Let $N_1 = \{1\}$, let $N_2 = \{2, \ldots, n\}$, $n \geq 3$, and let the market game v be defined by

$$v(S) = \min\{|S \cap N_1|, |S \cap N_2|\} \text{ for all } S \subseteq N$$

where $N = N_1 \cup N_2$ (see Example 2.2.2). To make the example more vivid we shall assume that player 1 is a landowner and that the members of N_2 are workers. The Shapley value of the landowner is $\phi^1(N, v) = (n-1)/n$. The workers cannot object to the payoff distribution $\phi(N, v)$ because $v(N_2) = 0$; that is, N_2 is unable to improve upon $\phi(N, v)$ by forming its own subgame. However, the workers may form a labor union and thereby "reduce" the game (N, v) to the two-person symmetric game $(\widehat{N}, \widehat{v})$ where $\widehat{N} = \{1, N_2\}$, $\widehat{v}(\emptyset) = \widehat{v}(\{1\}) = \widehat{v}(\{N_2\}) = 0$, and $\widehat{v}(\widehat{N}) = 1$. Clearly, $\phi^1(\widehat{N}, \widehat{v}) = \phi^{N_2}(\widehat{N}, \widehat{v}) = 1/2$. Thus, by forming a labor union the workers obtain a better bargaining position (and a larger payoff).

Now we shall present the Owen value. Let $\mathcal{U}$ be a set of players and let $\Delta_{\mathcal{U}}$ be the set of all games with coalition structures and with players in $\mathcal{U}$.

Now, let $G = (N, v, \mathcal{R}) \in \Delta_{\mathcal{U}}$. In the current section we consider G as a game with a *priori unions*. Thus, if $\mathcal{R} = \{\{i\} | i \in N\}$, then nontrivial a priori unions are absent. Unlike for games with coalition structures, the game (N, v), that is, a game without a priori unions, may be identified with $(N, v, \{\{i\} | i \in N\})$ in the present context.

Consequently $x \in \mathbb{R}^N$ is *feasible* for G, if x is feasible for (N, v), that is, if $x \in X^*(N, v)$. A *solution* on a set $\Delta \subseteq \Delta_{\mathcal{U}}$ of games with a priori unions assigns to each $(N, v, \mathcal{R}) \in \Delta$ a subset of $X^*(N, v)$. Note that this definition of a solution is in contrast to the definition of a solution for Δ when Δ is considered to be a set of games with coalition structures (see Definition 3.8.1).

Definition 8.8.2. *The* **intermediate game** $(\mathcal{R}, v_{\mathcal{R}})$ *is the game whose players are the coalitions of* $\mathcal{R}$ *and whose coalition function* $v_{\mathcal{R}}$ *is given by*

$$v_{\mathcal{R}}(\mathcal{T}) = v\left(\bigcup_{Q \in \mathcal{T}} Q\right) \quad \text{for all } \mathcal{T} \subseteq \mathcal{R} \tag{8.8.1}$$

(see Definition 2.13 in Maschler and Peleg (1967)).

Definition 8.8.3. *A single-valued solution* σ *on* $\Delta_{\mathcal{U}}$ *has the* **intermediate game property** *(*IGP*) if for every* $(N, v, \mathcal{R}) \in \Delta_{\mathcal{U}}$ *and all substitutes* $R, R' \in \mathcal{R}$ *(see Definition 3.8.5) of* $(\mathcal{R}, v_{\mathcal{R}})$*,*

$$\sum_{j \in R} \sigma^j(N, v, \mathcal{R}) = \sum_{j \in R'} \sigma^j(N, v, \mathcal{R}). \tag{8.8.2}$$

Definition 8.8.4 (Owen (1977)). *A* **coalition structure value** *(*CS*-value) on* $\Delta_{\mathcal{U}}$ *is a single-valued solution on* $\Delta_{\mathcal{U}}$ *that satisfies*[1] PO, RETP, NP, ADD, *and* IGP *(see Definitions 2.3.6, 5.3.11, 4.1.17, 8.7.1, and 8.8.2, respectively).*

Theorem 8.8.5 (Owen (1977)). *There is a unique* CS*-value on* $\Delta_{\mathcal{U}}$.

Proof: Let N be a finite nonempty set of players and let $\mathcal{R}$, $|\mathcal{R}| = m$, be a coalition structure for N. A permutation $\pi \in \mathcal{SYM}_N$ (Definition 2.1.15) is *consistent with* $\mathcal{R}$ if the following condition is satisfied: If $i, j \in R \in \mathcal{R}$, $\ell \in N$, and $\pi(i) < \pi(\ell) < \pi(j)$, then $\ell \in R$. Denote by $\mathcal{SYM}_N(\mathcal{R})$ the set of all permutations of N consistent with $\mathcal{R}$. Notice that $|\mathcal{SYM}_N(\mathcal{R})| = m! \prod_{R \in \mathcal{R}} |R|!$. Now we define the Owen value of a game $(N, v, \mathcal{R})$ by

$$\overline{\phi}^i(N, v, \mathcal{R}) = \sum_{\pi \in \mathcal{SYM}_N(\mathcal{R})} \frac{a^i_\pi(v)}{|\mathcal{SYM}_N(\mathcal{R})|} \quad \text{for every } i \in N \tag{8.8.3}$$

[1] If $(N, v, \mathcal{R})$ is a game with a priori unions, then $X(N, v)$, the set of preimputations of the underlying game (N, v), is the set of Pareto optimal payoffs of $(N, v, \mathcal{R})$.

(see (8.1.3)). As the reader may easily verify, $\overline{\phi}$ satisfies PO, RETP, NP, ADD, and IGP. Thus, it remains to prove the uniqueness part of the theorem.

Let σ be a CS-value on $\Delta_{\mathcal{U}}$. As σ is additive, by Lemma 8.1.4, it suffices to prove that for every unanimity game with coalition structure $(N, u_T, \mathcal{R}) \in \Delta_{\mathcal{U}}$, and all $\alpha \in \mathbb{R}$,

$$\sigma(N, \alpha u_T, \mathcal{R}) = \overline{\phi}(N, \alpha u_T, \mathcal{R}). \tag{8.8.4}$$

Thus, let $N \subseteq \mathcal{U}$ be a coalition, let $\mathcal{R}$ be a coalition structure for N, let $\emptyset \neq T \subseteq N$, and let $\alpha \in \mathbb{R}$. Let $\mathcal{T}_{\mathcal{R}} = \{R \in \mathcal{R} \mid R \cap T \neq \emptyset\}$. Furthermore, we abbreviate $u = u_T$. First consider the intermediate game $(\mathcal{R}, \alpha u_{\mathcal{R}})$. The coalition function $u_{\mathcal{R}}$ is given by

$$u_{\mathcal{R}}(\mathcal{Q}) = \begin{cases} 1 \text{ , if } \mathcal{T}_{\mathcal{R}} \subseteq \mathcal{Q}, \\ 0 \text{ , otherwise,} \end{cases}$$

for every $\mathcal{Q} \subseteq \mathcal{R}$. Thus, $u_{\mathcal{R}}$ is the unanimity game on $\mathcal{T}_{\mathcal{R}}$. By PO, NP, and IGP,

$$\sum_{j \in R} \sigma^j (N, \alpha u, \mathcal{R}) = \begin{cases} \alpha/|\mathcal{T}_{\mathcal{R}}| \text{ , if } R \in \mathcal{T}_{\mathcal{R}}, \\ \quad 0 \quad \text{ , if } R \in \mathcal{R} \setminus \mathcal{T}_{\mathcal{R}}. \end{cases} \tag{8.8.5}$$

Again by NP, $\sigma^i(N, \alpha u, \mathcal{R}) = 0$ for all $i \in N \setminus T$. If $i \in T$, let $R \in \mathcal{R}$ such that $i \in R$. Then $R \in \mathcal{T}_{\mathcal{R}}$. Therefore, by RETP and (8.8.5),

$$\sigma^i(N, \alpha u, \mathcal{R}) = \frac{\alpha}{|\mathcal{T}_{\mathcal{R}}||R \cap T|}.$$

We conclude that σ is uniquely determined on $(N, \alpha u, \mathcal{R})$. As $\overline{\phi}$ is also a CS-value, (8.8.4) has been proved. **q.e.d.**

The Owen value $\overline{\phi}$ of a game with a priori unions $G = (N, v, \mathcal{R}) \in \Delta_{\mathcal{U}}$ defined by (8.8.3) is related to the "classical" Shapley value of the intermediate game $(\mathcal{R}, v_{\mathcal{R}})$. Indeed, note that for every $R \in \mathcal{R}$,

$$\sum_{j \in R} \overline{\phi}(G) = \phi(\mathcal{R}, v_{\mathcal{R}}).$$

Remark 8.8.6. Each of the axioms RETP, NP, IGP, and ADD are independent of the remaining axioms in the foregoing theorem, provided that $|\mathcal{U}| \geq 2$. Analogously to Section 8.1 it can be shown that PO is not needed. Indeed, let σ be a single-valued solution on $\Delta_{\mathcal{U}}$ that satisfies RETP, NP, ADD, and IGP, and let $G = (N, v, \mathcal{R}) \in \Delta_{\mathcal{U}}$. By ADD and NP, $\sigma(G) + \sigma(N, -v, \mathcal{R}) = 0 \in \mathbb{R}^N$. Thus, PO is implied by feasibility.

We shall now introduce the consistency property which corresponds to the CS-value. Let σ be a single-valued solution on $\Delta_{\mathcal{U}}$, $(N, v, \mathcal{R}) \in \Delta_{\mathcal{U}}$, and $\emptyset \neq S \subseteq R$ for some $R \in \mathcal{R}$. The *second σ-reduced game* $(S, \overline{v}^{\mathcal{R}}_{S,\sigma})$ is given by

$$\overline{v}^{\mathcal{R}}_{S,\sigma}(T) = v((N \setminus S) \cup T) - \sum_{i \in N \setminus S} \sigma^i \left((N \setminus S) \cup T, v, \mathcal{R}_{|(N \setminus S) \cup T}\right) \qquad (8.8.6)$$

for all $\emptyset \neq T \subseteq S$, and $\overline{v}^{\mathcal{R}}_{S,\sigma}(\emptyset) = 0$. (Here $\left((N \setminus S) \cup T, v, \mathcal{R}_{|(N \setminus S) \cup T}\right)$ is a subgame of (N, v) with the coalition structure $\mathcal{R}_{|(N \setminus S) \cup T}$ as defined in Section 3.8.)

Definition 8.8.7. *A solution σ on $\Delta_{\mathcal{U}}$ is* **consistent** *(*CON2*) with respect to (8.8.6) if the following condition is satisfied. If $(N, v, \mathcal{R}) \in \Delta_{\mathcal{U}}$ and $\emptyset \neq S \subseteq R \in \mathcal{R}$, then*

$$\sigma^j(N, v, \mathcal{R}) = \sigma^j \left(S, \overline{v}^{\mathcal{R}}_{S,\sigma}\right) \text{ for all } j \in S. \qquad (8.8.7)$$

The proof of the following theorem is left to the reader (Exercise 8.8.3).

Theorem 8.8.8 (Winter (1988))**.** *There is a unique single-valued solution on $\Delta_{\mathcal{U}}$ that satisfies* PO, COV, RETP, IGP, *and* CON2, *and it is the* CS*-value.*

Exercises

Exercise 8.8.1. Prove the remaining statements of Remark 8.8.6 by means of examples.

Exercise 8.8.2. A single-valued solution σ on $\Delta_{\mathcal{U}}$ satisfies the

- *strong null player property* (SNP) if σ satisfies NP and if for all $(N, v, \mathcal{R}) \in \Delta_{\mathcal{U}}$ with $|N| \geq 2$ and for any null player $i \in N$, $\sigma\left(N \setminus \{i\}, v, \mathcal{R}_{|N \setminus \{i\}}\right) = \sigma^{N \setminus \{i\}}(N, v, \mathcal{R})$;
- *null intermediate game property* (NIGP) if for every $(N, v, \mathcal{R}) \in \Delta_{\mathcal{U}}$ such that $v_{\mathcal{R}}(\mathcal{T}) = 0$ for all $\mathcal{T} \subseteq \mathcal{R}$, (8.8.2) is true for all $R, R' \in \mathcal{R}$.

Prove that if $\mathcal{U}$ is infinite, then there exists a unique single-valued solution on $\Delta_{\mathcal{U}}$ that satisfies PO, AN, SNP, ADD, and NIGP (see Hart and Kurz (1983)).

Exercise 8.8.3. Prove Theorem 8.8.8.

8.9 Multilinear Extensions of Games

Throughout this section let $G = (N, v)$ be an n-person game. Denote by I^N the unit n-cube

$$I^N = \{x \in \mathbb{R}^N \mid 0 \leq x^i \leq 1 \text{ for all } i \in N\}.$$

The extreme points of I^N are the vectors χ_S, $S \subseteq N$ (see Notation 3.1.1). So, we notice that v determines a real function $\overline{v}$ on the corners of I^N by

$$\overline{v}(\chi_S) = v(S) \text{ for all } S \subseteq N. \tag{8.9.1}$$

Hence, $\overline{v}$ may be extended to I^N by

$$\overline{v}(x) = \sum_{T \subseteq N} \left(\prod_{i \in T} x^i \prod_{i \in N \setminus T} (1 - x^i) \right) v(T) \quad \text{for all } x \in I^N. \tag{8.9.2}$$

Owen (1972) has proved the following result.

Theorem 8.9.1. *There exists a unique multilinear function on I^N which satisfies (8.9.1).*

Proof: Clearly $\overline{v}$ is multilinear. Hence we have only to prove the uniqueness part of the theorem. Let $f : I^N \to \mathbb{R}$ be a multilinear function that satisfies $f(\chi_S) = v(S)$ for all $S \subseteq N$. As f is multilinear, there exist $c_T \in \mathbb{R},\ T \subseteq N$, such that

$$f(x) = \sum_{T \subseteq N} c_T \prod_{i \in T} x^i \text{ for all } x \in I^N.$$

Thus, in particular,

$$f(\chi_S) = v(S) = \sum_{T \subseteq S} c_T \text{ for every } S \subseteq N. \tag{8.9.3}$$

The corresponding homogeneous system of linear equations has a full rank; thus (8.9.3) has at most one solution. It has at least one solution, because $\overline{v}$ is multilinear. Therefore, $f = \overline{v}$. **q.e.d.**

Definition 8.9.2. *Let (N, v) be a coalitional game. The* **multilinear extension** *(*MLE*) of (N, v) is the function $\overline{v} : I^N \to \mathbb{R}$ defined by (8.9.2).*

Example 8.9.3. Let $N = \{1, 2, 3\}$ and $G \widehat{=} (3; 2, 1, 1)$. The MLE of G is given by

$$\overline{v}(x) = x^1x^2(1 - x^3) + x^1x^3(1 - x^2) + x^1x^2x^3 = x^1x^2 + x^1x^3 - x^1x^2x^3.$$

Let (N, v) be a game and let $\Omega = \{0, 1\}$. Then $\Omega^N = \{\chi_S \mid S \subseteq N\}$ is the set of extreme points of I^N. Every $x \in I^N$ defines a product probability on Ω^N by

$$Pr(\chi_S) = \prod_{i \in S} x^i \prod_{i \in N \setminus S} (1 - x^i) \text{ for all } S \subseteq N. \tag{8.9.4}$$

$Pr(\chi_S)$ may be considered as the probability of the formation of the (random) coalition $\widetilde{S}$ according to x. Now, (8.9.1) defines a random variable $\overline{v}$ on Ω^N. By (8.9.4), (8.9.2) is precisely the expected worth of the random coalition, that is,

$$\overline{v}(x) = E(\overline{v}(\chi_S)) = E(v(\widetilde{S})).$$

Now we shall find the relationship between the Shapley value of a game and its MLE. We shall assume that $N = \{1, \ldots, n\}$.

Theorem 8.9.4. *Let $G = (N, v)$ be an n-person game and let $\overline{v}$ be its* MLE. *Then*

$$\phi^j(N,v) = \int_0^1 \frac{\partial}{\partial x^j}\overline{v}(t,\dots,t)dt \quad \text{for all } j \in N. \tag{8.9.5}$$

Proof: By (8.9.2),

$$\begin{aligned}
&\tfrac{\partial \overline{v}}{\partial x^j}(x) \\
&= \textstyle\sum_{T\subseteq N\setminus\{j\}} \left(\prod_{i\in T} x^i \prod_{i\in(N\setminus T)\setminus\{j\}}(1-x^i)\right) v(T\cup\{j\}) \\
&\quad - \textstyle\sum_{T\subseteq N\setminus\{j\}} \left(\prod_{i\in T} x^i \prod_{i\in(N\setminus T)\setminus\{j\}}(1-x^i)\right) v(T) \\
&= \textstyle\sum_{T\subseteq N\setminus\{j\}} \left(\prod_{i\in T} x^i \prod_{i\in(N\setminus T)\setminus\{j\}}(1-x^i)\right) (v(T\cup\{j\}) - v(T)).
\end{aligned}$$

Hence, if $x = (t, \dots, t)$, then

$$\frac{\partial \overline{v}}{\partial x^j}(x) = \sum_{T\subseteq N\setminus\{j\}} t^{|T|}(1-t)^{n-|T|-1}(v(T\cup\{j\}) - v(T)). \tag{8.9.6}$$

Integrating (8.9.6) yields

$$\begin{aligned}
&\int_0^1 \frac{\partial \overline{v}}{\partial x^j}(t,\dots,t)dt \\
&= \sum_{T\subseteq N\setminus\{j\}} \int_0^1 t^{|T|}(1-t)^{n-|T|-1}(v(T\cup\{j\}) - v(T))dt \\
&= \sum_{T\subseteq N\setminus\{j\}} \frac{|T|!(n-|T|-1)!}{n!}(v(T\cup\{j\}) - v(T)) \\
&= \phi^j(N,v)
\end{aligned}$$

(see Exercise 8.9.1 which proves the second equality). **q.e.d.**

Example 8.9.5. Let $G \widehat{=} (3; 2, 1, 1)$ (see Example 8.9.3). Then

$$\frac{\partial \overline{v}}{\partial x^1}(x) = \frac{\partial}{\partial x^1}(x^1x^2 + x^1x^3 - x^1x^2x^3) = x^2 + x^3 - x^2x^3.$$

Hence $(\partial v/\partial x^1)(t,t,t) = 2t - t^2$ and

$$\phi^1(G) = \int_0^1 (2t - t^2)dt = [t^2 - \frac{t^3}{3}]_0^1 = \frac{2}{3}.$$

By PO and ETP, $\phi^2(G) = \phi^3(G) = 1/6$.

Remark 8.9.6. Formulae (8.9.5) and (8.9.6) are very useful in the computation of the Shapley value of large games (see Owen (1975) and (1972)).

Exercises

Exercise 8.9.1. Prove that

$$\int_0^1 t^k(1-t)^{n-k-1}dt = \frac{k!(n-k-1)!}{n!} \text{ for all } k = 0, \ldots, n-1.$$

Let (N, v) be a game. The **Banzhaf (power) index** $\eta(N, v) \in \mathbb{R}^N$ of (N, v), introduced by Banzhaf (1965) (see also Coleman (1971)), is defined by

$$\eta^i(N, v) = \frac{1}{2^{|N|-1}} \sum_{S \subseteq N \setminus \{i\}} v(S \cup \{i\}) - v(S) \text{ for all } i \in N.$$

Exercise 8.9.2. Let (N, v) be a game. Show that

$$\eta^i(N, v) = \frac{\partial \overline{v}}{\partial x^i}\left(\frac{1}{2}, \ldots, \frac{1}{2}\right) \text{ for all } i \in N.$$

8.10 Notes and Comments

The Shapley value is extensively studied in the literature and has many extensions and applications. We shall mention only some of them.

The Shapley value is proposed as a predominant solution to cost-sharing problems (see, e.g., Straffin and Heaney (1981), Suzuki and Nakayama (1976), and Littlechild and Owen (1973)), when a game-theoretic approach is presented. Note that in the case of airport games (see Example 2.2.4), the nucleolus and the modiclus as well yield reasonable cost-shares which may easily be computed recursively (see Potters and Sudhölter (1999)).

The Shapley value may be extended to TU games with a countable set of players. Two different approaches are proposed by Shapley (1962b) and Artstein (1971) on the one hand and by Pallaschke and Rosenmüller (1997) on the other.

Moreover, Aumann and Shapley (1974) extended the Shapley value to TU games with a continuum of players. In many papers (see, e.g., Billera, Heath, and Raanan (1978)) the solution is then proposed as a cost-sharing rule known as Aumann-Shapley pricing. Characterizations of Aumann-Shapley pricing by simple properties that allow for an economic interpretation are contained in Billera and Heath (1982) and Mirman and Tauman (1982).

For further discussions and applications of the Shapley value see Winter (2002).

8.11 A Summary of Some Properties of the Main Solutions

Table 8.11.1 presents a summary of some results on axiomatizations of the most important TU game solutions, the core and its relatives, the prekernel and prenucleolus, and the Shapley value.

Indeed, "+" means that the axiom is satisfied on suitable classes of games, "−" means that the axiom is not satisfied in general, and "⊕" means that the axiom occurs in at least one axiomatization of the corresponding solution. In the context of the core the axiomatizations are stated in Theorems 3.6.1 and 3.7.1 and in the final sentence of (2) in Section 3.9, whereas for the prekernel and the prenucleolus, Theorems 5.4.2, 6.3.1, and Remark 6.3.4 have to be mentioned. Finally, concerning the Shapley value we refer to Theorems 8.1.3, 8.1.7, and 8.2.6.

Note that the table is not exhaustive, neither in regard to the collection of properties nor in regard to the collection of properties employed in some axiomatizations.

Table 8.11.1. Solutions and Properties

Axioms \ Solutions	Core	Prekernel	Prenucleolus	Shapley value
Pareto optimality	+	⊕	+	⊕
Covariance	⊕	⊕	⊕	⊕
Anonymity	⊕	+	⊕	+
Equal treatment property	−	⊕	⊕	⊕
Individual rationality	⊕	−	−	−
Reasonableness	⊕	+	+	+
Additivity	−	−	−	⊕
Superadditivity	⊕	−	−	+
Reduced game property (RGP)	⊕	⊕	⊕	−
Reconfirmation property	⊕	−	+	−
Converse RGP	⊕	⊕	−	−
Aggregate monotonicity	+	−	−	+
Strong monotonicity	−	−	−	⊕
Null player property	+	+	+	⊕
HM consistency	−	−	−	⊕

+: Axiom is satisfied on suitable sets of games.
⊕: Axiom is used in at least one axiomatization.
−: Axiom is not satisfied in general.

9

Continuity Properties of Solutions

In this chapter we shall define and investigate continuity properties of the core, the unconstrained bargaining set, the prekernel, and the prenucleolus. The Shapley value, which is a linear function of the coalition function, is obviously continuous. We consider "minimal" continuity as a necessary "technical" condition for a solution to be acceptable. By "minimal" continuity we mean upper hemicontinuity (see Definition 9.1.5) plus closed-valuedness. Fortunately, all the foregoing solutions are "minimally" continuous. Furthermore, the core and the prenucleolus are actually continuous. (A set-valued solution is continuous if it is both upper hemicontinuous and lower hemicontinuous (see Definition 9.2.1).)

9.1 Upper Hemicontinuity of Solutions

Let X and Y be two metric spaces. A *set-valued* function from X to Y is a mapping φ that assigns to each $x \in X$ a **nonempty** subset $\varphi(x)$ of Y. (Note that this definition is in contrast to the definition of a set-valued solution, which may specify the **empty** set for some games under consideration.) If φ is a set-valued function from X to Y, then we abbreviate $\varphi : X \to 2^Y$ to $\varphi : X \rightrightarrows Y$.

Example 9.1.1. Let N be a (fixed) nonempty and finite set of players, let $n = |N|$, and let Γ_N be the set of all games (N, v). Then Γ_N may be identified with the set $\mathcal{V}$ (see Section 8.1) of coalition functions $v : 2^N \to \mathbb{R}$, $v(\emptyset) = 0$ and, thus, with the Euclidean space $\mathbb{R}^{2^n-1}$. The prekernel on Γ_N can be considered as a function $\mathcal{PK} : \mathbb{R}^{2^n-1} \to 2^{\mathbb{R}^N}$. By Corollary 5.1.18, $\mathcal{PK}$ is a set-valued function from $\mathbb{R}^{2^n-1}$ to $\mathbb{R}^N$.

The graph of a set-valued function is defined in the following (non-standard) way.

Definition 9.1.2. *Let $\varphi : X \rightrightarrows Y$ be a set-valued function. The* **graph** *of φ, $Gr(\varphi)$, is defined by*

$$Gr(\varphi) = \{(x, y) \mid x \in X \text{ and } y \in \varphi(x)\}.$$

Notice that $Gr(\varphi) \subseteq X \times Y$ (and that it is not a subset of $X \times 2^Y$).

Example 9.1.1 (continued). The graph of $\mathcal{PK}$ is given by

$$Gr(\mathcal{PK}) = \left\{ (v, x) \,\middle|\, \begin{array}{c} v \in \mathcal{V}, x \in X(N, v), \text{ and} \\ s_{k\ell}(x, v) = s_{\ell k}(x, v) \text{ for all } k, \ell \in N, k \neq \ell \end{array} \right\}$$

(see Definition 5.1.1).

Definition 9.1.3. *A set-valued function $\varphi : X \rightrightarrows Y$ is* **closed** *if $Gr(\varphi)$ is a closed subset of $X \times Y$.*

Example 9.1.1 (continued). The set-valued function $\mathcal{PK}$ is closed, because $s_{k\ell}(\cdot, \cdot) : \mathcal{V} \times \mathbb{R}^N \to \mathbb{R}$ is a continuous function for all $k, \ell \in N$, $k \neq \ell$.

We proceed with the following definition.

Definition 9.1.4. *A set-valued function $\varphi : X \rightrightarrows Y$ is* **bounded** *if for each compact subset B of X the* **image** *of B, $\varphi(B) = \bigcup_{x \in B} \varphi(x)$, is a bounded subset of Y.*

Example 9.1.1 (continued). The prekernel satisfies PO (by Definition 5.1.1) and REBE (see Theorem 5.5.5). Hence, if $B \subseteq \mathcal{V}$ is bounded, then $\mathcal{PK}(B)$ is bounded.

We now introduce one of the two main continuity concepts for set-valued functions.

Definition 9.1.5. *Let $\varphi : X \rightrightarrows Y$ be a set-valued function. Then φ is* **upper hemicontinuous** *(***uhc***) at $x \in X$ if for every open subset U of Y such that $U \supseteq \varphi(x)$ there exists an open subset V of X such that $x \in V$ and $\varphi(z) \subseteq U$ for every $z \in V$. Moreover, φ is* **uhc***, if it is* uhc *at each $x \in X$.*

Now we are ready for the following result.

Lemma 9.1.6. *Let X be a metric space and let $Y = \mathbb{R}^M$ for some nonempty and finite set M. If $\varphi : X \rightrightarrows Y$ is a closed and bounded set-valued function, then φ is* uhc.

Proof: Let $x \in X$ and let U be an open subset of Y such that $U \supseteq \varphi(x)$. Assume, on the contrary, that there exists no open subset V of X such that $x \in V$ and $\varphi(V) = \bigcup_{z \in V} \varphi(z) \subseteq U$. Then there exist sequences $(x_j)_{j \in \mathbb{N}}$ and $(y_j)_{j \in \mathbb{N}}$ such that $x_j \in X$, $y_j \in \varphi(x_j)$, $\lim_{i \to \infty} x_i = x$, and $y_j \in Y \setminus U$ for all $j \in \mathbb{N}$. In view of the boundedness of φ, there exists a convergent subsequence

$(y_{j_k})_{k\in\mathbb{N}}$ of $(y_j)_{j\in\mathbb{N}}$. Let $y = \lim_{k\to\infty} y_{j_k}$. Then $y \notin U$, because U is open. However, $y \in \varphi(x)$, because φ is closed, and the desired contradiction has been reached.

q.e.d.

Henceforth, N is a finite nonempty set of players and

$$\Gamma_N = \{v : 2^N \to \mathbb{R} \mid v(\emptyset) = 0\}.$$

Theorem 9.1.7. *The prekernel,* $\mathcal{PK} : \Gamma_N \rightrightarrows \mathbb{R}^N$, *is* uhc.

Theorem 9.1.7 is implied by the discussion in Example 9.1.1 and Lemma 9.1.6.

We shall now prove that the unconstrained bargaining set (see Definition 4.1.4) is uhc.

Theorem 9.1.8. *The unconstrained bargaining set,* $\mathcal{PM} : \Gamma_N \rightrightarrows \mathbb{R}^N$, *is* uhc.

Proof: We shall first prove that $\mathcal{PM}$ is a closed set-valued function. Let $v \in \Gamma_N$, $x \in X(v)$, and $k, \ell \in N$, $k \neq \ell$. If k has a justified objection against ℓ at x with respect to v, then we shall write $k \succ^{\mathcal{M}}_{(x,v)} \ell$ (see Definition 4.2.1). A careful examination of the proof of Lemma 4.2.2 shows that the set $\{(v, x) \mid v \in \Gamma_N, x \in X(v),$ and $k \succ^{\mathcal{M}}_{(x,v)} \ell\}$ is an open subset of the set $\{(v, x) \mid v \in \Gamma_N$ and $x \in X(v)\}$. Hence the graph of $\mathcal{PM}$,

$$Gr(\mathcal{PM}) = \{(v, x) \mid v \in \Gamma_N, x \in X(v) \text{ and } k \preceq^{\mathcal{M}}_{(x,v)} \ell \ \forall k, \ell \in N, k \neq \ell\}$$

(where $\preceq^{\mathcal{M}}_{(x,v)}$ is the negation of $\succ^{\mathcal{M}}_{(x,v)}$), is closed.

We shall now prove that $\mathcal{PM}$ is bounded. Let $B \subseteq \Gamma_N$ be a bounded set. As $\mathcal{PM}$ satisfies COV, we may assume that each $v \in B$ is monotonic and strictly positive (i.e., $v(S) > 0$ for all $\emptyset \neq S \subseteq N$; see Exercise 2.1.4). If $v \in B$ and $x \in \mathcal{PM}(v)$, then $x \in X(v)$. Hence it is sufficient to prove that, if $v \in B$ and $x \in \mathcal{PM}(v)$, then $x^i \leq v(N)$ for all $i \in N$. Assume, on the contrary, that $v \in B$, $x \in \mathcal{PM}(v)$, and $x^\ell > v(N)$ for some $\ell \in N$. Let $P = \{i \in N \mid x^i < 0\}$. Then $P \neq \emptyset$ and $\ell \notin P$, because $x^\ell > v(N) = x(N) > 0$. Let $y \in \mathbb{R}^P$ be given by $y^i = v(P)/|P|$ for all $i \in P$ and let $k \in P$. We claim that (P, y) is a justified objection of k against ℓ at x. Indeed, let $Q \in \mathcal{T}_{\ell k}$. Then

$$v(Q) - y(Q \cap P) - x(Q \setminus P) \leq v(Q) - x^\ell < v(Q) - v(N) \leq 0;$$

hence ℓ cannot counterobject (P, y) via Q.

q.e.d.

We conclude this section with the proof of upper hemicontinuity of the core. Let

$$\Gamma_N^{\mathcal{C}} = \{v \in \Gamma_N \mid \mathcal{C}(v) \neq 0\}.$$

Theorem 9.1.9. *The core correspondence* $\mathcal{C} : \Gamma_N^{\mathcal{C}} \rightrightarrows \mathbb{R}^N$ *is* uhc.

Proof: The graph of $\mathcal{C}$,

$$Gr(\mathcal{C}) = \{(v, x) \in \Gamma_N^{\mathcal{C}} \times \mathbb{R}^N \mid x(N) = v(N) \text{ and } x(S) \geq v(S)\ \forall S \subseteq N\},$$

is a closed subset of $\Gamma_N^{\mathcal{C}} \times \mathbb{R}^N$. The core correspondence is also bounded by PO and IR. Thus, the core is uhc by Lemma 9.1.6. **q.e.d.**

Exercises

Exercise 9.1.1. Give an example of a closed set-valued function that is not uhc.

Exercise 9.1.2. Let X and Y be two metric spaces and let $\varphi : X \rightrightarrows Y$ be a set-valued function.

(1) If Y is compact and φ is closed, then φ is uhc.

(2) If φ is uhc and *closed-valued* (that is, $\varphi(x)$ is a closed subset of Y for every $x \in X$), then φ is closed.

Exercise 9.1.3. Let $v \in \Gamma_N$ and $\varepsilon_0(v) = \min_{x \in X(v)} \max_{\emptyset \neq S \subsetneq N} e(S, x, v)$ (see (7.1.1)). Prove that $\varepsilon_0(\cdot) : \Gamma_N \to \mathbb{R}$ is continuous. (Hint: Show that there exists $\widehat{v} \in \Gamma_N$ which is strategically equivalent to v, such that $\varepsilon_0(\widehat{v})$ is the value of the matrix game, the matrix, A, of which is defined by $A = (A_{Sj})_{\substack{\emptyset \neq S \subsetneq N \\ j \in N}}$, where each row $a_{S\cdot}$ is given by $a_{S\cdot} = \widehat{v}(S)\chi_N - \chi_S$.)

Exercise 9.1.4. Use Exercise 9.1.3 to prove that the least-core $\mathcal{LC} : \Gamma_N \rightrightarrows \mathbb{R}^N$ (see Definition 7.1.2) is uhc.

9.2 Lower Hemicontinuity of Solutions

We shall start with the definition of lower hemicontinuity of set-valued functions. Later in this section we shall prove that the core is lower hemicontinuous, whereas the prekernel is not lower hemicontinuous. The discussion of the lower hemicontinuity of the unconstrained bargaining set and the least-core will be left to the reader as an exercise.

Let X and Y be metric spaces.

Definition 9.2.1. *Let $\varphi : X \rightrightarrows Y$ be a set-valued function. Then φ is* **lower hemicontinuous** *(***lhc***) at $x \in X$ if for every open set $U \subseteq Y$ such that $\varphi(x) \cap U \neq \emptyset$, there exists an open set $V \subseteq X$ such that $x \in V$ and $\varphi(z) \cap U \neq \emptyset$ for every $z \in V$. Moreover, φ is* **lhc**, *if it is* lhc *at each $x \in X$.*

The following lemma is very helpful for the analysis of lower hemicontinuity.

Lemma 9.2.2. *Let X be a convex polyhedral set in $\mathbb{R}^n$, $X \neq \emptyset, \mathbb{R}^n$, and let Y be a bounded set in $\mathbb{R}^m$ ($m, n \in \mathbb{N}$). If $\varphi : X \rightrightarrows Y$ is a set-valued function such that the graph of φ, $Gr(\varphi)$, is convex, then φ is* lhc.

Proof: Let $x_0 \in X$ and let U be an open subset of Y such that

$$U \cap \varphi(x_0) \neq \emptyset.$$

Let $y_0 \in U \cap \varphi(x_0)$. Then, for some $0 < \delta < 1$,

$$U \supseteq \{y \in Y \mid \|y - y_0\| < \delta\}$$

($\|\cdot\|$ denotes the Euclidean norm). As Y is bounded, for some $M > 1$,

$$\|y - y_0\| < M \text{ for all } y \in Y.$$

Define $f : X \to \mathbb{R}^n$ by

$$f(x) = \frac{M}{\delta}x + \left(1 - \frac{M}{\delta}\right) x_0 \text{ for all } x \in X.$$

Claim: There exists an open set $V \subseteq X$ such that $x_0 \in V$ and $f(x) \in X$ for all $x \in V$.

As X is a convex polyhedral subset of $\mathbb{R}^n$, it has a representation of the form $X = \{x \in \mathbb{R}^n \mid Ax \leq b\}$, where A is a $k \times n$ matrix and $b \in \mathbb{R}^k$. As $X \neq \emptyset, \mathbb{R}^n$, $A \neq 0$. Let $\|A\| = \max\{\|Ax\| \mid x \in \mathbb{R}^n, \|x\| \leq 1\}$. Then $\|A\| > 0$. Now define $\varepsilon > 0$ by distinguishing the following cases. If $Ax_0 = b$, let $\varepsilon = 1$. Otherwise let

$$\varepsilon = \frac{\delta}{M\|A\|} \min\{b^j - (Ax_0)^j \mid j \in \{1, \ldots, k\}, b^j - (Ax_0)^j > 0\}.$$

Define $V = \{x \in X \mid \|x - x_0\| < \varepsilon\}$ and let $x \in V$. Then

$$Af(x) = Ax_0 + \frac{M}{\delta}A(x - x_0).$$

Thus, if $(Ax_0)^j = b^j$, then $(Af(x))^j \leq b^j$, because $(Ax)^j \leq b^j$. If $(Ax_0)^j < b^j$, then $b^j - (Ax_0)^j \geq \frac{\varepsilon M \|A\|}{\delta}$ and, hence,

$$(Af(x))^j = (Ax_0)^j + \frac{M}{\delta}(A(x - x_0))^j \leq b^j - \frac{\varepsilon M\|A\|}{\delta} + \frac{M}{\delta}\|A\|\|x - x_0\| \leq b^j.$$

Thus, the claim has been proved.

Now the proof can be completed. Let $x \in V$. We shall show that $\varphi(x) \cap U \neq \emptyset$. By our claim, $f(x) \in X$. Also,

$$x = \frac{\delta}{M}f(x) + \left(1 - \frac{\delta}{M}\right) x_0.$$

Let $y_1 \in \varphi(f(x))$. Then, by convexity of $Gr(\varphi)$,

$$y = \frac{\delta}{M} y_1 + \left(1 - \frac{\delta}{M}\right) y_0 \in \varphi(x).$$

Also, $\|y - y_0\| = \frac{\delta}{M}\|y_1 - y_0\| < \delta$, and therefore $y \in U$. **q.e.d.**

Lemma 9.2.2 has the following immediate corollary.

Corollary 9.2.3. *Let X be a convex polyhedral subset of $\mathbb{R}^n$, $X \neq \emptyset$, and let $Y \subseteq \mathbb{R}^m$. If $\varphi : X \rightrightarrows Y$ is a bounded set-valued function with a convex graph, then φ is* lhc.

Let N be a fixed coalition, let $\mathcal{V} = \mathcal{V}(N) = \{v : 2^N \to \mathbb{R} \mid v(\emptyset) = 0\}$ and $\mathcal{V}_{\mathcal{C}} = \{v \in \mathcal{V} \mid \mathcal{C}(v) \neq \emptyset\}$.

Theorem 9.2.4. *The core correspondence $\mathcal{C} : \mathcal{V}_{\mathcal{C}} \rightrightarrows \mathbb{R}^N$ is* lhc.

Proof: The set $\mathcal{V}_{\mathcal{C}}$ is a polyhedral convex set by Theorem 3.1.10. Also, $\mathcal{C}$ is bounded and its graph is convex. By Corollary 9.2.3, $\mathcal{C}$ is lhc. **q.e.d.**

We shall now prove by an example that the prekernel is not lhc.

Example 9.2.5. Let $0 \leq \varepsilon < 1$ and $N = \{1, 2, 3, 4, 5\}$. Define $v_\varepsilon \in \mathcal{V}(N)$ by

$$v_\varepsilon(S) = \begin{cases} 1 & \text{, if } |S| \geq 4 \text{ or } S \in \left\{ \begin{matrix} \{1,2,4\}, \{1,3,4\}, \{2,3,4\}, \\ \{1,2,5\}, \{2,3,5\} \end{matrix} \right\}, \\ 1 - \varepsilon & \text{, if } S = \{1,3,5\}, \\ 0 & \text{, otherwise.} \end{cases}$$

Let $\varepsilon > 0$ and let $x \in \mathcal{PK}(N, v_\varepsilon)$. By Theorem 5.3.5 the prekernel preserves the desirability relation. Hence, $x^1 = x^3$, $x^2 \geq x^1$, and $x^4 \geq x^5$. By 0-monotonicity of the game Theorem 5.5.6 yields $x \geq 0$. We now claim that $x^2 > 0$. Assume, on the contrary, that $x^2 = 0$. Then $x^1 = x^3 = 0$. Also, $s_{45}(x, v_\varepsilon)$ and $s_{54}(x, v_\varepsilon)$ are attained by $\{1,2,4\}$ and $\{1,2,5\}$, respectively. Hence $x^4 = x^5 = 1/2$ by Pareto optimality. A contradiction is derived by the observation that

$$s_{25}(x, v_\varepsilon) = e(\{1,2,4\}, x, v_\varepsilon) > e(\{1,3,5\}, x, v_\varepsilon) = s_{52}(x, v_\varepsilon).$$

Hence $x^2 > 0$. Let $\mu = \max_{S \subseteq N} e(S, x, v_\varepsilon)$. Then $\mu > 0$. Let

$$\mathcal{D}(x) = \{S \subseteq N \mid e(S, x, v_\varepsilon) = \mu\}.$$

By Lemma 5.5.3, $\mathcal{D}(x)$ is separating. Thus, for every $i \in N$, there exists $S \in \mathcal{D}(x)$ satisfying $i \in S$. Applying this fact to $i = 2$, we conclude that $\{1,2,5\}, \{2,3,5\} \in \mathcal{D}(x)$. As $\mu = s_{54}(x, v_\varepsilon) = s_{45}(x, v_\varepsilon)$, $\{1,3,4\} \in \mathcal{D}(x)$. Similarly, by $\mu = s_{14}(x, v_\varepsilon) = s_{41}(x, v_\varepsilon)$, we conclude that $\{2,3,4\} \in \mathcal{D}(x)$. Hence $x^1 = x^2 = x^3$ and $x^4 = x^5$. Thus, $\{1,3,5\} \notin \mathcal{D}(x)$. As $\mu = s_{25}(x, v_\varepsilon) = s_{52}(x, v_\varepsilon)$, $\{1,3,4,5\} \in \mathcal{D}(x)$. Hence $x = (\frac{1}{3}, \frac{1}{3}, \frac{1}{3}, 0, 0)$. On the other hand, it can easily be verified that $(0,0,0,\frac{1}{2},\frac{1}{2}) \in \mathcal{PK}(N,v_0)$. (In fact it can be shown that

$$\mathcal{PK}(N,v_0) = \{(\alpha,\alpha,\alpha,\beta,\beta) \mid 3\alpha + 2\beta = 1, \alpha,\beta \geq 0\}.)$$

Hence $\mathcal{PK} : \mathcal{V}(N) \rightrightarrows \mathbb{R}^N$ is not lhc at v_0.

Remark 9.2.6. Let v_ε, $0 \leq \varepsilon < 1$, be defined as in Example 9.2.5. By Theorem 5.5.10, $\mathcal{K}(N,v_\varepsilon) = \mathcal{PK}(N,v_\varepsilon)$; hence the kernel, when restricted to 0-monotonic games, is not lhc at v_0. The same example may be used to show that the prebargaining set is not lhc at v_0. Indeed, $\mathcal{PM}(v_\varepsilon) = \mathcal{PK}(v_\varepsilon)$ for all $0 \leq \varepsilon < 1$ (see Exercise 9.2.3).

Exercises

Exercise 9.2.1. Let X and Y be metric spaces, let $f : X \to X$ be a continuous mapping, and let $\varphi : X \rightrightarrows Y$ be a set-valued function. Show that if φ is lhc (uhc), then the composition $\varphi \circ f$ is lhc (uhc).

Exercise 9.2.2. Prove that the least-core $\mathcal{LC} : \mathcal{V}(N) \rightrightarrows \mathbb{R}^N$ is lhc. (Hint: Use Exercise 9.1.3 and Theorem 9.2.4.)

Exercise 9.2.3. Prove that $\mathcal{PK}(N,v_\varepsilon) = \mathcal{PM}(N,v_\varepsilon)$ for all $0 \leq \varepsilon < 1$, where (N,v_ε) is defined as in Example 9.2.5 (see Stearns (1965)).

Exercise 9.2.4. Let $X = \{x \in \mathbb{R}^2 \mid \|x\| \leq 1\}$ and let $Y = [0,1]$. Give an example of a set-valued function $\varphi : X \rightrightarrows Y$ with a convex and closed graph that is not lhc. (Hint: It suffices to describe a suitable graph.)

9.3 Continuity of the Prenucleolus

Let N be a fixed coalition and let $\mathcal{V}(N) = \{v : 2^N \to \mathbb{R} \mid v(\emptyset) = 0\}$. We shall now prove that the prenucleolus is a continuous function on $\mathcal{V}(N)$. Let $v \in \mathcal{V}(N)$. Recall that the unique member of $\mathcal{PN}(N,v)$ is denoted by $\nu(v)$.

Theorem 9.3.1. *The prenucleolus $\nu : \mathcal{V}(N) \to \mathbb{R}^N$ is continuous.*

Proof: Let $v_k \in \mathcal{V}(N)$, $k \in \mathbb{N}$, such that $\lim_{k\to\infty} v_k = v$. For each $k \in \mathbb{N}$ let $x_k = \nu(v_k)$. Then the sequence $(x_k)_{k\in\mathbb{N}}$ is bounded, because $x_k \in \mathcal{PK}(N,v_k)$ for each $k \in \mathbb{N}$. Let $(x_{k_j})_j \in \mathbb{N}$ be a convergent subsequence and let $x = \lim_{j\to\infty} x_{k_j}$. We have to prove that $x = \nu(v)$. Clearly, $x \in X(N,v)$. Now let $\alpha \in \mathbb{R}$ such that $\mathcal{D}(\alpha,x,v) \neq \emptyset$ (see Definition 5.2.5). We shall show that $\mathcal{D}(\alpha,x,v)$ is balanced. Choose $\varepsilon > 0$ such that $\mathcal{D}(\alpha,x,v) = \mathcal{D}(\alpha - 2\varepsilon,x,v)$. As $\lim_{j\to\infty} x_{k_j} = x$ and $\lim_{j\to\infty} v_{k_j} = v$, there exists $J \in \mathbb{N}$ such that $e(S,x,v) - \varepsilon < e(S,x_{k_j},v_{k_j}) < e(S,x,v) + \varepsilon$ for each $S \subseteq N$ and all $j > J$. Therefore $\mathcal{D}(\alpha,x,v) = \mathcal{D}(\alpha - \varepsilon, x_{k_j}, v_{k_j})$ for all $j > J$. Now $x_k = \nu(v_k)$.

Thus, by Theorem 6.1.1, $\mathcal{D}(\alpha, x, v)$ is balanced. The same theorem implies that $x = \nu(v)$. **q.e.d.**

Exercises

Exercise 9.3.1. Use Theorem 6.6.8 to show that the modiclus $\psi : \mathcal{V}(N) \rightrightarrows \mathbb{R}^N$ is a continuous mapping.

Exercise 9.3.2. Show that the positive core $\mathcal{C}_+ : \mathcal{V}(N) \rightrightarrows \mathbb{R}^N$ (see Definition 6.3.11) is continuous, that is, both uhc and lhc. (Hint: Let (N, v) be a game and let $\nu = \nu(N, v)$. Show that $\mathcal{C}_+(N, v)$ is the core of the minimum of the games (N, v) and (N, ν).)

9.4 Notes and Comments

There are several papers which are related to this chapter. Example 9.2.5 and Exercise 9.2.3 are due to Stearns (1965) (see also Example 1 of Stearns (1968)). The continuity of the (pre)nucleolus is due to Schmeidler (1969) (see also Kohlberg (1971)). Finally, the continuity of the least-core was first discussed in Lucchetti, Patrone, Tijs, and Torre (1987).

10

Dynamic Bargaining Procedures for the Kernel and the Bargaining Set

Up to now we have only considered the merits of various solutions. We have not explicitly asked how the players may arrive at (a point of) a particular solution of a given game. One possibility is that the players meet and discuss which solution to adopt. If a single-valued solution is chosen, then it is implemented. However, if a multi-valued solution is adopted, then an additional round of negotiation is necessary in order to completely determine the final payoff distribution.

In this chapter we consider a more realistic model. The players start at some initial preimputation and proceed to make bilateral demands at each period of time $t = 0, 1, \ldots$. More precisely, at time t some pair $(i(t), j(t))$ of players such that $i(t)$ has a justified objection against $j(t)$ (with respect to a given bargaining set), agree upon a transfer from $j(t)$ to $i(t)$ that (partially) meets $i(t)$'s demand. If there is no justified objection at time t, then the process stops. This procedure is repeated ad infinitum. We provide sufficient conditions for such dynamic processes to converge to the underlying bargaining set.

We now review the contents of this chapter. In Section 10.1 we define a dynamic system $\varphi_{\mathcal{K}}$ which leads to the kernel of a game. The dynamic subsystems of $\varphi_{\mathcal{K}}$ lead to many known bargaining sets. The study of stable sets and stable points of $\varphi_{\mathcal{K}}$ and its subsystems is the topic of Section 10.2. Section 10.3 is devoted to the analysis of (local) asymptotic stability. It is shown that the nucleolus of a game is an asymptotically stable point of an lhc subsystem φ of $\varphi_{\mathcal{K}}$ iff it is an isolated critical point of φ. No other preimputation can be asymptotically stable.

10.1 Dynamic Systems for the Kernel and the Bargaining Set

Let X be a metric space. A *dynamic system* on X is a set-valued function $\varphi : X \rightrightarrows X$. A *$\varphi$-sequence from* $x_0 \in X$ is a sequence $(x_t)_{t\in\mathbb{N}_0}$ such that $x_{t+1} \in \varphi(x_t)$ for all $t = 0, 1, \ldots$. A point $x \in X$ is called an *endpoint* of φ if $\varphi(x) = \{x\}$. (Endpoints of φ are also called *critical points* or *rest-points.*) The set of all endpoints of φ is denoted by E_φ.

Example 10.1.1. Let (N, v) be a 0-1-normalized game (that is, $v(N) = 1$, $v(\{i\}) = 0$ for all $i \in N$; hence $|N| \geq 2$ is implicitly assumed), and let

$$X = I(N, v) = \{x \in \mathbb{R}^N \mid x(N) = 1, x^i \geq 0 \text{ for all } i \in N\}.$$

For $i, j \in N,\ i \neq j$, and $x \in X$ let

$$k_{ij} = \left(\min\left\{x^j, \frac{1}{2}\Big(s_{ij}(x, v) - s_{ji}(x, v)\Big)\right\}\right)_+ .$$

We now define a dynamic system $\varphi_{\mathcal{K}}$ for the kernel of (N, v) by the following rule. Let $x \in X$. Then $y \in \varphi_{\mathcal{K}}(x)$ iff there exist $i, j \in N$, $i \neq j$, and $0 \leq \alpha \leq k_{ij}(x)$ such that

$$y = x + \alpha\left(\chi_{\{i\}} - \chi_{\{j\}}\right) \tag{10.1.1}$$

(that is, $y^i = x^i + \alpha$, $y^j = x^j - \alpha$, and $y^\ell = x^\ell$ for all $\ell \in N \setminus \{i, j\}$).

Remark 10.1.2. Let (N, v), X and $\varphi_{\mathcal{K}}$ be defined as in Example 10.1.1. Note that $x \in \varphi_{\mathcal{K}}(x)$ for all $x \in X$. Also, $\varphi_{\mathcal{K}}(x) \subseteq X$ for all $x \in X$ and $E_{\varphi_{\mathcal{K}}} = \mathcal{K}(N, v)$. Furthermore, $\varphi_{\mathcal{K}}$ is lhc (by the continuity of the $s_{ij}(\cdot, v) : X \to \mathbb{R}$ the proof is straightforward).

Let X be a metric space, let $\varphi : X \rightrightarrows X$ be a dynamic system, and let $d(\cdot,\cdot) : X \times X \to \mathbb{R}$ be a metric for X. It is convenient at this point to introduce the generalized real-valued function $f : X \to \mathbb{R}_+ \cup \{\infty\}$ defined by

$$f(x) = \sup\left\{\sum_{t=0}^{\infty} d(x_t, x_{t+1}) \,\middle|\, (x_t)_{t\in N_0} \text{ is a } \varphi\text{-sequence and } x_0 = x\right\}. \tag{10.1.2}$$

Thus, $f(x)$ is the supremum of the lengths of all trajectories (i.e., φ-sequences) that start at x.

Remark 10.1.3. If X is complete, $x \in X$, and $f(x) < \infty$, then every φ-sequence converges (to a point in X).

The following result is very useful.

Lemma 10.1.4. *The function f is bounded on X iff there exists a bounded real-valued function $\Psi : X \to \mathbb{R}$ such that*

$$y \in \varphi(x) \Rightarrow \Psi(x) - \Psi(y) \geq d(x, y) \text{ for all } x, y \in X. \tag{10.1.3}$$

Proof: Sufficiency. Let $\Psi : X \to \mathbb{R}$ satisfy (10.1.3), let $x_0 \in X$, and let $(x_t)_{t \in \mathbb{N}_0}$ be a φ-sequence from x_0. Then, for each $T \in \mathbb{N}$,

$$\sum_{t=0}^{T-1} d(x_t, x_{t+1}) \leq \sum_{t=0}^{T-1} (\Psi(x_t) - \Psi(x_{t+1})) = \Psi(x_0) - \Psi(x_T) \leq 2M,$$

where M is a bound of Ψ (i.e., $|\Psi(x)| \leq M$ for all $x \in X$). Thus $f(x_0) \leq 2M$.

Necessity. If f is bounded, we may choose $\Psi = f$. **q.e.d.**

Definition 10.1.5. *A* **valuation** *for φ is a continuous function $\Psi : X \to \mathbb{R}$ that satisfies (10.1.3).*

Remark 10.1.6. If φ has a valuation and X is compact, then f is bounded.

A function $g : X \to \mathbb{R} \cup \{\infty, -\infty\}$ is *lower semicontinuous* (lsc) at $x \in X$ if for each $\alpha \in \mathbb{R}$, $\alpha < g(x)$, there exists $\delta > 0$ such that $g(y) > \alpha$ for all $y \in X$ satisfying $d(y, x) < \delta$.

We now show that the function $\rho : X \to \mathbb{R} \cup \{\infty\}$, defined by

$$\rho(x) = \sup\{d(x, y) \mid y \in \varphi(x)\}, \tag{10.1.4}$$

is lsc if φ is lhc.

Lemma 10.1.7. *If φ is* lhc, *then ρ is* lsc.

Proof: Let $x \in X$ and let $\rho(x) > \alpha$. Then there exists $y \in \varphi(x)$ such that $d(x, y) > \alpha$. Choose $0 < \delta < d(x, y) - \alpha$ and denote

$$V = \{z \in X \mid d(z, y) < \delta/2\}.$$

As φ is lhc, there exists $0 < \varepsilon < \delta/2$ such that if $x^* \in X$ and $d(x, x^*) < \varepsilon$, then $\varphi(x^*) \cap V \neq \emptyset$. Now let $x^* \in X$, $d(x, x^*) < \varepsilon$, and choose any $y^* \in \varphi(x^*) \cap V$. Then

$$d(x, y) \leq d(x, x^*) + d(x^*, y^*) + d(y^*, y).$$

Hence $\rho(x^*) \geq d(x^*, y^*) > \alpha + \delta - \delta/2 - \delta/2 = \alpha$. Thus, ρ is lsc. **q.e.d.**

Let $x \in X$, $\alpha \in \mathbb{R}$, $\alpha > 0$, and $y \in \varphi(x)$. The transition from x to y is *α-maximal* if $d(x, y) \geq \alpha\rho(x)$. A φ-sequence $(x_t)_{t \in \mathbb{N}_0}$ is *maximal* if there exist $\alpha > 0$ and a subsequence $\left(x_{t_j}\right)_{j \in N}$ such that

$$d\left(x_{t_j}, x_{t_j+1}\right) \geq \alpha\rho\left(x_{t_j}\right) \text{ for all } j \in N, \tag{10.1.5}$$

that is, infinitely many α-maximal transitions occur. We are now ready to prove the following lemma.

Lemma 10.1.8. *Let φ be* lhc*, let $(x_t)_{t\in\mathbb{N}_0}$ be a maximal φ-sequence, and let $x \in X$. If $\lim_{t\to\infty} x_t = x$, then $x \in E_\varphi$.*

Proof: Assume, on the contrary, that $\rho(x) > 0$. Choose $0 < \alpha < 1$ and a subsequence $\left(x_{t_j}\right)_{j\in\mathbb{N}}$ satisfying (10.1.5). By Lemma 10.1.7 there exists $J \in \mathbb{N}$ such that $\rho(x_{t_j}) \geq \frac{\rho(x)}{2}$ for all $j > J$. Moreover,

$$d\left(x_{t_j}, x_{t_j+1}\right) \geq \alpha\rho(x_{t_j}) \geq \alpha\frac{\rho(x)}{2} \text{ for all } j > J,$$

which is the desired contradiction. **q.e.d.**

Corollary 10.1.9. *Let X be a compact metric space and let $\varphi : X \rightrightarrows X$ be an* lhc *dynamic system. If φ has a valuation, then*

(1) every φ-sequence converges to some point in X;

(2) every maximal φ-sequence converges to an endpoint of φ.

The following theorem shows that Corollary 10.1.9 may be applied to the dynamic system $\varphi_{\mathcal{K}}$ defined in Example 10.1.1.

Theorem 10.1.10. *Let (N, v) be a 0-1-normalized game. Then the dynamic system for $\mathcal{K}(N, v)$, $\varphi_{\mathcal{K}}$, has a valuation.*

We postpone the proof of Theorem 10.1.10 and shall now introduce some notation, which will also be used in subsequent sections. We start with the following simple remark.

Remark 10.1.11. Let $q \in \mathbb{N}$ and let $a, b \in \mathbb{R}^q$ such that $a^1 \geq \cdots \geq a^q \geq 0$ and $b^1 \geq \cdots \geq b^q$. Then, for every $1 \leq p \leq q$ and every permutation π of $\{1, \ldots, q\}$,

$$\sum_{i=1}^{p} a^i b^i \geq \sum_{i=1}^{p} a^i b^{\pi(i)}.$$

Definition 10.1.12. *Let X be a metric space, let $\varphi : X \rightrightarrows X$ be a dynamic system, and let $G = (G^1, \ldots, G^m) : X \to \mathbb{R}^m$. The function G is* **φ-monotone** *if*

$$\Big(x \in X \text{ and } y \in \varphi(x)\Big) \Rightarrow G^i(x) \geq G^i(y)\ \forall\, i = 1, \ldots, m.$$

Proof of Theorem 10.1.10: For $x \in X = I(N, v)$ let

$$\theta(x) = \left(\theta^1(x), \ldots, \theta^{2^n}(x)\right) \in \mathbb{R}^{2^n}$$

be the vector of excesses, $e(S, x, v)$, $S \subseteq N$, arranged in a non-increasing order (where $n = |N|$). Define $\Psi : X \to \mathbb{R}^{2^n}$ by the formula

$$\Psi^k(x) = \sum_{\ell=1}^{k} 2^{k-\ell}\theta^\ell(x). \tag{10.1.6}$$

We shall prove that Ψ is $\varphi_{\mathcal{K}}$-monotone and deduce that Ψ^{2^n} is a valuation.

Let $x \in X$ and let $y \in \varphi_{\mathcal{K}}(x)$. Then there exist $i, j \in N$, $i \neq j$, and $0 \leq \alpha \leq k_{ij}(x)$ such that (10.1.1) is satisfied. The case $\alpha = 0$ is trivial. Thus, we assume that $\alpha > 0$. Then $\alpha \leq \frac{1}{2}(s_{ij}(x,v) - s_{ji}(x,v))$. Let $\{S_1, \ldots, S_{2^n}\} = 2^N$ be such that $\theta(y) = (e(S_1, y, v), \ldots, e(S_{2^n}, x, v))$ and let $s_{ij}(x,v) = e(S_p, x, v)$. We may assume that $e(S_q, y, v) > e(S_p, y, v)$ for all $1 \leq q < p$. As $s_{ij}(y,v) = e(S_p, y, v) \geq s_{ji}(y,v)$, $S_q \notin \mathcal{T}_{ij} \cup \mathcal{T}_{ji}$ for $1 \leq q < p$. Hence $e(S_q, x, v) = e(S_q, y, v)$ for all $1 \leq q < p$. By Remark 10.1.11,

$$\Psi^q(x) \geq \sum_{\ell=1}^{q} 2^{q-\ell} e(S_\ell, x, v) = \sum_{\ell=1}^{q} 2^{q-\ell} e(S_\ell, y, v) = \Psi^q(y) \text{ for } 1 \leq q < p.$$

For $q \geq p$, by the same remark,

$$\begin{aligned}
\Psi^q(x) - \Psi^q(y) &\geq \textstyle\sum_{\ell=1}^{q} 2^{q-\ell}(e(S_\ell, x, v) - e(S_\ell, y, v)) \\
&= 2^{q-p}(e(S_p, x, v) - e(S_p, y, v)) \\
&\quad + \textstyle\sum_{\ell=p+1}^{q} 2^{q-\ell}(e(S_\ell, x, v) - e(S_\ell, y, v)) \\
&\geq 2^{q-p}\alpha - \textstyle\sum_{\ell=p+1}^{q} 2^{q-\ell}\alpha = \alpha = \|x - y\|_\infty,
\end{aligned}$$

where $\|z\|_\infty = \max_{i \in N} |z^i|$ for $z \in \mathbb{R}^N$. Clearly, all component functions Ψ^q, $q = 1, \ldots, 2^n$, are continuous. Hence Ψ^{2^n} is a valuation of $\varphi_{\mathcal{K}}$. **q.e.d.**

Note that in the foregoing proof only Ψ^{2^n} is relevant. However, in the next section we shall use the fact that Ψ is $\varphi_{\mathcal{K}}$-monotone.

Let X be a metric space and let $\varphi : X \rightrightarrows X$ be a dynamic system. A set-valued function $\varphi^* : X \rightrightarrows X$ is a *subsystem* of φ, if $\varphi^*(x) \subseteq \varphi(x)$ for all $x \in X$. If Ψ is a valuation of φ and φ^* is a subsystem of φ , then Ψ is a valuation of φ^*. We shall now prove that a natural dynamic system for the bargaining set $\mathcal{M}$ is a subsystem of $\varphi_{\mathcal{K}}$.

Example 10.1.13. Let (N, v) be a 0-1-normalized game and let $X = I(N, v)$. For $i, j \in N$, $i \neq j$, let $M_{ij} = \{x \in X \mid i \preceq_x^{\mathcal{M}} j\}$, that is, M_{ij} is the set of imputations where i has no justified objection against j (see Definition 4.2.1). Let $m_{ij}(x) = \min\{\alpha \geq 0 \mid x + \alpha(\chi_{\{i\}} - \chi_{\{j\}}) \in M_{ij}\}$. Let $x \in X$. Clearly, $m_{ij}(x) \leq x^j$. Also, if $y \in X$ and $s_{ij}(y,v) \leq s_{ji}(y,v)$, then $y \in M_{ij}$. Thus, $m_{ij}(x) \leq \frac{1}{2}(s_{ij}(x,v) - s_{ji}(x,v))_+$. Hence, $m_{ij}(x) \leq k_{ij}(x)$.

Now we define the dynamic system $\varphi_{\mathcal{M}}$ for the bargaining set by the following rule. Let $x \in X$. Then $y \in \varphi_{\mathcal{M}}(x)$ iff there exist $i, j \in N$, $i \neq j$, and $0 \leq \alpha \leq m_{ij}(x)$ such that y satisfies (10.1.1). Note that $x \in \varphi_{\mathcal{M}}(x)$ and $\varphi_{\mathcal{M}}(x) \subseteq X$ for all $x \in X$. Also, $E_{\varphi_{\mathcal{M}}} = \mathcal{M}(N, v)$ and $\varphi_{\mathcal{M}}$ is lhc. As $\varphi_{\mathcal{M}}$ is a subsystem of $\varphi_{\mathcal{K}}$, $\varphi_{\mathcal{M}}$ has a valuation. Thus every $\varphi_{\mathcal{M}}$-sequence converges to an imputation and every maximal $\varphi_{\mathcal{M}}$-sequence converges to a point in the bargaining set of (N, v).

We shall now find an explicit bound to the lengths of trajectories in $\varphi_{\mathcal{K}}$.

Example 10.1.14. Let (N, v) be a 0-1-normalized game and let $X = I(N, v)$. For $x \in X$ let $\theta_*(x) \in \mathbb{R}^{n(n-1)}$ be the vector of the $n(n-1)$ maximal surpluses $s_{ij}(x, v)$, $i \in N$, $j \in N \setminus \{i\}$, arranged in a non-increasing order. Define $\Psi_* : X \to \mathbb{R}^{n(n-1)}$ by

$$\Psi_*^k(x) = \sum_{\ell=1}^{k} 2^{k-\ell}\theta_*^\ell(x) \text{ for all } 1 \le k \le n(n-1).$$

Then Ψ_* is $\varphi_{\mathcal{K}}$-monotone and $\Psi_*^{n(n-1)}$ is a valuation of $\varphi_{\mathcal{K}}$. (The proof of this fact is the same as the proof of Theorem 10.1.10.) We shall now find upper and lower bounds for $\Psi_*^{n(n-1)}$. Let $i, j \in N$, $i \ne j$. Denote $v_{ij} = \sup\{v(S) \mid S \in \mathcal{T}_{ij}\}$. Then $v_{ij} - 1 \le s_{ij}(x, v) \le v_{ij}$ for all $x \in X$. Let $b \in \mathbb{R}^{n(n-1)}$ be the vector of the v_{ij}, $i \in N$, $j \in N \setminus \{i\}$, arranged in a non-increasing order, and let $b^\ell = v_{i_\ell j_\ell}$, $\ell = 1, \ldots, n(n-1)$. Then, for all $x \in X$, Remark 10.1.11 yields

$$\Psi_*^{n(n-1)}(x) = \textstyle\sum_{\ell=1}^{n(n-1)} 2^{n(n-1)-\ell}\theta_*^\ell(x)$$
$$\ge \textstyle\sum_{\ell=1}^{n(n-1)} 2^{n(n-1)-\ell} s_{i_\ell j_\ell}(x, v) \ge \sum_{\ell=1}^{n(n-1)} 2^{n(n-1)-\ell}(b^\ell - 1).$$

Moreover,

$$\Psi_*^{n(n-1)}(x) = \sum_{\ell=1}^{n(n-1)} 2^{n(n-1)-\ell} s_{i_{\pi(\ell)} j_{\pi(\ell)}},$$

where π is a suitable permutation of $1, \ldots, n(n-1)$. Hence, again by Remark 10.1.11,

$$\Psi_*^{n(n-1)}(x) \le \sum_{\ell=1}^{n(n-1)} 2^{n(n-1)-\ell} b^{\pi(\ell)} \le \sum_{\ell=1}^{n(n-1)} 2^{n(n-1)-\ell} b^\ell.$$

Thus, if $(x_t)_{t \in \mathbb{N}_0}$ is a $\varphi_{\mathcal{K}}$-sequence, then

$$\sum_{t=0}^{\infty} d(x_t, x_{t+1}) \le \sum_{\ell=1}^{n(n-1)} 2^{n(n-1)-\ell}(b^\ell - (b^\ell - 1)) = 2^{n(n-1)} - 1,$$

where $d(x, y) = \|x - y\|_\infty$ for all $x, y \in X$.

Exercises

Exercise 10.1.1. Let X be a metric space and let $\varphi : X \rightrightarrows X$ be an lhc set-valued function. Prove that (1) the function $f : X \to \mathbb{R} \cup \{\infty\}$ given by (10.1.2) is lsc and (2) E_φ is a closed set.

Exercise 10.1.2. Let X be a compact metric space, let $\varphi : X \rightrightarrows X$ be an lhc set-valued function, and let $U \subseteq X$, $U \supseteq E_\varphi$, be an open set. Assume further that the function f defined by (10.1.2) is bounded on X. Prove that for every $\alpha > 0$ there exists $q(\alpha) \in \mathbb{N}$ such that the number of α-maximal transitions in $X \setminus U$ of any φ-sequence is less than $q(\alpha)$.

Exercise 10.1.3. Let (N, v) be a game, let $n \geq 2$, and let $X = X(N, v)$. For $i, j \in N$, $i \neq j$, and $x \in X$ let $r_{ij}(x) = \frac{1}{2}(s_{ij}(x, v) - s_{ji}(x, v))_+$. Define the dynamic system $\varphi_{\mathcal{PK}} : X \rightrightarrows X$ as follows. Let $x \in X$. Then $y \in \varphi_{\mathcal{PK}}(x)$ iff there exist $i, j \in N$, $i \neq j$, and $0 \leq \alpha \leq r_{ij}(x)$ such that y satisfies (10.1.1). Show that
(1) $E_{\varphi_{\mathcal{PK}}} = \mathcal{PK}(N, v)$;
(2) $\varphi_{\mathcal{PK}}$ is lhc and has a valuation;
(3) the assertions of Corollary 10.1.9 are valid for $\varphi = \varphi_{\mathcal{PK}}$.

10.2 Stable Sets of the Kernel and the Bargaining Set

Let φ be a dynamic system on a metric space X.

Definition 10.2.1. *Let $Q \subseteq X$, $Q \neq \emptyset$. The set Q is* **stable** *with respect to φ if for every open set U, $X \supseteq U \supseteq Q$, there exists an open set V, $X \supseteq V \supseteq Q$, such that the following condition is satisfied: If $x_0 \in V$ and $(x_t)_{t \in \mathbb{N}_0}$ is a φ-sequence, then $x_t \in U$ for all $t \in \mathbb{N}$. A point $x \in X$ is* **stable** *if $\{x\}$ is stable.*

Remark 10.2.2. If $Q \subseteq X$ is stable, then $\varphi(Q) \subseteq Q$, that is, Q is an invariant set. If $x \in X$ is stable, then $x \in E_\varphi$.

Let $G : X \to \mathbb{R}^m$. A point $a \in \mathbb{R}^m$ is *Pareto-minimal* with respect to G if a is a Pareto-minimal point of $G(X)$, that is, if $b \leq a$ implies $b = a$ for every $b \in G(X)$.

Definition 10.2.3. *Let $A \subseteq X$. The set A is a (generalized)* **nucleolus** *of G if there exists a Pareto-minimal point a with respect to G such that $A = G^{-1}(a) = \{x \in X \mid G(x) = a\}$.*

The function $G : X \to \mathbb{R}^m$ is *strictly* φ-monotone if G is φ-monotone and satisfies, in addition, the following property for every $x \in X$:

$$\Big(y \in \varphi(x) \text{ and } y \neq x\Big) \Rightarrow G^k(x) > G^k(y) \text{ for some } k, 1 \leq k \leq m.$$

Strictly φ-monotone vectorial functions will serve as Lyapunov functions.

Remark 10.2.4. If G is strictly φ-monotone, then every nucleolus of G is a subset of E_φ.

Our first result in this section establishes, essentially, the stability of nucleoli of φ-monotone lsc vector functions.

Theorem 10.2.5. *Let X be a compact metric space, let $G : X \to \mathbb{R}^m$, and let $\varphi : X \rightrightarrows X$ be a dynamic system. Further, let $A = G^{-1}(a)$ be a nucleolus of G. If*

(1) G is φ-monotone,

(2) G^i is lsc *on X for $i = 1, \ldots, m$, and*

(3) G^i is continuous on A for $i = 1, \ldots, m$,

then A is stable with respect to φ.

Remark 10.2.6. As $A = \{x \in X \mid G^i(x) \leq a^i,\ i = 1, \ldots, m\}$, it follows from Condition (2) that A is closed.

Proof of Theorem 10.2.5: Let $U \supseteq A$, $U \subseteq X$, be an open set in X. Then $B = X \setminus U$ is compact. If $y \in B$, then there exist $1 \leq k = k(y) \leq m$ and $r = r(y) \in \mathbb{N}$ such that $G^k(y) > a^k + 1/r$. Hence $y \in U_{k,r}$, where

$$U_{k,r} = \left\{ x \in X \,\middle|\, G^k(x) > a^k + \frac{1}{r} \right\}.$$

As G^k is lsc, $U_{k,r}$ is open. As B is compact, B is covered by finitely many of the sets $U_{k(y),r(y)}$, $y \in B$. Hence, there exist $S \subseteq \{1, \ldots, m\}$ and $r \in \mathbb{N}$ such that $B \subseteq \bigcup_{k \in S} U_{k,r}$. For $k \in S$ let

$$\widetilde{V}_k = \left\{ x \in X \,\middle|\, G^k(x) < a^k + \frac{1}{r} \right\}.$$

Then $\widetilde{V} = \bigcap_{k \in S} \widetilde{V}_k \subseteq U$. As G^k is continuous on A for all $k \in S$, $\widetilde{V}_k$ contains an open set $V_k \supseteq A$. Let $V = \bigcap_{k \in S} V_k$. Then V is open and $A \subseteq V \subseteq U$. Let $x_0 \in V$ and let $(x_t)_{t \in \mathbb{N}_0}$ be a φ-sequence. By Condition (1), $G^k(x_t) < a^k + 1/r$ for all $k \in S$ and $t \in \mathbb{N}$. Hence $x_t \in U$ for all $t \in \mathbb{N}$. **q.e.d.**

Theorem 10.2.5 has the following important corollary. By Remark 5.1.12 the nucleolus $\mathcal{N}(N, v)$ of a 0-1-normalized game (N, v) is a single point, which is denoted by $\bar{\nu}(N, v)$.

Corollary 10.2.7. *Let (N, v) be a* 0-1*-normalized game and let φ be a subsystem of $\varphi_{\mathcal{K}}$. Then $\bar{\nu}(N, v)$ is a stable point of φ.*

Proof: Let $X = I(N, v)$ and $\Psi : X \to \mathbb{R}^{2^n}$ be given by (10.1.6). Then $G = \Psi$ satisfies Conditions (1)-(3) of Theorem 10.2.5 and $\mathcal{N}(N, v)$ is a generalized nucleolus of G. **q.e.d.**

We shall now discuss the Lyapunov function $G = \Psi_*$ defined in Example 10.1.14. The following definition and lemma are needed.

Definition 10.2.8. *Let X be a metric space, let $G : X \to \mathbb{R}^m$, and let $\varphi : X \rightrightarrows X$ be a dynamic system. The vectorial function G is* **strongly** *φ-monotone if it is φ-monotone and satisfies, in addition,*

$$\Big(x \in X \text{ and } y \in \varphi(x)\Big) \Rightarrow G^k(x) - G^k(y) \geq d(x,y) \text{ for some } 1 \leq k \leq m.$$

Lemma 10.2.9. *Let φ be a dynamic system on a compact metric space X, let $G : X \to \mathbb{R}^m$ be a strongly φ-monotone function, and let $A = G^{-1}(a)$ be a generalized nucleolus of G. If G satisfies (1)-(3) of Theorem 10.2.5, then each point $\xi \in A$ is a stable point of φ.*

Proof: Let $\xi \in A$ and define $\widehat{G} : X \to \mathbb{R}^{m+1}$ by $\widehat{G}^i = G^i$ for $i = 1, \ldots, m$, and

$$\widehat{G}^{m+1}(x) = \sum_{\ell=1}^{m} G^\ell(x) + d(x, \xi) \text{ for } x \in X.$$

The assumptions on G guarantee that $\widehat{G}$ satisfies (1) and (2). Define $\widehat{a} \in \mathbb{R}^{m+1}$ by $\widehat{a}^\ell = a^\ell$ for $1 \leq \ell \leq m$ and $\widehat{a}^{m+1} = \sum_{\ell=1}^m a^\ell$. Then $\widehat{a}$ is Pareto-minimal with respect to $\widehat{G}$ and $\{\xi\} = \widehat{G}^{-1}(\widehat{a})$. By Theorem 10.2.5, ξ is a stable point of φ.

q.e.d.

Now we are ready for the following example.

Example 10.2.10. Let (N, v) be a 0-1-normalized game, let $X = I(N, v)$, and let φ be a subsystem of $\varphi_{\mathcal{K}}$. Consider the Lyapunov function Ψ_* defined in Example 10.1.14. Define the *lexicographic kernel* of (N, v), $\mathcal{LK}(N, v)$, by

$$\mathcal{LK}(N, v) = \{x \in X \mid \theta_*(y) \geq_{lex} \theta_*(x) \text{ for all } y \in X\}.$$

Then $\mathcal{LK}(N, v)$ is a generalized nucleolus of Ψ_*. Hence, by Lemma 10.2.9, each point of $\mathcal{LK}(N, v)$ is a stable point of φ.

Exercises

Exercise 10.2.1. Give an example of a 0-1-normalized game (N, v) such that $\bar{\nu}(N, v) \notin \mathcal{LK}(N, v)$ (see Example 4.14 in Maschler and Peleg (1976)).

Exercise 10.2.2. Let (N, v) be a game, let $n \geq 2$, and let $\varphi_{\mathcal{PK}}$ be the dynamic system for $\mathcal{PK}(N, v)$ (see Exercise 10.1.3). Prove that every nonempty ε-core (see Definition 7.1.1) is a stable set of $\varphi_{\mathcal{PK}}$.

Exercise 10.2.3. Let (N, v) be a 0-1-normalized game, let $x \in I(N, v)$ and $y \in \varphi_{\mathcal{K}}(x)$. Prove that (1) $\theta(y) \leq_{lex} \theta(x)$ and (2) $\theta_*(y) \leq_{lex} \theta_*(x)$ (see Example 10.1.14).

Exercise 10.2.4. Let (N, v) be a game, $n \geq 2$, and let φ be a subsystem of $\varphi_{\mathcal{PK}}$ (see Exercise 10.1.3). Show that $\nu(N, v)$ (the unique member of $\mathcal{PN}(N, v)$) is a stable point of φ.

10.3 Asymptotic Stability of the Nucleolus

Let X be a metric space and let $\varphi : X \rightrightarrows X$ be a dynamic system.

Definition 10.3.1. *A point $x \in E_\varphi$ is called (locally)* **asymptotically** *stable (with respect to φ) if (i) x is a stable point of φ and (ii) there exists an open set U, $x \in U \subseteq X$, such that for each $x_0 \in U$ and each maximal φ-sequence $(x_t)_{t \in \mathbb{N}_0}$, $\lim_{t\to\infty} x_t = x$.*

Remark 10.3.2. If $x \in E_\varphi$ is asymptotically stable, then x is an *isolated* point of E_φ, that is, there exists an open set $V \subseteq X$ such that $V \cap E_\varphi = \{x\}$.

The following lemma shows that the nucleolus is the only candidate for local asymptotic stability with respect to $\varphi_{\mathcal{K}}$ and its subsystems.

Lemma 10.3.3. *Let (N, v) be a 0-1-normalized game and let φ be a subsystem of $\varphi_{\mathcal{K}}$. If $x \in E_\varphi$ and $x \neq \mathcal{N}(N, v)$, then x is not asymptotically stable with respect to φ.*

Proof: Denote $\bar{\nu} = \bar{\nu}(N, v)$ and let $x_\beta = \beta\bar{\nu} + (1 - \beta)x$ for all $0 < \beta < 1$. By Lemma 5.1.9,

$$\theta(x_\beta) \leq_{lex} \beta\theta(\bar{\nu}) + (1 - \beta)\theta(x) <_{lex} \theta(x) \text{ for all } 0 < \beta < 1.$$

Let $0 < \beta < 1$ and $(x_t)_{t \in \mathbb{N}_0}$ be a φ-sequence such that $x_0 = x_\beta$. Let Ψ be defined by (10.1.6). Then there exists $1 \leq k \leq 2^n$ such that $\Psi^k(x) > \Psi^k(x_\beta)$. As Ψ^k is continuous and φ-monotone, the sequence $(x_t)_{t \in \mathbb{N}_0}$ does not converge to x. **q.e.d.**

We are now ready for the following theorem.

Theorem 10.3.4. *Let (N, v) be a 0-1-normalized game and let φ be an* lhc *subsystem of $\varphi_{\mathcal{K}}$. If $\bar{\nu} = \bar{\nu}(N, v)$ is an isolated point of E_φ, then $\bar{\nu}$ is an asymptotically stable point of φ.*

Proof: By Corollary 10.2.7, $\bar{\nu}$ is a stable point of φ. Let U, $\bar{\nu} \in U \subseteq X$, be an open set such that $\mathrm{cl}(U) \cap E_\varphi = \{\bar{\nu}\}$ ("cl" denotes "closure"). As $\bar{\nu}$ is stable, there exists an open set $V \subseteq U$ such that if $x_0 \in V$ and $(x_t)_{t \in \mathbb{N}_0}$ is a φ-sequence, then $x_t \in U$ for all $t \in \mathbb{N}$. If, in addition, $(x_t)_{t \in \mathbb{N}_0}$ is maximal, then $\lim_{t\to\infty} x_t \in \mathrm{cl}(U) \cap E_\varphi = \{\bar{\nu}\}$. **q.e.d.**

Exercises

Exercise 10.3.1. Let (N, v) be a 0-1-normalized game, let $\bar{\nu} = \bar{\nu}(N, v)$, and let φ be an lhc subsystem of $\varphi_{\mathcal{K}}$. Prove that if $\bar{\nu}$ is an isolated point of E_φ, then $\{\bar{\nu}\} = \mathcal{LK}(N, v)$. (Hint: Show that if $x \in E_\varphi \setminus \mathcal{LK}(N, v)$, then x is not asymptotically stable.)

Exercise 10.3.2. Let (N, v) be a 0-1-normalized game and let φ be an lhc subsystem of $\varphi_{\mathcal{K}}$. Prove that if $x \in E_{\varphi} \setminus \mathcal{N}(N, v)$ is an isolated point of E_{φ}, then x is not a stable point of φ.

Exercise 10.3.3. Let (N, v) be a game, $n \geq 2$, let $\nu = \nu(N, v)$, and let φ be a subsystem of $\varphi_{\mathcal{PK}}$. Prove the following assertions: (1) If x is an asymptotically stable point of φ, then $x = \nu$. (2) If φ is lhc and ν is an isolated point of E_{φ}, then ν is an asymptotically stable point of φ.

10.4 Notes and Comments

The dynamic system $\varphi_{\mathcal{K}}$ and its lhc subsystems were first introduced in Stearns (1968). Stearns also proved that if φ is an lhc subsystem of $\varphi_{\mathcal{K}}$, then the maximal φ-sequences converge to the points of some (generalized) bargaining set E_{φ}. Our proofs follow Maschler and Peleg (1976).

The lhc subsystems of $\varphi_{\mathcal{K}}$ leading to the reactive bargaining set and to the semi-reactive bargaining set are explicitly described in Granot and Maschler (1997) and Sudhölter and Potters (2001).

The results on stability in Section 10.2 are due to Kalai, Maschler, and Owen (1975). Again, our approach is due to Maschler and Peleg (1976). The asymptotic stability of the nucleolus is due to Kalai, Maschler, and Owen (1975). Some of the techniques in Maschler and Peleg (1976) appeared first in Justman (1977).

Remark 10.4.1. A continuous version, that is, a system of differential equations, of each of the lhc subsystems of $\varphi_{\mathcal{K}}$ was introduced in Billera (1972). As shown in Billera and Wu (1977), the continuous dynamic system leading to the kernel allows us to partition the set of games on a fixed player set into finitely many equivalence classes. Results on stability of the continuous versions are contained in Kalai, Maschler, and Owen (1975).

Remark 10.4.2. There are dynamic approaches leading to other solutions. We mention three of them. Justman (1977) presents a discrete dynamic system such that the maximal trajectories converge to the nucleolus and Wu (1977) describes a discrete dynamic system leading to the core, provided that the core is nonempty. Continuous dynamic systems leading to the core, the Shapley value, and the nucleolus, are contained in Grotte (1976).

Part II

NTU Games

11

Cooperative Games in Strategic and Coalitional Form

This chapter is devoted to an analysis of the relationship between a cooperative game in strategic form with nontransferable utility (NTU), and its coalition functions. Section 11.1 introduces the concept of cooperative game in strategic form by means of a definition and three examples. The main example is a game in strategic form, where correlated strategies (of any coalition) may be supported by binding agreements.

Then, in Section 11.2, we proceed to define the main coalition functions of a cooperative NTU game in strategic form, namely, the α and β NTU coalition functions. The α coalition function describes what each coalition can guarantee for itself. The β approach defines for each coalition the set of payoff vectors (to the members of the coalition) which the complement cannot block. The main properties of the α and β coalition functions are verified under various assumptions. The analysis motivates our axiomatic approach to coalitional NTU games in Section 11.3.

We conclude in Section 11.4 with a discussion of cooperative games with side payments but without transferable utilities. We show that the core of TU games is closely related to the core of NTU games with side payments and concave utilities (for money).

11.1 Cooperative Games in Strategic Form

In this section we shall briefly discuss cooperation in strategic games. This will enable us to define some coalition functions of a cooperative game in strategic form. These examples will serve as the basis for the axiomatic treatment of coalitional games without transferable utilities or NTU games for short.

Definition 11.1.1. *A* **cooperative game in strategic form** *is a triple* $\left(N, (\Sigma(S))_{\emptyset \neq S \subseteq N}, \left(u^i\right)_{i \in N}\right)$ *that has the following properties:*

(1) N is a finite nonempty set (the set of players).

(2) For each coalition $\emptyset \neq S \subseteq N$, $\Sigma(S)$ is a nonempty set (the set of strategies of S).

(3) If $\emptyset \neq S, T \subseteq N$, $S \cap T = \emptyset$, then $\Sigma(S \cup T) \supseteq \Sigma(S) \times \Sigma(T)$.

(4) For every $i \in N$, $u^i : \Sigma(N) \to \mathbb{R}$ is a function (i's payoff function).

Intuitively, every $\sigma^S \in \Sigma(S)$, $|S| \geq 2$, can be supported by a binding agreement between the members of S. We now present some examples of cooperative games in strategic form.

Example 11.1.2. Let N be a finite set of players and let

$$g = \left(N, \left(A^i\right)_{i\in N}, \left(h^i\right)_{i\in N}\right)$$

be a finite game in strategic form, that is, $h^i : \prod_{j\in N} A^j \to \mathbb{R}$ for all $i \in N$ and the strategy sets A^i, $i \in N$, are finite (and nonempty). Let $\emptyset \neq S \subseteq N$. A *correlated* strategy for S is a probability distribution on $A^S = \prod_{i\in S} A^i$. Let $\Sigma(S)$ be the set of all correlated strategies for S. If for each coalition $\emptyset \neq S \subseteq N$, $|S| \geq 2$, every $\sigma^S \in \Sigma(S)$ can be supported by a binding agreement between the members of S, then the situation may be modelled by the triple $\left(N, (\Sigma(S))_{\emptyset\neq S\subseteq N}, \left(u^i\right)_{i\in N}\right)$, where

$$u^i(\sigma^N) = \sum_{a\in A^N} \sigma^N(a)\; h^i(a) \;\text{ for all } \sigma^N \in \Sigma(N) \text{ and all } i \in N,$$

which is a cooperative game in strategic form (Exercise 11.1.1).

The following example is very simple.

Example 11.1.3. Let $g = \left(N, \left(A^i\right)_{i\in N}, \left(h^i\right)_{i\in N}\right)$ be a game in strategic form. For any $\emptyset \neq S \subseteq N$, let $\Sigma(S) = A^S$. Again, if binding agreements are possible for multi-player coalitions, then we may consider

$$G(g) = \left(N, (\Sigma(S))_{\emptyset\neq S\subseteq N}, \left(h^i\right)_{i\in N}\right),$$

which is a cooperative game in strategic form (Exercise 11.1.2), as a description of the situation.

We shall now introduce NTU-markets. A quadruple $(N, \mathbb{R}^m_+, A, W)$ is an *NTU-market* if $N, \mathbb{R}^m_+$, and A are given as in Subsection 2.2.1, and $W = (w^i)_{i\in N}$ is an indexed collection of continuous real functions on $\mathbb{R}^m_+$ (the utility functions). Hence an NTU-market is a market as defined in Subsection 2.2.1 except that the utility functions may not be concave.

Example 11.1.4. Let $(N, \mathbb{R}^m_+, A, W)$ be an NTU-market. For each $\emptyset \neq S \subseteq N$ define

$$\Sigma(S) = X^S = \left\{ (x^i)_{i \in S} \,\middle|\, x^i \in \mathbb{R}^m_+ \text{ for all } i \in S \text{ and } \sum_{i \in S} x^i = \sum_{i \in S} a^i \right\},$$

that is, $\Sigma(S)$ is the set of feasible S-allocations. For any $x_N = (x^j)_{j \in N} \in \Sigma(N)$ and each $i \in N$ let $u^i(x_N) = w^i(x^i)$. Then

$$\left(N, (\Sigma(S))_{\emptyset \neq S \subseteq N}, (u^i)_{i \in N}\right)$$

is a cooperative game in strategic form (Exercise 11.1.1).

Exercises

Exercise 11.1.1. Show that the triples $\left(N, (\Sigma(S))_{\emptyset \neq S \subseteq N}, (u^i)_{i \in N}\right)$ as defined in Examples 11.1.2 and 11.1.4 are a cooperative game in strategic form.

Exercise 11.1.2. Verify that $\left(N, (\Sigma(S))_{\emptyset \neq S \subseteq N}, (h^i)_{i \in N}\right)$ defined in Example 11.1.3 is a cooperative game in strategic form.

11.2 α- and β-Effectiveness

Let $G = \left(N, (\Sigma(S))_{\emptyset \neq S \subseteq N}, (u^i)_{i \in N}\right)$ be a cooperative game in strategic form. For every $\emptyset \neq S \subseteq N$, every $\sigma^S \in \Sigma(S)$ can be adopted by a binding agreement. (In economic situations binding agreements are mainly agreements whose violation entails high monetary penalties which deter potential violators.) We shall now see which payoff vectors a coalition S can guarantee when every $\sigma^S \in \Sigma(S)$ is available.

Definition 11.2.1. *Let $G = \left(N, (\Sigma(S))_{\emptyset \neq S \subseteq N}, (u^i)_{i \in N}\right)$ be a cooperative game in strategic form, let $\emptyset \neq S \subseteq N$, and let $x^S \in \mathbb{R}^S$. The coalition S is α-**effective** for x^S if there exists $\sigma^S \in \Sigma(S)$ such that for every $\sigma^{N \setminus S} \in \Sigma(N \setminus S)$, $u^i(\sigma^S, \sigma^{N \setminus S}) \geq x^i$ for all $i \in S$. (Here we assume as a convention that $\Sigma(\emptyset)$ is a singleton and that $\Sigma(S) \times \Sigma(\emptyset)$ is identified with $\Sigma(S)$ for all $S \subseteq N$.)*

Clearly, if a coalition S is α-effective for some vector $x^S \in \mathbb{R}^S$, then S can, indeed, guarantee the utility x^i for each $i \in S$.

Let $G = \left(N, (\Sigma(S))_{\emptyset \neq S \subseteq N}, (u^i)_{i \in N}\right)$ be a cooperative game in strategic form and let $\emptyset \neq S \subseteq N$. We denote

$$V_\alpha(S,G) = V_\alpha(S) = \{x^S \in \mathbb{R}^S \mid S \text{ is } \alpha\text{-effective for } x^S\}. \tag{11.2.1}$$

Moreover, let $V_\alpha(\emptyset) = \emptyset$.

Remark 11.2.2. The mapping V_α has the following two properties:

(1) For every $S \subseteq N$, $V_\alpha(S)$ is *comprehensive*, that is, if $x^S \in V_\alpha(S)$, $y^S \in \mathbb{R}^S$, and $y^S \leq x^S$, then $y^S \in V_\alpha(S)$.

(2) The mapping V_α is *superadditive*, that is, if $S, T \subseteq N$ and $S \cap T = \emptyset$, then

$$V_\alpha(S \cup T) \supseteq V_\alpha(S) \times V_\alpha(T).$$

Under the standard assumptions of compactness of the strategy sets and continuity of the payoff functions, V_α has closed and bounded (from above) values. More precisely, the following lemma is valid.

Lemma 11.2.3. *Let* $G = \left(N, (\Sigma(S))_{\emptyset \neq S \subseteq N}, (u^i)_{i \in N}\right)$ *be a cooperative game in strategic form. Assume that* $\Sigma(S)$ *is a compact metric space for every* $\emptyset \neq S \subseteq N$, *and that* $u^i : \Sigma(N) \to \mathbb{R}$ *is continuous for every* $i \in N$. *Then*

(1) $V_\alpha(S) \neq \emptyset$ *for each* $S \in 2^N \setminus \{\emptyset\}$;

(2) $V_\alpha(S)$ *is a closed subset of* $\mathbb{R}^S$ *for each* $S \subseteq N$;

(3) for each $S \subseteq N$ *there exists* $y^S \in \mathbb{R}^S$ *such that* $x^S \leq y^S$ *for every* $x^S \in V_\alpha(S)$.

Proof: In order to prove (1), choose $\sigma \in \Sigma(N)$. Then $u(\sigma) = (u^i(\sigma))_{i \in N}$ is a member of $V_\alpha(N)$. Now let $\emptyset \neq S \subsetneqq N$. Choose $\sigma^S \in \Sigma(S)$ and define, for $i \in S$, $x^i = \inf_{\sigma^{N \setminus S} \in \Sigma(N \setminus S)} u^i(\sigma^S, \sigma^{N \setminus S})$. By continuity of u^i and by the compactness of $\Sigma^{N \setminus S}$, $x^i \in \mathbb{R}$. Let $x^S = (x^i)_{i \in S}$. Then $x^S \in V_\alpha(S)$.

The proof of (3) is left to the reader (Exercise 11.2.1). In order to prove (2), let $\emptyset \neq S \subseteq N$, let $\left(x^S(k)\right)_{k \in \mathbb{N}}$ be a convergent sequence such that $x^S(k) \in V_\alpha(S)$ for all $k \in \mathbb{N}$ and let $y^S = \lim_{k \to \infty} x^S(k)$. For each $k \in \mathbb{N}$ choose $\sigma^S(k) \in \Sigma(S)$ such that $u^i(\sigma^S(k), \sigma^{N \setminus S}) \geq x^i(k)$ for all $i \in S$ and all $\sigma^{N \setminus S} \in \Sigma(N \setminus S)$. As $\Sigma(S)$ is compact, we may assume that there is a subsequence $\left(\sigma^S(k_j)\right)_{j \in \mathbb{N}}$ and $\tau^S \in \Sigma(S)$ such that $\lim_{j \to \infty} \sigma^S(k_j) = \tau^S$. Let $\sigma^{N \setminus S} \in \Sigma(N \setminus S)$ and $i \in S$. Then

$$u^i(\tau^S, \sigma^{N \setminus S}) = \lim_{j \to \infty} u^i(\sigma^S(k_j), \sigma^{N \setminus S}) \geq \lim_{j \to \infty} x^i(k_j) = y^i.$$

Hence, $y^S \in V_\alpha(S)$. **q.e.d.**

We remark that Examples 11.1.2 and 11.1.4 satisfy the conditions of Lemma 11.2.3. The next lemma shows that, if $\Sigma(S)$ is convex for every $\emptyset \neq S \subseteq N$, and u^i, $i \in N$, are concave on $\Sigma(N)$, then V_α is convex-valued.

Lemma 11.2.4. *Let $G = \left(N, (\Sigma(S))_{\emptyset \neq S \subseteq N}, (u^i)_{i \in N}\right)$ be a cooperative game in strategic form. If $\Sigma(S)$, $\emptyset \neq S \subseteq N$, is a convex subset of some real vector space, and if u^i is concave on $\Sigma(N)$ for each $i \in N$, then $V_\alpha(S)$ is convex for each $S \subseteq N$.*

Proof: Let $\emptyset \neq S \subseteq N$, $x^S, y^S \in V_\alpha(S)$, $0 < \vartheta < 1$, and $\sigma^{N \setminus S} \in \Sigma(N \setminus S)$. There exist $\sigma_x^S, \sigma_y^S \in \Sigma(S)$ such that $u^i(\sigma_x^S, \tau^{N\setminus S}) \geq x^i$ and $u^i(\sigma_y^S, \tau^{N\setminus S}) \geq y^i$ for all $i \in S$ and all $\tau^{N \setminus S} \in \Sigma(N \setminus S)$. Let $\sigma^S = \vartheta \sigma_x^S + (1-\vartheta)\sigma_y^S$ and $i \in S$. Then

$$u^i(\sigma^S, \sigma^{N\setminus S}) \geq \vartheta u^i(\sigma_x^S, \sigma^{N\setminus S}) + (1-\vartheta) u^i(\sigma_y^S, \sigma^{N\setminus S}) \geq \vartheta x^i + (1-\vartheta) y^i.$$

Thus, $\vartheta x^S + (1-\vartheta) y^S \in V_\alpha(S)$. **q.e.d.**

Let $G = \left(N, (\Sigma(S))_{\emptyset \neq S \subseteq N}, (u^i)_{i \in N}\right)$ be a cooperative game in strategic form, let $\emptyset \neq S \subseteq N$, and let $x^S \in \mathbb{R}^S$. We shall now define when $N \setminus S$ cannot prevent S from getting x^S.

Definition 11.2.5. *Let $G = \left(N, (\Sigma(S))_{\emptyset \neq S \subseteq N}, (u^i)_{i \in N}\right)$ be a cooperative game in strategic form, let $\emptyset \neq S \subseteq N$, and let $x^S \in \mathbb{R}^S$. The coalition S is* β-**effective** *for x^S if for each $\sigma^{N\setminus S} \in \Sigma(N \setminus S)$ there exists $\sigma^S \in \Sigma(S)$ such that $u^i(\sigma^S, \sigma^{N\setminus S}) \geq x^i$ for all $i \in S$.*

Let $G = \left(N, (\Sigma(S))_{\emptyset \neq S \subseteq N}, (u^i)_{i \in N}\right)$ be a cooperative game in strategic form and let $\emptyset \neq S \subseteq N$. We denote

$$V_\beta(S, G) = V_\beta(S) = \{x^S \in \mathbb{R}^S \mid S \text{ is } \beta\text{-effective for } x^S\}. \tag{11.2.2}$$

We also define $V_\beta(\emptyset) = \emptyset$. Clearly, $V_a(S) \subseteq V_\beta(S)$ for all $S \subseteq N$.

Example 11.2.6 (Aumann (1961)). Let $g = \left(N, (A^i)_{i \in N}, (h^i)_{i \in N}\right)$ be the game in strategic form (partially) given by $N = \{1, 2, 3\}$, $A^1 = \{s_1^1\}$, and $A^j = \{s_1^j, s_2^j\}$ for $j = 2, 3$. Moreover, let $T = \{1, 2\}$, let $A^T = A^1 \times A^2 = \{s_1^T, s_2^T\}$, and let $h^T = (h^1, h^2)$ be given by the following matrix:

	s_1^3	s_2^3
s_1^T	$(1, -1)$	$(0, 0)$
s_2^T	$(0, 0)$	$(-1, 1)$

Assume that correlated strategies can be used (see Example 11.1.2). Then $(0, 0) \in V_\beta(T) \setminus V_\alpha(T)$.

Remark 11.2.7. Let $G = \left(N, (\Sigma(S))_{\emptyset \neq S \subseteq N}, (u^i)_{i \in N}\right)$ be a cooperative game in strategic form. Then the following assertions are valid:

(1) The set $V_\beta(S)$ is comprehensive for every $S \subseteq N$.

(2) If, for each $\emptyset \neq S \subseteq N$, $\Sigma(S)$ is a compact metric space, and if, for every $i \in N$, $u^i : \Sigma(N) \to \mathbb{R}$ is continuous, then $V_\beta(S)$ is closed and bounded from above for each $S \subseteq N$.

(3) If each $\Sigma(S)$, $\emptyset \neq S \subseteq N$, is a convex subset of some real vector space and if each $u^i : \Sigma(N) \to \mathbb{R}$, $i \in N$, is concave, then $V_\beta(S)$ is convex for every $S \subseteq N$.

Example 11.2.8 (Example 11.1.4 continued). Let G be the cooperative game in strategic form which is derived from an NTU-market. Then $V_\alpha(\cdot, G) = V_\beta(\cdot, G)$. Also, $V_\alpha(\cdot, G)$ has closed values which are bounded from above.

A cooperative game in strategic form is *tight* if $V_\alpha(\cdot, G) = V_\beta(\cdot, G)$. Note that $V_\beta(\cdot, G)$ may not be superadditive (see Exercise 11.2.2). We now shall give sufficient conditions on G for the superadditivity of $V_\beta(G)$.

Theorem 11.2.9. *Let* $G = \left(N, (\Sigma(S))_{\emptyset \neq S \subseteq N}, \left(u^i\right)_{i \in N}\right)$ *be a cooperative game in strategic form. Assume that for each* $\emptyset \neq S \subseteq N$, $\Sigma(S)$ *is a convex and compact subset of some Euclidean space, and that* $u^i : \Sigma(N) \to \mathbb{R}$ *is continuous and quasi-concave for each* $i \in N$. *Then* $V_\beta(\cdot, G)$ *is superadditive.*

Proof: Let $S, T \subseteq N$, $S, T \neq \emptyset$, $S \cap T = \emptyset$, $x^S \in V_\beta(S)$, and $x^T \in V_\beta(T)$. If $\sigma^{N \setminus (S \cup T)} \in \Sigma(N \setminus (S \cup T))$, then we have to prove that there exists $\mu^{S \cup T} \in \Sigma(S \cup T)$ such that $u^i\left(\mu^{S \cup T}, \sigma^{N \setminus (S \cup T)}\right) \geq x^i$ for all $i \in S \cup T$. For $\sigma^S \in \Sigma(S)$ let

$$B(\sigma^S) = \left\{\sigma^T \in \Sigma(T) \,\middle|\, u^i\left(\sigma^S, \sigma^T, \sigma^{N \setminus (S \cup T)}\right) \geq x^i \text{ for all } i \in T\right\}.$$

Then $B(\sigma^S)$ is (nonempty and) convex, because u^i, $i \in T$, are quasi-concave. Similarly, let

$$B(\sigma^T) = \left\{\sigma^S \in \Sigma(S) \,\middle|\, u^i\left(\sigma^S, \sigma^T, \sigma^{N \setminus (S \cup T)}\right) \geq x^i \text{ for all } i \in S\right\}$$

for each $\sigma^T \in \Sigma(T)$. Then $B(\sigma^T)$ is convex. Thus, we may define a set-valued mapping $\varphi : \Sigma(S) \times \Sigma(T) \rightrightarrows \Sigma(S) \times \Sigma(T)$ by

$$\varphi(\sigma^S, \sigma^T) = B(\sigma^T) \times B(\sigma^S) \;\text{ for all } \sigma^S \in \Sigma(S) \text{ and all } \sigma^T \in \Sigma(T).$$

Thus φ is convex-valued. Also, as the functions u^i, $i \in S \cup T$, are continuous, φ is upper hemicontinuous. By Kakutani's fixed-point theorem there exist $\mu^S \in \Sigma(S)$ and $\mu^T \in \Sigma(T)$ such that $\mu^{S \cup T} = (\mu^S, \mu^T) \in B(\mu^T) \times B(\mu^S)$. Thus

$$u^i\left(\mu^{S \cup T}, \sigma^{N \setminus (S \cup T)}\right) \geq x^i \;\text{ for all } i \in S \cup T.$$

As $\Sigma(S \cup T) \supseteq \Sigma(S) \times \Sigma(T)$, $\mu^{S \cup T} \in \Sigma(S \cup T)$. Hence, $(x^S, x^T) \in V_\beta(S \cup T)$.

q.e.d.

We conclude this section with a continuation of Example 11.1.2.

Example 11.2.10 (Example 11.1.2 continued). Let

$$g = \left(N, \left(A^i\right)_{i\in N}, \left(h^i\right)_{i\in N}\right)$$

be a finite game in strategic form, let $\Sigma(S)$, $\emptyset \neq S \subseteq N$, be the set of all correlated strategies of S and let $G = \left(N, (\Sigma(S))_{\emptyset\neq S\subseteq N}, \left(u^i\right)_{i\in N}\right)$ be the resulting cooperative game in strategic form. Then, for each $S \subseteq N$, $V_\alpha(S,G)$ and $V_\beta(S,G)$ are comprehensive, closed, bounded from above, and convex. Also, both $V_\alpha(\cdot,G)$ and $V_\beta(\cdot,G)$ are superadditive.

Exercises

Exercise 11.2.1. Prove Assertion (3) of Lemma 11.2.3.

Exercise 11.2.2. Find a two-person game g in strategic form such that $V_\beta(\cdot,G)$ is not superadditive, where $G = G(g)$ is defined as in Example 11.1.3.

Exercise 11.2.3 (Example 11.1.2 continued). Let

$$g = \left(N, \left(A^i\right)_{i\in N}, \left(h^i\right)_{i\in N}\right)$$

be a finite game in strategic form, let $\Sigma(S)$, $\emptyset \neq S \subseteq N$, be the set of all correlated strategies of S and let $G = \left(N, (\Sigma(S))_{\emptyset\neq S\subseteq N}, \left(u^i\right)_{i\in N}\right)$ be the resulting cooperative game in strategic form. Prove the following assertions:

(1) $V_\alpha(N,G) = V_\beta(N,G)$.

(2) $V_\alpha(\{i\},G) = V_\beta(\{i\},G)$ for all $i \in N$.

(3) $V_\alpha(S,G)$ is polyhedral for each $S \subseteq N$.

(4) $V_\beta(S,G)$ may not be polyhedral.

(Concerning (3) and (4) see Aumann (1961).)

11.3 Coalitional Games with Nontransferable Utility

Let $G = \left(N, (\Sigma(S))_{\emptyset\neq S\subseteq N}, \left(u^i\right)_{i\in N}\right)$ be a cooperative game in strategic form. The mappings $V_\alpha(\cdot,G)$ and $V_\beta(\cdot,G)$ are two examples of NTU coalition functions. The set $V_\alpha(S,G)$, $S \subseteq N$, consists of the set of payoff vectors that S can guarantee for its members. On the other hand, $V_\beta(S,G)$, $S \subseteq N$, consists of the set of payoff vectors that $N \setminus S$ cannot prevent S from getting. The examples in Section 11.2 motivate the following definition.

Definition 11.3.1. *A (coalitional)* **NTU game** *is a pair (N, V) where N is a coalition and V is a function which associates with each coalition $S \subseteq N$ a subset $V(S)$ of $\mathbb{R}^S$ such that*

$$V(S) \neq \emptyset, \text{ if } S \neq \emptyset, \text{ and } V(\emptyset) = \emptyset; \tag{11.3.1}$$
$$V(S) \text{ is comprehensive;} \tag{11.3.2}$$
$$V(S) \text{ is closed;} \tag{11.3.3}$$
$$V(S) \cap (x^S + \mathbb{R}^S_+) \text{ is bounded for every } x^S \in \mathbb{R}^S. \tag{11.3.4}$$

Clearly, Definition 11.3.1 is compatible with our examples $V = V_\alpha(\cdot, G)$ and $V = V_\beta(\cdot, G)$ of Section 11.2, provided that G satisfies some standard properties (see Lemma 11.2.3 and Remark 11.2.7). Also, with every TU game (N, v) we associate the NTU game (N, V_v) defined by

$$V_v(S) = \left\{ x^S \in \mathbb{R}^S \,\middle|\, \sum_{i \in S} x^i \leq v(S) \right\} \quad \text{for every } \emptyset \neq S \subseteq N. \tag{11.3.5}$$

Exercises

Exercise 11.3.1. Let (N, v) be a TU game. Prove that (N, V_v) is superadditive iff (N, v) is superadditive.

Exercise 11.3.2. Let $(N, \mathbb{R}^2_+, A, W)$ be the market defined by $N = \{1, 2\}$, $a^1 = (1, 0)$, $a^2 = (0, 1)$, and $u^i(x_1, x_2) = \sqrt{x_1 x_2}$ for $i = 1, 2$, and $(x_1, x_2) \in \mathbb{R}^2_+$. (As u^i, $i = 1, 2$, is a Cobb-Douglas function, it is concave.) Let (N, v) be the corresponding (TU) market game and let G be the cooperative game in strategic form defined in Example 11.1.4. Provide sketches of the NTU coalition functions V_v and $V_\alpha(\cdot, G)$.

11.4 Cooperative Games with Side Payments but Without Transferable Utility

We shall now introduce NTU pregames.

Definition 11.4.1. *An* **NTU pregame** *is a pair $(N, \widehat{V})$ where N is a coalition and $\widehat{V}$ is a function which associates with every coalition $S \subseteq N$ a subset $\widehat{V}(S)$ of $\mathbb{R}^S$ such that (11.3.1) and (11.3.2) are satisfied.*

We proceed now to define the core of an NTU pregame.

Definition 11.4.2. *Let $(N,\widehat{V})$ be an* NTU *pregame. The* **core** *of $(N,\widehat{V})$, $\mathcal{C}(N,\widehat{V})$, is given by*

$$\mathcal{C}(N,\widehat{V}) = \left\{ x \in \widehat{V}(N) \,\middle|\, \begin{array}{c} \textit{for every } \emptyset \neq S \subseteq N \textit{ and every } y^S \in \widehat{V}(S), \\ \textit{there exists } i \in S \textit{ such that } x^i \geq y^i \end{array} \right\}.$$

In Chapter 12 we shall investigate the core of NTU games. Here we shall only find a connection between the core of TU games and the core of a special class of NTU games.

Let (N,v) be a TU game where $v(S)$, $\emptyset \neq S \subseteq N$, is expressed in some monetary unit (say, dollars). Assume that the von Neumann-Morgenstern utility functions of the players for money, u^i, $i \in N$, are continuous and strictly increasing (on $\mathbb{R}$). Then the triple $\left(N, v, \left(u^i\right)_{i\in N}\right)$ yields an NTU pregame $(N,\widehat{V})$ in the following way:

$$\widehat{V}(S) = \operatorname{convh} \left\{ x^S \in \mathbb{R}^S \,\middle|\, \begin{array}{c} x^i \leq u^i(y^i),\ i \in S, \\ \text{for some } y^S \in \mathbb{R}^S \\ \text{such that } \sum_{i\in S} y^i = v(S) \end{array} \right\}. \tag{11.4.1}$$

(We recall (see Remark 2.1.4) that the members of S may use lotteries in order to distribute $v(S)$, $\emptyset \neq S \subseteq N$.) If the utility functions u^i, $i \in N$, are concave (and strictly monotonic), then

$$\widehat{V}(S) = \left\{ x^S \in \mathbb{R}^S \,\middle|\, \begin{array}{c} x^i \leq u^i(y^i),\ i \in S, \text{ for some } y^S \in \mathbb{R}^S \\ \text{such that } \sum_{i\in S} y^i = v(S) \end{array} \right\}$$

and $\widehat{V}(S)$ is closed for all $\emptyset \neq S \subseteq N$. Also, $(N,\widehat{V})$ satisfies (11.3.4). The straightforward proof of this fact is left to the reader as an exercise (see Exercise 11.4.1). Hence, $(N,\widehat{V}) = (N,V)$ is an NTU game, provided that the utilities are concave. Furthermore, there is a simple relationship between $\mathcal{C}(N,v)$ and $\mathcal{C}(N,V)$. Let $u_* : \mathbb{R}^N \to \mathbb{R}^N$ be defined by $u^i_*(x) = u^i(x^i)$ for all $x \in \mathbb{R}^N$ and all $i \in N$. Then $\mathcal{C}(N,V) = u_*(\mathcal{C}(N,v))$, because the functions u^i, $i \in N$, are strictly increasing. Thus, the results on the core of TU games which were obtained in Chapters 2 and 3 may be applied to games with side payments and concave utilities.

A similar remark holds for the bargaining set $\mathcal{M}$. However, it is deferred to Chapter 16.

Exercises

Exercise 11.4.1. Let (N,v) be a TU game, let $u^i : \mathbb{R} \to \mathbb{R}$, $i \in N$, be concave and strictly increasing functions, let $\emptyset \neq S \subseteq N$, and let $(N,\widehat{V})$ be given by (11.4.1). Show that $\widehat{V}(S)$ satisfies (11.3.4).

Exercise 11.4.2. Find a triple $\left(N, v, (u^i)_{i\in N}\right)$, where (N,v) is a TU game, and $u^i : \mathbb{R} \to \mathbb{R}$, $i \in N$, are continuous, strictly increasing, and bounded, such that the pregame $(N,\widehat{V})$ defined by (11.4.1) is not closed-valued.

11.5 Notes and Comments

As far as we know, Definition 11.1.1 is new. However, it is clearly motivated by the examples in Section 11.1. Section 11.2 is based on Aumann and Peleg (1960) and Aumann (1961). Section 11.4 follows Aumann (1967).

A proof of Kakutani's fixed-point theorem, which is used in the proof of Theorem 11.2.9, is contained in Klein (1973).

12

The Core of NTU Games

The concept of the core is naturally generalized to NTU games. However, the analysis of the core when there are no side payments is much more difficult than in the TU case. In this chapter we report some basic results on the core of NTU games.

A precise definition of the core must also specify the set of payoff vectors that are potential candidates for core membership. Aumann (1961) introduced the core of an arbitrary set of payoff vectors. In Section 12.1 we consider the core under individual rationality, (weak) Pareto optimality, and the conjunction of these properties. It turns out that for superadditive games all the resulting cores coincide.

The basic existence result of the core of NTU games was proved by Scarf (1967). In Section 12.2 we prove a generalization due to Billera (1970b) using the proof of Shapley and Vohra (1991). In Section 12.3 we prove the nonemptiness of the core of ordinal convex games (due to Greenberg (1985)) and of the core of cardinal convex games (due to Sharkey (1981)).

Section 12.4 is devoted to an axiomatization of the core. Peleg (1985) showed that the core may be characterized by nonemptiness, individual rationality, and the reduced game property (suitably extended to the NTU case).

In Section 12.5 further remarkable properties are described that may even be used to characterize the core, if some of the considered NTU games have an empty core.

12.1 Individual Rationality, Pareto Optimality, and the Core

Let (N, V) be an NTU game and let $x, y \in \mathbb{R}^N$. The vector y *dominates* x *via a coalition* S ($\emptyset \neq S \subseteq N$) if $y^S \in V(S)$ and $y^S \gg x^S$ (see Notation 4.4.5). Moreover, y *dominates* x if there exists a coalition S such that y dominates x via S. By Definition 11.4.2, the core of (N, V), $\mathcal{C}(N, V)$, is the set of all members of $V(N)$ that are not dominated by any vector in $\mathbb{R}^N$. Historically, the core was defined somewhat differently, and we shall now investigate some variants of the core.

Definition 12.1.1. *Let $A \subseteq \mathbb{R}^N$. The* **core of A with respect to** (N, V), *$\mathcal{C}(N, V, A)$, is the set of members of A that are not dominated by any member of A.*

Given the NTU game (N, V) we shall be interested in the cores of four subsets A of $\mathbb{R}^N$ with respect to (N, V). Let us denote

$$V(N)_e = \left\{ x \in V(N) \,\middle|\, \begin{array}{c} \text{for all } y \in V(N) \text{ there exists} \\ i \in N \text{ such that } x^i \geq y^i \end{array} \right\}. \tag{12.1.1}$$

The set $V(N)_e$ is the set of weakly Pareto optimal (*efficient*) elements of $V(N)$. Note that $V(N)_e$ is the boundary of $V(N)$, because $V(N)$ is closed and comprehensive.

We shall restrict our attention to superadditive NTU games (see Remark 11.2.2 (2)).

Lemma 12.1.2. *If (N, V) is superadditive, then*

$$\mathcal{C}(N, V) = \mathcal{C}(N, V, V(N)_e).$$

Proof: Clearly, $\mathcal{C}(N, V) \subseteq V(N)_e$. Hence $\mathcal{C}(N, V) \subseteq \mathcal{C}(N, V, V(N)_e)$. In order to show the opposite inclusion, assume, on the contrary, that there exists $x \in \mathcal{C}(N, V, V(N)_e) \setminus \mathcal{C}(N, V)$. Then there exist $y \in \mathbb{R}^N$ and $\emptyset \neq S \subsetneqq N$ such that $y^S \in V(S)$ and $y^S \gg x^S$. By nonemptiness of $V(N \setminus S)$, there exists $z^{N \setminus S} \in V(N \setminus S)$. By superadditivity of (N, V), $(y^S, z^{N \setminus S}) \in V(N)$. Clearly, $(y^S, z^{N \setminus S})$ dominates x via S and, hence, we may assume that $y \in V(N)$.

Define $A = \{z \in V(N) \mid z^S = y^S \text{ and } z^{N \setminus S} \geq y^{N \setminus S}\}$ and observe that $A \neq \emptyset$, because $y \in A$. Also, each member of A dominates x via S. By (11.3.3) and (11.3.4), A is compact. Choose $\bar{y} \in A$ such that $\bar{y}(N \setminus S)$ is maximal. Then $\bar{y} \in V(N)_e$ and the desired contradiction has been obtained. **q.e.d.**

Let (N, V) be an NTU game. For every $i \in N$ let

$$v^i = \max\{x^i \mid x^{\{i\}} \in V(\{i\})\}.$$

Then $V(\{i\}) = \{x^{\{i\}} \in \mathbb{R}^{\{i\}} \mid x^i \leq v^i\}$. (Frequently, $x^{\{i\}} \in \mathbb{R}^{\{i\}}$, a vector with one component, is identified with the scalar x^i.)

Definition 12.1.3. *A vector $x \in \mathbb{R}^N$ is* **individually rational** *(with respect to the NTU game (N, V)) if $x^i \geq v^i$ for all $i \in N$. We denote*

$$V(N)_{ir} = \{x \in V(N) \mid x \text{ is individually rational}\}.$$

Lemma 12.1.4. *If (N, V) is superadditive, then*

$$\mathcal{C}(N, V) = \mathcal{C}(N, V, V(N)_{ir}).$$

Proof: Clearly, $\mathcal{C}(N, V) \subseteq V(N)_{ir}$. Hence $\mathcal{C}(N, V) \subseteq \mathcal{C}(N, V, V(N)_{ir})$. In order to show the opposite inclusion, assume, on the contrary, that there exists $x \in \mathcal{C}(N, V, V(N)_{ir}) \setminus \mathcal{C}(N, V)$. Then there exist $S \subseteq N$, $|S| \geq 2$, and $y^S \in V(S)$ such that $y^S \gg x^S$. Let $z = (y^S, v^{N \setminus S})$. Then, by superadditivity of (N, V), $z \in V(N)$. Clearly, $z \in V(N)_{ir}$, and the desired contradiction has been obtained.

q.e.d.

Finally, let $V(N)_{e,ir}$ denote the set of weakly Pareto optimal and individually rational feasible payoff vectors, i.e., $V(N)_{e,ir} = V(N)_e \cap V(N)_{ir}$. The main result of this section is the following theorem.

Theorem 12.1.5. *If (N, V) is a superadditive NTU game, then*

$$\mathcal{C}(N, V) = \mathcal{C}(N, V, V(N)_e) = \mathcal{C}(N, V, V(N)_{ir}) = \mathcal{C}(N, V, V(N)_{e,ir}).$$

Clearly, it suffices to show that $\mathcal{C}(N, V, V(N)_e) = \mathcal{C}(N, V, V(N)_{e,ir})$. The proof of this assertion may easily be deduced from the proofs of Lemmata 12.1.2 and 12.1.4. Hence it is left as an exercise (Exercise 12.1.1).

Remark 12.1.6. An NTU game (N, V) is *superadditive at N* if, for every partition $\mathcal{P}$ of N, $\prod_{S \in \mathcal{P}} V(S) \subseteq V(N)$. We remark that in all results of this section "superadditivity" may be replaced by "superadditivity at N".

Exercises

Exercise 12.1.1. Prove Theorem 12.1.5.

Exercise 12.1.2. Show that the superadditivity condition is needed in Theorem 12.1.5 by providing games (N_k, V_k), $k = 1, 2$, satisfying

$$\mathcal{C}(N_1, V_1) \neq \mathcal{C}(N_1, V_1, V_1(N)_e) \text{ and } \mathcal{C}(N_2, V_2) \neq \mathcal{C}(N_2, V_2, V_2(N)_{ir}).$$

12.2 Balanced NTU Games

Let N be a (nonempty) finite set of players. A family $(\pi_S)_{S \in 2^N \setminus \{\emptyset\}}$ of vectors is *permissible* (for N) if, for all $\emptyset \neq S \subseteq N$, $\pi_S \in \mathbb{R}^N$, $\pi_S > 0$ (see Notation

4.4.5), $\pi_S^{N\setminus S} = 0 \in \mathbb{R}^{N\setminus S}$, and $\pi_N \gg 0$. Let π be a permissible family. A collection of coalitions $\mathcal{B}$ is π-*balanced* if there exist $\lambda_S > 0$, $S \in \mathcal{B}$, such that $\sum_{S\in\mathcal{B}} \lambda_S \pi_S = \pi_N$. Thus $\mathcal{B}$ is balanced (see Definition 3.1.2), iff $\mathcal{B}$ is π-balanced with respect to the family $(\pi_S)_{S\in 2^N\setminus\{\emptyset\}}$ defined by $\pi_S = \chi_S$ for all $\emptyset \neq S \subseteq N$.

Remark 12.2.1. Let π be a permissible family and let $\mathcal{B}$ be a collection of coalitions. Denote

$$\widehat{\pi}_S = \pi_S/\pi_S(S). \tag{12.2.1}$$

Then $\mathcal{B}$ is π-balanced iff $\widehat{\pi}_N$ is a convex combination of the $\widehat{\pi}_S$, $S \in \mathcal{B}$.

Let (N, V) be an NTU game. For every $S \subseteq N$ let $V_S = V(S) \times \mathbb{R}^{N\setminus S}$. The game (N, V) is *balanced* if

$$\bigcap_{S\in\mathcal{B}} V_S \subseteq V(N) \tag{12.2.2}$$

for every balanced collection $\mathcal{B}$ of coalitions. Moreover, (N, V) is π-*balanced* if (12.2.2) holds for every π-balanced collection $\mathcal{B}$.

Scarf (1967) proved that every balanced NTU game has a nonempty core. We shall prove that every π-balanced NTU game has a nonempty core. This more general result is due to Billera (1970b). First we consider the following example.

Example 12.2.2. Let $N = \{1, 2, 3\}$ and let

$$\begin{aligned}
V(\{i\}) &= \{x^i \in \mathbb{R}^{\{i\}} \mid x^i \leq 0\}, i \in N;\\
V(\{1,2\}) &= \{x^{\{1,2\}} \in \mathbb{R}^{\{1,2\}} \mid x^1 \leq 4, x^2 \leq 3\};\\
V(\{2,3\}) &= \{x^{\{2,3\}} \in \mathbb{R}^{\{2,3\}} \mid x^2 \leq 4, x^3 \leq 3\};\\
V(\{1,3\}) &= \{x^{\{1,3\}} \in \mathbb{R}^{\{1,3\}} \mid x^1 \leq 2, x^3 \leq 5\};\\
V(N) &= \left\{ x \in \mathbb{R}^N \,\middle|\, \begin{array}{l} (x^1 \leq 4, x^2 \leq 3, x^3 \leq 0) \text{ or} \\ (x^1 \leq 2, x^2 \leq 4, x^3 \leq 3) \end{array} \right\}.
\end{aligned}$$

Then (N, V) is not balanced, because $(\{2\}), \{1, 3\})$ is a balanced collection and $(2, 0, 5) \in (V_{\{2\}} \cap V_{\{1,3\}}) \setminus V(N)$. However, if $\delta = (1, 1/2, 1/3)$, $\pi_S = (\delta^S, 0^{N\setminus S})$ for all $\emptyset \neq S \subsetneq N$, and $\pi_N = \chi_N$, then (N, V) is π-balanced (see Exercise 12.2.1).

We may now formulate the main result of this section.

Theorem 12.2.3. *Let (N, V) be an NTU game and let π be a permissible family of vectors. If (N, V) is π-balanced, then $\mathcal{C}(N, V) \neq \emptyset$.*

We postpone the proof of Theorem 12.2.3 and start with the following observation. Let $b \in \mathbb{R}^N$ and let the game $(N, V + b)$ be defined by $(V + b)(S) = V(S) + b^S$ for all $S \subseteq N$. Then $(N, V + b)$ has a nonempty core (is π-balanced), iff (N, V) has a nonempty core (is π-balanced). Thus, in the proof of Theorem 12.2.3, we may assume that

$$v^i > 0 \text{ for all } i \in N. \tag{12.2.3}$$

Also, by (11.3.4), there exists $q > 0$ such that for every $S \subseteq N$,

$$x^S \in V(S) \cap \mathbb{R}^S_+ \Rightarrow x^i < q \text{ for all } i \in S.$$

Denote $Q = \{x \in \mathbb{R}^N \mid x \leq q\chi_N\}$ and define

$$Z = \left(\bigcup_{S \in 2^N} V_S \right) \cap Q.$$

As $V(S)$ and hence V_S are comprehensive, Z is comprehensive, that is, $Z = Z - \mathbb{R}^N_+$. Let ∂Z denote the boundary of Z. Then

$$(z \in \partial Z \text{ and } y \gg z) \Rightarrow y \notin Z. \tag{12.2.4}$$

Notice that for every $i \in N$, $v^i \chi_{\{i\}} + q\chi_{N \setminus \{i\}} \in Z$. Hence, by (12.2.4),

$$(z \in \partial Z \text{ and } z^j = 0 \text{ for some } j \in N) \Rightarrow \max_{i \in N} z^i = q. \tag{12.2.5}$$

We shall now prove that there exist a π-balanced collection $\mathcal{B}$ and $\widehat{z} \in \partial Z$ such that $\widehat{z} \in \bigcap_{S \in \mathcal{B}} V_S$. Finally, we shall prove that $\widehat{z} \in \mathcal{C}(N, V)$.

Proof of Theorem 12.2.3: Let $\Delta = \{x \in \mathbb{R}^N_+ \mid x(N) = 1\}$ and define, for every $x \in \Delta$,

$$f(x) = \{y \in \partial Z \mid y = tx \text{ for some } t \geq 0\}.$$

Notice that for every $x \in \Delta$, $n(q+1)x \notin Z$, whereas $0 \in \text{int } Z$ ("int" denotes the "interior of"). Hence, $f(x) \neq \emptyset$ for every $x \in \Delta$. We conclude that $f : \Delta \rightrightarrows \partial Z$ is closed and bounded, and thus, by Lemma 9.1.6, uhc. We claim that f is single-valued and, hence, continuous. Assume, on the contrary, that f is not single-valued. Let $x \in \Delta$ and $y, \widehat{y} \in f(x)$, $y = tx$, $\widehat{y} = \widehat{t}x$, where $\widehat{t} > t$. If $x \gg 0$, then $\widehat{y} \gg y$, contradicting (12.2.4). Thus, $K = \{k \in N \mid x^k = 0\} \neq \emptyset$. Now, for $j \in N \setminus K$, $y^j < \widehat{y}^j \leq q$, and $y^K = 0$, which contradicts (12.2.5). Hence, f is continuous.

With the help of f we define the set-valued mapping $g : \Delta \rightrightarrows \Delta$ by

$$g(x) = \{\widehat{\pi}_S \mid S \in 2^N \setminus \{\emptyset\} \text{ and } f(x) \in V_S\}$$

(see Equation (12.2.1)). Indeed, $g(x) \neq \emptyset$ for all $x \in \Delta$, because $f(x)$ is a nonempty subset of $\bigcup_{S \subseteq N} V_S$. Moreover, as the reader may verify, g is uhc (see Exercise 12.2.2). Thus, the set-valued function $\widetilde{g} : \Delta \times \Delta \rightrightarrows \Delta$, defined by

$\widetilde{g}(x,y) = \text{convh}\ g(x)$ for all $x, y \in \Delta$, is uhc and convex-valued. We proceed to define a continuous function $h : \Delta \times \Delta \to \Delta$ by

$$h^i(x,y) = \frac{x^i + \left(y^i - \widehat{\pi}^i_N\right)_+}{1 + \sum_{j \in N} \left(y^j - \widehat{\pi}^j_N\right)_+} \text{ for all } i \in N, x, y \in \Delta.$$

Finally, $h \times \widetilde{g} : \Delta \times \Delta \rightrightarrows \Delta \times \Delta$ satisfies the conditions of Kakutani's fixed-point theorem. Hence, there exists $(\widehat{x}, \widehat{y}) \in \Delta \times \Delta$ such that $\widehat{x} = h(\widehat{x}, \widehat{y})$ and $\widehat{y} \in \text{convh}\ g(\widehat{x})$. Thus,

$$\widehat{x}^i \left(\sum_{j \in N} \left(\widehat{y}^j - \widehat{\pi}^j_N\right)_+ \right) = \left(\widehat{y}^i - \widehat{\pi}^i_N\right)_+ \text{ for all } i \in N. \tag{12.2.6}$$

We claim that $\widehat{y} = \widehat{\pi}_N$. Assume, on the contrary, that

$$\sum_{j \in N} \left(\widehat{y}^j - \widehat{\pi}^j_N\right)_+ > 0.$$

Let $K_+ = \{i \in N \mid \widehat{x}^i > 0\}$ and $K_0 = N \setminus K_+$. If $i \in K_+$, then $\widehat{y}^i > \widehat{\pi}^i_N > 0$ by (12.2.6). Thus $K_0 \neq \emptyset$, because $\widehat{y} \in \Delta$. As $\widehat{y} \in \text{convh}\ g(\widehat{x})$, for each $i \in K_+$ there exists $S \subseteq N$ such that $i \in S$ and $f(\widehat{x}) \in V_S$. Hence, the choice of q and $f(\widehat{x}) \geq 0$ imply $f^i(\widehat{x}) < q$ for all $i \in K_+$. As $f^j(\widehat{x}) = 0$ for all $j \in K_0$ and $K_0 \neq \emptyset$, (12.2.5) is violated. Thus, $\widehat{y} = \widehat{\pi}_N$.

Let now $\mathcal{B} = \{S \subseteq N \mid f(\widehat{x}) \in V_S\}$. Then

$$\widehat{\pi}_N \in \text{convh}\ \{\widehat{\pi}_S \mid S \in \mathcal{B}\}.$$

By Remark 12.2.1, $\mathcal{B}$ is balanced. Let $\widehat{z} = f(\widehat{x})$. Then $\widehat{z} \in \bigcap_{S \in \mathcal{B}} V_S$. As (N, V) is π-balanced, $\widehat{z} \in V(N)$. We conclude the proof by showing that $\widehat{z} \in \mathcal{C}(N, V)$. By definition, $\widehat{z} \geq 0$. Hence $\widehat{z} \ll q\chi_N$. Assume now, on the contrary, that $\widehat{z} \notin \mathcal{C}(N, V)$. Then there exist $\emptyset \neq S \subseteq N$ and $x \in V_S$ such that $x^S \gg \widehat{z}^S$. We may assume that $x^i = q$ for all $i \in N \setminus S$. Then $x \in Z$ and $x \gg \widehat{z} \in \partial Z$, contradicting (12.2.4). **q.e.d.**

In view of Theorem 12.2.3, π-balancedness for some permissible π is a sufficient condition for the nonemptiness of the core. The following example shows that it is not a necessary condition.

Example 12.2.4. Let $N = \{1, 2, 3\}$ and let

$$\begin{aligned} V(N) &= \{x \in \mathbb{R}^N \mid x^1 \leq \tfrac{1}{2}, x^2 \leq \tfrac{1}{2}, \text{ and } x^3 \leq 0\}; \\ V(\{1,2\}) &= \{x^{\{1,2\}} \in \mathbb{R}^{\{1,2\}} \mid x^1 + x^2 \leq 1\}; \\ V(S) &= \{x^S \in \mathbb{R}^S \mid x^S \leq 0\} = -\mathbb{R}^S_+ \text{ otherwise.} \end{aligned}$$

Then $(\frac{1}{2}, \frac{1}{2}, 0) \in \mathcal{C}(N, V)$. We claim that there does not exist any permissible family π such that (N, V) is π-balanced. Indeed, if π is permissible on N, then at least one of the following collections is π-balanced:

$$\{\{1,2\},\{3\}\}, \{\{1,2\},\{2\},\{3\}\}, \{\{1,2\},\{1\},\{3\}\}.$$

If $\mathcal{B}$ is one of these collections of coalitions, then either $(1, 0, 0)$ or $(0, 1, 0)$ is a member of $\bigcap_{S \in \mathcal{B}} V_S$. But neither $(1, 0, 0)$ nor $(0, 1, 0)$ belongs to $V_N = V(N)$.

Theorem 12.2.3, Example 12.2.2, and Example 12.2.4 are due to Billera (1970b). However, the present proof of Theorem 12.2.3 is due to Shapley and Vohra (1991).

We conclude this section with a result of Scarf (1967) on the nonemptiness of the core of market games. If $(N, \mathbb{R}^m_+, A, W)$ is an NTU-market (see Section 11.1), then (N, V), defined by

$$V(S) = \left\{ u^S \in \mathbb{R}^S \;\middle|\; \begin{array}{c} \text{there exists } x_S \in X^S \text{ such that} \\ w^i(x^i_S) \geq u^i \text{ for all } i \in S \end{array} \right\} \tag{12.2.7}$$

for all $S \in 2^N \setminus \{\emptyset\}$ and $V(\emptyset) = \emptyset$, is the *derived* NTU game (see Example 11.1.4 for the definition of X^S).

Theorem 12.2.5. *Let $(N, \mathbb{R}^m_+, A, W)$ be an NTU-market and let (N, V) be the derived NTU game. If w^i is quasi-concave for each $i \in N$, then $\mathcal{C}(N, V) \neq \emptyset$.*

Proof: We shall prove that the derived game (N, V) is balanced. Let $\mathcal{B}$ be a balanced collection, let $(\delta_S)_{S \in \mathcal{B}}$ be a system of balancing weights (see Definition 3.1.2) and let $u \in \bigcap_{S \in \mathcal{B}} V_S$. Then for each $S \in \mathcal{B}$ there exists $x_S \in X^S$ such that $w^i(x^i_S) \geq u^i$ for all $i \in S$. Let $z = (z^i)_{i \in N} \in (\mathbb{R}^m_+)^N$ be defined by

$$z^i = \sum_{S: S \in \mathcal{B}, S \ni i} \delta_S x^i_S.$$

The quasi-concavity of w^i implies that $w^i(z^i) \geq u^i$ for all $i \in N$. It remains to prove that $\sum_{i \in N} z^i = \sum_{i \in N} a^i$. The observation that

$$\begin{aligned} \textstyle\sum_{i \in N} z^i &= \textstyle\sum_{i \in N} \sum_{S: i \in S \in \mathcal{B}} \delta_S x^i_S = \sum_{S \in \mathcal{B}} \delta_S \sum_{i \in S} x^i_S \\ = \textstyle\sum_{S \in \mathcal{B}} \delta_S \sum_{i \in S} a^i &= \textstyle\sum_{i \in N} a^i \sum_{S: i \in S \in \mathcal{B}} \delta_S = \sum_{i \in N} a^i \end{aligned}$$

shows that $u \in V(N)$. **q.e.d**

Exercises

Exercise 12.2.1. Prove that the NTU game defined in Example 12.2.2 is π-balanced (with respect to the family $(\pi_S)_{\emptyset \neq S \subseteq N}$ which is specified therein).

Exercise 12.2.2. Prove that $g : \Delta \rightrightarrows \Delta$, defined in the proof of Theorem 12.2.3, is uhc.

Exercise 12.2.3. Let (N, v) be a TU game and let (N, V_v) be defined by (11.3.5). Prove that (N, v) is balanced iff (N, V_v) is balanced.

12.3 Ordinal and Cardinal Convex Games

In this section we shall investigate the core of convex NTU games.

12.3.1 Ordinal Convex Games

In this subsection we shall prove that the core of an ordinal convex NTU game is nonempty. We start with the following definition.

Definition 12.3.1. *A pair $(N, \overline{V})$ is a* **quasi-game** *if $(N, \overline{V})$ is a pregame (see Definition 11.4.1) that satisfies (11.3.4) and if $\overline{V}(N)$ is closed.*

The following definition, due to Vilkov (1977), is a first generalization of Shapley's notion of "convex game" to NTU games.

Definition 12.3.2. *A quasi-game $(N, \overline{V})$ is an* **ordinal convex quasi-game** *if for all $S, T \subseteq N$*

$$\overline{V}_S \cap \overline{V}_T \subseteq \overline{V}_{S\cap T} \cup \overline{V}_{S\cup T} \tag{12.3.1}$$

where $\overline{V}_Q = \overline{V}(Q) \times \mathbb{R}^{N\setminus Q}$ for all $Q \subseteq N$ (see Section 12.2).

We shall now prove the following result.

Theorem 12.3.3. *The core of an ordinal convex quasi-game is nonempty.*

Proof: Let $(N, \overline{V})$ be an ordinal convex quasi-game. By induction on $n = |N|$ we shall prove that $\mathcal{C}(N, \overline{V}) \neq \emptyset$. The case $n = 1$ follows from the assumptions that $\overline{V}(N)$ satisfies (11.3.1), (11.3.3), and (11.3.4). Now let $n \geq 2$, let $i \in N$, let $M = N \setminus \{i\}$, and let $\alpha = \sup \overline{V}(\{i\})$. Clearly, $\alpha < \infty$. We consider α as a member of $\mathbb{R}^{\{i\}}$. Let (M, V) be the restriction of $(N, \overline{V})$ to the player set M, i.e., $V(S) = \overline{V}(S)$ for all $S \subseteq M$. Note that (M, V) is a pregame that satisfies (11.3.4). Define the pregame (M, W) by

$$W(S) = \{x^S \in \mathbb{R}^S \mid (x^S, \beta) \in \overline{V}(S \cup \{i\}) \text{ for some } \beta \in \mathbb{R}^{\{i\}}, \beta > \alpha\}$$

for all $S \in 2^M \setminus \{\emptyset\}$ and $W(\emptyset) = \emptyset$.

We now define the quasi-game (M, U) by

$$U(M) = \{x^M \in \mathbb{R}^M \mid (x^M, \alpha) \in \overline{V}(N)\},\ U(\emptyset) = \emptyset$$

and

$$U(S) = V(S) \cup W(S) \text{ for all } \emptyset \neq S \subsetneqq M.$$

Clearly, $U(M)$ is closed. We claim that $U(M) \supseteq V(M)$. Indeed, let $x^M \in V(M)$ and let $\alpha_t \in \mathbb{R}^{\{i\}}$, $\alpha_t < \alpha$, $t \in \mathbb{N}$, satisfy $\lim_{t\to\infty} \alpha_t = \alpha$. By (12.3.1), $(x^M, \alpha_t) \in \overline{V}(N)$ for $t \in \mathbb{N}$ and, thus, $(x^M, \alpha) \in \overline{V}(N)$.

Clearly, (M, U) is a quasi-game. We shall now prove that (M, U) is ordinal convex. Let $S, T \subseteq M$ and $x \in U_S \cap U_T$. Clearly, we may assume that $S \neq M \neq T$ and that $S \cup T \neq \emptyset$. We distinguish the following possible cases:

(1) $x \in V_S \cap V_T$. Then $(x, \alpha) \in \overline{V}_S \cap \overline{V}_T$. (Note that $V_Q \subseteq \mathbb{R}^M$ for every $Q \subseteq M$.) Thus, by (12.3.1),

$$(x, \alpha) \in \overline{V}_{S\cap T} \cup \overline{V}_{S\cup T}.$$

Hence,

$$x \in V_{S\cap T} \cup V_{S\cup T} \subseteq U_{S\cap T} \cup U_{S\cup T}.$$

(If $S \cup T = M$, then the above inclusion follows from the fact that $U_M \supseteq V_M$.)

(2) $x \in V_S \cap W_T$. Let $\beta \in \mathbb{R}^{\{i\}}$, $\beta > \alpha$, such that $(x, \beta) \in \overline{V}_{T\cup\{i\}}$. Then $(x, \beta) \in \overline{V}_S \cap \overline{V}_{T\cup\{i\}}$. Hence, by ordinal convexity of $(N, \overline{V})$,

$$(x, \beta) \in \overline{V}_{S\cap T} \cup \overline{V}_{S\cup T\cup\{i\}}.$$

Hence, $x \in V_{S\cap T} \cup W_{S\cup T} \subseteq U_{S\cap T} \cup U_{S\cup T}$. (If $S \cup T = M$, then the above inclusion follows from comprehensiveness of $\overline{V}(N)$.)

(3) The case $x \in W_S \cap V_T$ may be treated analogously to the preceding case.

(4) $x \in W_S \cap W_T$. By comprehensiveness there exists $\beta \in \mathbb{R}^{\{i\}}$, $\beta > \alpha$, such that $(x, \beta) \in \overline{V}_{S\cup\{i\}} \cap \overline{V}_{T\cup\{i\}}$. Hence,

$$(x, \beta) \in \overline{V}_{S\cup\{i\}} \cap \overline{V}_{T\cup\{i\}} \subseteq \overline{V}_{(S\cap T)\cup\{i\}} \cup \overline{V}_{S\cup T\cup\{i\}}.$$

If $(x, \beta) \in \overline{V}_{S\cup T\cup\{i\}}$, then $x \in W_{S\cup T} \subseteq U_{S\cup T}$. Finally, if $(x, \beta) \notin \overline{V}_{S\cup T\cup\{i\}}$, then $(x, \beta) \in \overline{V}_{(S\cap T)\cup\{i\}}$. As $\beta > \alpha$, $S \cap T \neq \emptyset$. Hence, $x \in W_{S\cap T} \subseteq U_{S\cap T}$.

By the induction hypothesis there exists $x \in \mathcal{C}(M, U)$. We claim that y, defined by $y = (x, \alpha)$, is an element of $\mathcal{C}(N, \overline{V})$. The vector y is weakly Pareto optimal in $\overline{V}(N)$, because x is weakly Pareto optimal in $U(M)$. Assume, on the contrary, that $y \notin \mathcal{C}(N, \overline{V})$. Then there exist $\emptyset \neq S \subsetneqq N$, $S \neq M$, and $z \in \overline{V}_S$ such that z dominates x via S. If $i \notin S$, then $z^M \in V_S \subseteq U_S$ and $z^S \gg x^S$, contradicting $x \in \mathcal{C}(M, U)$. If $i \in S$, then $z^i > \alpha$. Hence $z^M \in W_{\widehat{S}} \subseteq U_{\widehat{S}}$ and $z^{\widehat{S}} \gg x^{\widehat{S}}$, where $\widehat{S} = S \setminus \{i\}$, which is impossible. **q.e.d.**

Theorem 12.3.3 is due to Greenberg (1985).

12.3.2 Cardinal Convex Games

Let (N, V) be an NTU game. We denote

$$V_0(S) = V(S) \times \{0^{N\setminus S}\} \subseteq \mathbb{R}^N \text{ for all } S \in 2^N \setminus \{\emptyset\} \tag{12.3.2}$$

and $V_0(\{\emptyset\}) = \{0\}$. This notation enables us to define cardinal balanced games.

Definition 12.3.4. *An NTU game (N, V) is* **cardinal balanced** *if for every balanced collection of coalitions $\mathcal{B}$ with a system $(\delta_S)_{S\in\mathcal{B}}$ of balancing coefficients,*

$$\sum_{S\in\mathcal{B}} \delta_S V_0(S) \subseteq V(N). \tag{12.3.3}$$

A cardinal balanced game is balanced (see Exercise 12.3.2). Thus, the core of a cardinal balanced game is nonempty.

Let (N, v) be a TU game and let (N, V_v) be the associated NTU game defined by (11.3.5). Clearly, (N, v) is balanced if and only if (N, V_v) is cardinal balanced.

Let $(N, \mathbb{R}^m_+, A, W)$ be an NTU-market and (N, V) be the derived NTU game (defined by (12.2.7)). If the utility functions w^i, $i \in N$, are concave, then (N, V) is cardinal balanced (see Exercise 12.3.3).

In this section we shall prove that a cardinal convex game is cardinal balanced. We start with the following definition.

Definition 12.3.5. *An NTU game (N, V) is* **cardinal convex** *if*

$$V_0(S) + V_0(T) \subseteq V_0(S \cap T) + V_0(S \cup T) \text{ for all } S, T \in 2^N. \tag{12.3.4}$$

The following theorem is due to Sharkey (1981).

Theorem 12.3.6. *Let (N, V) be a cardinal convex NTU game. If $V(N)$ is convex, then (N, V) is cardinal balanced.*

Proof: As $V(N)$ is convex and in view of Corollary 3.1.9 it is sufficient to prove (12.3.3) for minimal balanced collections of coalitions. Let $\mathcal{B} = \{S_1, \ldots, S_\ell\}$ be a minimal balanced collection of coalitions and let $\lambda_1, \ldots, \lambda_\ell$ be its balancing coefficients. Then $\lambda_1, \ldots, \lambda_\ell$ are rational numbers. Hence, there exists $\delta \in \mathbb{N}$ such that $\delta_j = \delta\lambda_j \in \mathbb{N}$ for all $j = 1, \ldots, \ell$. We have to prove (12.3.3), that is,

$$\sum_{j=1}^{\ell} \delta_j V_0(S_j) \subseteq \delta V(N). \tag{12.3.5}$$

Let $p = \sum_{j=1}^{\ell} \delta_j$. Then (12.3.5) can be written as

$$\sum_{k=1}^{p} V_0(T_k) \subseteq \delta V(N), \tag{12.3.6}$$

where every coalition S_j, $j = 1, \ldots, \ell$, is repeated δ_j times in the sequence $T_1, \ldots, T_p$.

Each $i \in N$ appears in exactly δ coalitions in the sequence $T_1, \ldots, T_p$. A pair of coalitions (T_h, T_k) is *incomparable* if $T_k \setminus T_h \neq \emptyset \neq T_h \setminus T_k$. With the help of (12.3.4) we can reduce the number of incomparable pairs in the sequence $T_1, \ldots, T_p$. Indeed, let (T_h, T_k) be an incomparable pair. Then

$$\sum_{j=1}^{p} V_0(T_j) \subseteq V_0(T_h \cap T_k) + V_0(T_h \cup T_k) + \sum_{j \in \{1,\ldots,p\} \setminus \{h,k\}} V_0(T_j). \tag{12.3.7}$$

As the reader may easily verify, the number of incomparable pairs on the right-hand side of (12.3.7) (that is, in the sequence that arises from $T_1, \ldots, T_p$ by only replacing T_h and T_k by $T_h \cap T_k$ and $T_h \cup T_k$) is smaller than the number of incomparable pairs on the left-hand side (that is, in the original sequence $T_1, \ldots, T_p$). Continuing in this manner, we finally obtain coalitions $U_1 \subseteq \cdots \subseteq U_p$ such that $\sum_{j=1}^{p} V_0(T_j) \subseteq \sum_{j=1}^{p} V_0(U_j)$.

As each player appears in exactly δ coalitions in the sequence $U_1, \ldots, U_p$, it follows that $U_1 = \cdots = U_{p-\delta} = \emptyset$ and $U_{p-\delta+1} = \cdots = U_p = N$. Finally, as $V(N)$ is convex by assumption,

$$\sum_{j=1}^{p} V_0(U_j) = \sum_{j=p-\delta+1}^{p} V(N) \subseteq \delta V(N).$$

q.e.d.

Exercises

Exercise 12.3.1. Find an ordinal convex game which is not balanced (see Greenberg (1985)).

Exercise 12.3.2. Prove that any cardinal balanced NTU game is balanced.

Exercise 12.3.3. Let $(N, \mathbb{R}^m_+, A, W)$ be an NTU-market. Prove that if the utility functions w^i, $i \in N$, are concave, then the NTU game derived from the market is cardinal balanced.

Exercise 12.3.4. Give an example of an ordinal convex NTU game that is not cardinal convex.

Exercise 12.3.5. Give an example of a cardinal convex NTU game that is not ordinal convex (see Sharkey (1981)).

Exercise 12.3.6. Let (N, v) be a TU game and let (N, V_v) be the corresponding NTU game. Show that the following assertions are equivalent: (a) (N, v) is convex; (b) (N, V_v) is ordinal convex; (c) (N, V_v) is cardinal convex.

12.4 An Axiomatization of the Core

We now turn to the axiomatization of the core. As in the TU context, reduced games and their properties will play a central role. In the first subsection we shall define and study reduced games of NTU games. First, however, we introduce an additional property of NTU games. If $A \subseteq \mathbb{R}^N$ for some finite set N, then ∂A denotes the boundary of A, that is, the intersection of the closures of A and its complement.

Definition 12.4.1. *Let (N, V) be an NTU game. The game V is* **non-levelled** *if for all $S \subseteq N$, $V(S)$ is* **non-levelled**, *that is, for all $x^S, y^S \in \partial V(S)$, $x^S \geq y^S$ implies that $x = y$.*

Non-levelness is a standard property of NTU games. Now we assume it for the rest of this section. Also, we shall use it throughout the next chapter.

12.4.1 Reduced Games of NTU Games

The concept of a reduced game is now generalized to NTU games.

Definition 12.4.2. *Let (N, V) be an NTU game, $x \in V(N)$, and let $S \in 2^N \setminus \{\emptyset\}$. The* **reduced game** *(of (N, V)) with respect to S and x is the game $(S, V_{S,x})$ defined by*

$$V_{S,x}(S) = \{y^S \in \mathbb{R}^S \mid (y^S, x^{N\setminus S}) \in V(N)\}, \text{ and} \tag{12.4.1}$$

$$V_{S,x}(T) = \bigcup_{Q \subseteq N\setminus S} \{y^T \in \mathbb{R}^T \mid (y^T, x^Q) \in V(T \cup Q)\} \tag{12.4.2}$$

for all $T \in 2^S \setminus \{\emptyset, S\}$, and $V_{S,x}(\emptyset) = \emptyset$.

In the reduced game the players of S are allowed to choose only payoff vectors y^S that are compatible with $x^{N\setminus S}$, the fixed payoff distribution to the members of $N \setminus S$. On the other hand, proper subcoalitions T of S may count on the cooperation of any subset Q of $N \setminus S$, provided that in the resulting payoff vectors for $T \cup Q$ each member i of Q receives exactly x^i. (Hence, if T counts on the cooperation of some $Q \neq N \setminus S$, then T has to ensure the feasibility of x^Q but not that of $x^{N\setminus S}$.) Thus the reduced game $(S, V_{S,x})$ describes the following situation. Suppose that all the members of the grand coalition N

agree that the members of $N \setminus S$ will get $x^{N\setminus S}$. Further, assume that the members of $N \setminus S$ continue to cooperate with the members of S (subject to the foregoing agreement). Then $V_{S,x}$ describes the possible payoff vectors that various coalitions of members of S may obtain. However, it is assumed that S will choose some payoff vector in $V_{S,x}(S)$. Thus, the sets $V_{S,x}(T)$, $\emptyset \neq T \subsetneqq S$, serve only to determine the final choice in $V_{S,x}(S)$.

Reduced games of NTU games were used in several papers (e.g., Greenberg (1985)). The foregoing definition is due to Peleg (1985). As the reader may easily verify, a reduced game of an NTU game is an NTU game. We shall now prove that also non-levelness is hereditary.

Lemma 12.4.3. *Let (N, V) be a non-levelled NTU game, let $x \in V(N)$, and let $S \in 2^N \setminus \{\emptyset\}$. Then the reduced game $(S, V_{S,x})$ is non-levelled.*

Proof: Let $\emptyset \neq T \subseteq S$ and let $y^T, z^T \in \partial V_{S,x}(T)$. Assume, on the contrary, that $y^T > z^T$. Then there exists $Q \subseteq N \setminus S$ such that $(y^T, x^Q) \in V(T \cup Q)$. By comprehensiveness, $(z^T, x^Q) \in V(T \cup Q)$. By non-levelness of (N, V) there exists $w^{T\cup Q} \in V(T \cup Q)$ such that $w^{T\cup Q} \gg (z^T, x^Q)$. Clearly, $(w^T, x^Q) \in V(T \cup Q)$. Hence $w^T \in V_{S,x}(T)$ and $w^T \gg z^T$. Thus, $z^T \notin \partial V_{S,x}(T)$ and the desired contradiction has been obtained. **q.e.d.**

We now recall the following well-known definition.

Definition 12.4.4. *Let (N, V) be an NTU game and $x \in V(N)$. The vector x is* **Pareto optimal** *(in (N, V)) if there is no $y \in V(N)$ such that $y > x$.*

Let (N, V) be a non-levelled NTU game. A payoff vector $x \in V(N)$ is Pareto optimal if and only if $x \in \partial V(N)$. As $\mathcal{C}(N, V) \subseteq \partial V(N)$, every member of $\mathcal{C}(N, V)$ is Pareto optimal.

Pareto optimality is a hereditary property, that is, it is inherited by the restricted vectors in the corresponding reduced games. More precisely, we have the following lemma.

Lemma 12.4.5. *Let (N, V) be a non-levelled NTU game, let $x \in V(N)$, and let $S \in 2^N \setminus \{\emptyset\}$. Then x is Pareto optimal iff x^S is Pareto optimal in the reduced game $(S, V_{S,x})$.*

Proof: If x^S is not Pareto optimal, then there exists $y^S \in V_{S,x}(S)$ such that $y^S > x^S$. Hence $(y^S, x^{N\setminus S}) \in V(N)$ and $(y^S, x^{N\setminus S}) > x$. Therefore x is not Pareto optimal. Conversely, if x is not Pareto optimal, then, by the non-levelness of $V(N)$, there exists $y \in V(N)$ satisfying $y \gg x$. Clearly, $(y^S, x^{N\setminus S}) \in V(N)$. Therefore $y^S \in V_{S,x}(S)$ and x^S is not Pareto optimal. **q.e.d.**

12.4.2 Axioms for the Core

Let $\mathcal{U}$ be a set of players. Denote by $\widehat{\Gamma}_{\mathcal{U}}$ the set of non-levelled NTU games (N, V) such that $N \subseteq \mathcal{U}$ and let $\widehat{\Gamma} \subseteq \widehat{\Gamma}_{\mathcal{U}}$.

We shall be interested in the following properties of solutions on $\widehat{\Gamma}$.

A *solution* on $\widehat{\Gamma}$ is a mapping σ that assigns to each $(N, V) \in \widehat{\Gamma}$ a subset $\sigma(N, V)$ of $V(N)$. Let σ be a solution on $\widehat{\Gamma}$. Then σ satisfies *nonemptiness* (NE) if $\sigma(N, V) \neq \emptyset$ for every $(N, V) \in \widehat{\Gamma}$. Moreover, σ satisfies *Pareto optimality* (PO) if $\sigma(N, V) \subseteq \partial V(N)$ for every $(N, V) \in \widehat{\Gamma}$. Finally, σ satisfies *individual rationality* (IR) if for every game $(N, V) \in \widehat{\Gamma}$, every $x \in \sigma(N, V)$ is individually rational (see Definition 12.1.3).

We further denote $\widehat{\Gamma}^{\mathcal{C}} = \{(N, V) \in \widehat{\Gamma} \mid \mathcal{C}(N, v) \neq \emptyset\}$. Obviously, $\sigma = \mathcal{C}$ satisfies PO and IR on $\widehat{\Gamma}$. Also, it satisfies NE on $\widehat{\Gamma}^{\mathcal{C}}$. We proceed with the definition of the reduced game property.

Definition 12.4.6. *A solution σ on $\widehat{\Gamma}$ has the* **reduced game property** *(*RGP*) if it satisfies the following condition: If $(N, V) \in \widehat{\Gamma}$, $S \in 2^N \setminus \{\emptyset\}$, and $x \in \sigma(N, V)$, then $(S, V_{S,x}) \in \widehat{\Gamma}$ and $x^S \in \sigma(S, V_{S,x})$.*

RGP is a condition of *self-consistency*: If $(N, V) \in \widehat{\Gamma}$ and $x \in \sigma(N, V)$, that is, x is prescribed as a solution to (N, V), then RGP requires that, for every subcoalition S of N, x^S be a solution to the reduced game $(S, V_{S,x})$.

Note that, by Lemma 12.4.3, $\widehat{\Gamma}_{\mathcal{U}}$ is closed under reduction, that is, if $(N, V) \in \widehat{\Gamma}_{\mathcal{U}}$, $x \in V(N)$, and $\emptyset \neq S \subseteq N$, then $(S, V_{S,x}) \in \widehat{\Gamma}_{\mathcal{U}}$.

Lemma 12.4.7. *The core satisfies* RGP *on $\widehat{\Gamma}^{\mathcal{C}}_{\mathcal{U}}$.*

Proof: Let $(N, V) \in \widehat{\Gamma}^{\mathcal{C}}_{\mathcal{U}}$, $x \in \sigma(N, v)$, and $\emptyset \neq S \subseteq N$. Assume, on the contrary, that $x^S \notin \mathcal{C}(S, V_{S,x})$. Then there exist $\emptyset \neq T \subseteq S$ and $y^T \in V_{S,x}(T)$ such that $y^T \gg x^T$. Hence, there exists $Q \subseteq N \setminus S$ such that $(y^T, x^Q) \in V(T \cup Q)$ and $(y^T, x^Q) > x^{T \cup Q}$. As (N, V) is non-levelled, there exists $z^{T \cup Q} \in V(T \cup Q)$ such that $z^{T \cup Q} \gg x^{T \cup Q}$, contradicting the assumption that $x \in \mathcal{C}(N, V)$. **q.e.d.**

We proceed to formulate the main result of this section.

Theorem 12.4.8. *Assume that $\mathcal{U}$ is infinite. Then there exists a unique solution on $\widehat{\Gamma}^{\mathcal{C}}_{\mathcal{U}}$ that satisfies* NE, IR, *and* RGP, *and it is the core.*

The proof of Theorem 12.4.8 is postponed to Subsection 12.4.3. We still need some preliminary definitions and results. We recall that if N is a coalition, then $\mathcal{P}(N) = \{\{i, j\} \mid i, j \in N, i \neq j\}$.

Definition 12.4.9. *A solution σ on $\widehat{\Gamma}$ has the* **converse reduced game property** *(*CRGP*) if it satisfies the following condition: If $(N,V) \in \widehat{\Gamma}$, $|N| \geq 2$, $x \in \partial V(N)$, and if for every $S \in \mathcal{P}(N)$, $(S, V_{S,x}) \in \widehat{\Gamma}$ and $x^S \in \sigma(S, V_{S,x})$, then $x \in \sigma(N,V)$.*

The proofs of the two following lemmata are left as exercises.

Lemma 12.4.10. *The core satisfies* CRGP *on $\widehat{\Gamma}$.*

Moreover, the following lemma is needed.

Lemma 12.4.11. *Let σ be a solution on $\widehat{\Gamma}$ that satisfies* IR *and* RGP*. Then σ satisfies* PO.

12.4.3 Proof of Theorem 12.4.8

We have already shown that the core satisfies RGP on $\widehat{\Gamma}^{\mathcal{C}}_{\mathcal{U}}$. As mentioned before, it satisfies NE and IR. Thus, we only have to show uniqueness. First we prove the following lemma.

Lemma 12.4.12. *If σ is a solution on $\widehat{\Gamma}$ that satisfies* IR *and* RGP*, then $\sigma(N,V) \subseteq \mathcal{C}(N,V)$ for every $(N,V) \in \widehat{\Gamma}$.*

Proof: By Lemma 12.4.11, σ satisfies PO. Let $(N,V) \in \widehat{\Gamma}$ and let $n = |N|$. If $n \leq 2$, then IR and PO imply that $\sigma(N,V) \subseteq \mathcal{C}(N,V)$. Now let $n \geq 3$ and let $x \in \sigma(N,V)$. By RGP, $x^S \in \sigma(S, V_{S,x})$ for every $S \in \mathcal{P}(N)$. By PO, $x \in \partial V(N)$. By Lemma 12.4.10, the core satisfies CRGP. Thus, $x \in \mathcal{C}(N,V)$. **q.e.d.**

The following lemma plays a key role in the proof.

Lemma 12.4.13. *Let $(N,V) \in \widehat{\Gamma}^{\mathcal{C}}_{\mathcal{U}}$, let $\ell \in U \setminus N$, let $M = N \cup \{\ell\}$, and let $\bar{x} \in \mathcal{C}(N,V)$. Then there exists $(M,W) \in \widehat{\Gamma}^{\mathcal{C}}_{\mathcal{U}}$ that satisfies the following properties:*

$$\mathcal{C}(M,W) = \{z\} = \left\{\left(\bar{x}, 0^{\{\ell\}}\right)\right\}; \tag{12.4.3}$$

$$W_{N,z} = V. \tag{12.4.4}$$

We postpone the proof of Lemma 12.4.13 and use it first to prove Theorem 12.4.8.

Proof of Theorem 12.4.8: Let σ be a solution on $\widehat{\Gamma}^{\mathcal{C}}_{\mathcal{U}}$ that satisfies NE, IR, and RGP. By Lemma 12.4.12 we only have to prove that the core is a subsolution of σ. Let $(N,V) \in \widehat{\Gamma}^{\mathcal{C}}_{\mathcal{U}}$ and $\bar{x} \in \mathcal{C}(N,V)$. By our infinity assumption there exists $\ell \in U \setminus N$. Let $M = N \cup \{\ell\}$. By Lemma 12.4.13 there exists a game (M, W) that satisfies (12.4.3) and (12.4.4). By NE, $\{z\} = \sigma(M, W)$. By RGP, $z^N \in \sigma(N, W_{N,z})$. But $z^N = \bar{x}$ and $W_{N,z} = V$. **q.e.d.**

Given $(N, V) \in \widehat{\Gamma}^{\mathcal{C}}_{\mathcal{U}}$, $\ell \in \mathcal{U} \setminus N$, we now construct a game (M, W) satisfying (12.4.3) and (12.4.4).

Proof of Lemma 12.4.13: In order to define W, let $\emptyset \neq S \subseteq M$. We distinguish the cases (1) – (4) and complete the proof in Step (5).

(1) $\ell \notin S$ and $S \neq N$: In this case we define $W(S) = V(S)$.

(2) $S = N$: For $x \in \mathbb{R}^N$ and $i \in N$ let

$$f^i(x^i) = x^i + \frac{(x^i - \bar{x}^i)_+}{1 + (x^i - \bar{x}^i)_+}.$$

Then $f^i(\cdot)$ is a continuous and strictly increasing function of x^i. Let $f = (f^i)_{i \in N} : V(N) \to \mathbb{R}^N$ and define $W(N) = f(V(N))$.

Note that

$$x^i \leq f^i(x^i) \leq f^i(x^i) + 1 \text{ for all } x \in V(N) \text{ and } i \in N. \tag{12.4.5}$$

The proof that $W(N)$ is closed and nonempty is straightforward. Also, by (12.4.5), $W(N)$ satisfies (11.3.4). We shall now prove that $W(N)$ is comprehensive. Let $y \in W(N)$ and $u \in \mathbb{R}^N$, $u \leq y$. Then there exists $x \in V(N)$ such that $y = f(x)$. By the continuity of f^i, $i \in N$, and by (12.4.5), there exists $\widetilde{x}^i \leq x^i$ such that $f^i(\widetilde{x}^i) = u^i$ for all $i \in N$. By comprehensiveness of $V(N)$, $\widetilde{x} = (\widetilde{x}^i)_{i \in N} \in V(N)$. Hence, $W(N)$ satisfies (11.3.2). Finally, the non-levelness of $W(N)$ follows immediately from the strict monotonicity of f^i, $i \in N$.

We shall use the following properties (12.4.6) and (12.4.7) of $W(N)$:

$$\bar{x} \in \partial W(N). \tag{12.4.6}$$

$$\text{If } x \in \partial V(N) \text{ and } x \neq \bar{x}, \text{ then } x \notin \partial W(N). \tag{12.4.7}$$

In order to prove (12.4.6), let $y \in W(N)$. Then there exists $x \in V(N)$ such that $f(x) = y$. As $\bar{x} \in \partial V(N)$, there exists $j \in N$ such that $x^j \leq \bar{x}^j$. Hence, $y^j = x^j \leq \bar{x}^j$.

In order to prove (12.4.7), let $x \in \partial V(N)$, $x \neq \bar{x}$. Then there exists $j \in N$ such that $x^j > \bar{x}^j$. By the definition of f^j there exists $\varepsilon > 0$ such that $f^j(\widetilde{x}^j) > x^j$, where $\widetilde{x}^j = x^j - \varepsilon$. By non-levelness of $V(N)$, $\widetilde{x}^j$ is a component of some vector $\widetilde{x} \in V(N)$ satisfying $\widetilde{x}^k > x^k$ for all $k \in N \setminus \{j\}$. By the strict monotonicity of the f^k, $k \neq j$, $f(\widetilde{x}) \gg x$. Hence, $x \notin \partial W(N)$.

(3) $S = \{\ell\}$: We define $W(\{\ell\}) = \{x^{\{\ell\}} \in \mathbb{R}^{\{\ell\}} \mid x^\ell \leq 0\}$.

(4) $S = T \cup \{\ell\}$ for some $\emptyset \neq T \subseteq N$: Let $a^S \in \mathbb{R}^S$ be defined by $a^\ell = 1$ and $a^i = -1$ for all $i \in T$. Denote by L_S the line $\{ta^S \mid t \in \mathbb{R}\}$ and define $W(S)$ as the vectorial sum

$$W(S) = \left(V(T) \times \left\{0^{\{\ell\}}\right\}\right) + L_S. \tag{12.4.8}$$

As the reader may easily verify, $W(S)$ is nonempty, closed, and comprehensive. We now shall prove that $W(S)$ satisfies (11.3.4). Let $b^S \in \mathbb{R}^S$ and let $y^S \in (b^S + \mathbb{R}^S_+) \cap W(S)$. Then $y^S = (x^T, 0) + ta^S$ for some $x^T \in V(T)$ and some $t \geq b^\ell$. Thus, $x^i \geq b^i + t \geq b^i + b^\ell$ for every $i \in T$. Let $c^T \in \mathbb{R}^T$ be defined by $c^i = b^i + b^\ell$, $i \in T$. As $(c^T + \mathbb{R}^T_+) \cap V(T)$ is bounded, $(b^S + \mathbb{R}^S_+) \cap W(S)$ is bounded as well.

Finally, we shall prove that $W(S)$ is non-levelled. Let $y_1^S, y_2^S \in W(S)$ such that $y_1^S > y_2^S$. We have to prove that $y_2^S \notin \partial W(S)$. For $k = 1, 2$, $y_k^S = (x_k^T, 0) + t_k a^S$ for some $t_k \in \mathbb{R}$, $x_k^T \in V(T)$. Clearly, $t_1 \geq t_2$. If $t_1 = t_2$, then $x_1^T > x_2^T$. Hence there exists $x_3^T \in V(T)$ such that $x_3^T \gg x_2^T$. If $z^S = (x_3^T, 0) + ta^S$ where $(t - t_1) > 0$ is small enough, then $z^S \gg y_2^S$. Also, $z^S \in W(S)$ by construction. Now, if $t_1 > t_2$, then let $t_3 \in \mathbb{R}$ satisfy $t_1 < t_3 < t_2$. Then $z^S = (x_1^T, 0) + t_3 a^S \gg y_2^S$ and $z^S \in W(S)$.

We shall use the following obvious property of $W(S)$:

$$\left\{x^T \in \mathbb{R}^T \,\middle|\, \left(x^T, 0^{\{\ell\}}\right) \in W(S)\right\} = V(T). \tag{12.4.9}$$

(5) We now complete the proof: In order to show (12.4.3), let $y \in \mathcal{C}(M, W)$. Then $y = (x^N, 0) + ta^M$ for some $x^N \in V(N)$ and some $t \in \mathbb{R}$. By Case (3), $t \geq 0$. If $t > 0$, then $x^N \gg y^N$ and $x^N \in W(N)$. Thus, $t = 0$. By (12.4.7) and (12.4.9), $x^N = \bar{x}$. Additionally employing (12.4.6) yields $z = (\bar{x}, 0) \in \mathcal{C}(M, W)$. Finally, (12.4.4) is implied by (12.4.9) and Case (1).
q.e.d.

The following examples show that each of the axioms NE, IR, and RGP is logically independent of the remaining axioms, provided that $|\mathcal{U}| \geq 3$.

Example 12.4.14. Let $\sigma(N, V) = \emptyset$ for every $(N, V) \in \widehat{\Gamma}^{\mathcal{C}}_{\mathcal{U}}$. Then σ violates only NE.

Example 12.4.15. Let $(N, V) \in \widehat{\Gamma}^{\mathcal{C}}_{\mathcal{U}}$. If $|N| \leq 2$, let $\sigma(N, V) = \partial V(N)$. If $|N| \geq 3$, let $\sigma(N, V) = \mathcal{C}(N, V)$. Then σ violates only IR.

Example 12.4.16. For every $(N, V) \in \widehat{\Gamma}^{\mathcal{C}}_{\mathcal{U}}$ let

$$\sigma(N, V) = \{x \in \partial V(N) \mid x^i \geq v^i \text{ for all } i \in N\}.$$

That is, $\sigma(N, V)$ is the set of all Pareto optimal and individually rational payoff vectors of (N, V). Clearly, σ violates only RGP.

The results of this section are due to Peleg (1985).

Exercises

Exercise 12.4.1. Prove Lemma 12.4.10 and Lemma 12.4.11.

WRGP, the *weak* reduced game property (see Definition 2.3.17), and RCP, the *reconfirmation property* (see Definition 2.3.18), may be generalized to a solution on a set $\widehat{\Gamma}$ of non-levelled NTU games by replacing the set Γ of TU games and the TU coalition functions v and $v_{S,x}$ by $\widehat{\Gamma}$ and the NTU coalition functions V and $V_{S,x}$, respectively.

Exercise 12.4.2. Assume that $|\mathcal{U}| \geq 3$. Prove that there exists a unique solution on $\widehat{\Gamma}_{\mathcal{U}}^{\mathcal{C}}$ that satisfies NE, IR, WRGP, and CRGP, and it is the core. Also, show that CRGP is logically independent of NE, IR, and RGP in the foregoing characterization of the core, provided that $|\mathcal{U}| < \infty$.

Exercise 12.4.3. Assume that $|\mathcal{U}| \geq 3$. Use Lemma 12.4.13 to show that the core on $\widehat{\Gamma}_{\mathcal{U}}^{\mathcal{C}}$ does not satisfy RCP.

12.5 Additional Properties and Characterizations

Denote by $\widehat{\Gamma}_{\mathcal{U}}$ the set of non-levelled NTU games whose players are members of $\mathcal{U}$, let $\widehat{\Gamma} \subseteq \widehat{\Gamma}_{\mathcal{U}}$, and let $(N, V) \in \widehat{\Gamma}_{\mathcal{U}}$. For every $x \in V(N)$ and every $S \in 2^N \setminus \{\emptyset\}$ the *Moulin reduced game* with respect to S and x, $(S, V_{S,x}^M)$, is defined by

$$V_{S,x}^M(T) = \{y^T \in \mathbb{R}^T \mid (y^T, x^{N\setminus S}) \in V(T \cup (N \setminus S))\} \text{ for } \emptyset \neq T \subseteq S$$

and $V_{S,x}^M(\emptyset) = \emptyset$ (compare with Exercise 2.3.3). It is straightforward to show that $(S, V_{S,x}^M) \in \widehat{\Gamma}_{\mathcal{U}}$. Note that, if $V = V_v$ for some TU game (N, v), then $V_{S,x}^M = V_{v_{S,x}^M}$.

With the help of the foregoing definition, the reduced game property with respect to Moulin reduced games, RGP^M, may be generalized to solutions on $\widehat{\Gamma}$. Using RGP^M the core on $\widehat{\Gamma}_{\mathcal{U}}^{\mathcal{C}}$, the set of NTU games with a nonempty core, may be characterized.

Theorem 12.5.1. *Assume that $\mathcal{U}$ is infinite. Then there exists a unique solution on $\widehat{\Gamma}_{\mathcal{U}}^{\mathcal{C}}$ that satisfies* NE, IR, *and* RGP^M, *and it is the core.*

Proof: Clearly, the core satisfies NE on $\widehat{\Gamma}_{\mathcal{U}}^{\mathcal{C}}$. Also, it satisfies IR in general. The straightforward proof of RGP^M is left to the reader (Exercise 12.5.1).

Conversely, let σ be a solution on $\widehat{\Gamma}_{\mathcal{U}}^{\mathcal{C}}$ that satisfies NE, IR, and RGP^M. Then σ is PO. The proof is the simple Exercise 12.5.2.

Let $(N,V) \in \widehat{\Gamma}_{\mathcal{U}}^{\mathcal{C}}$. In order to show that $\sigma(N,V) \subseteq \mathcal{C}(N,V)$, let $x \in \sigma(N,V)$. Assume, on the contrary, that $x \notin \mathcal{C}(N,V)$. Then there exist a coalition $T \subseteq N$ and $y^T \in V(T)$ such that $y^T \gg x^T$. Let $i \in T$ and let $S = (N \setminus T) \cup \{i\}$. By RGP^M, $x^S \in \sigma(S,W)$, where $W = V^M_{S,x}$. However, $y^{\{i\}} \in W(\{i\})$ and $y^i > x^i$ contradict IR.

In order to show the opposite inclusion, let $z \in \mathcal{C}(N,V)$ and $\ell \in \mathcal{U} \setminus N$. Let $\widehat{N} = N \cup \{\ell\}$ and let $(\widehat{N},U)$ be defined as follows: Let

$$U(S) = \left\{ x^S \in \mathbb{R}^S \,\middle|\, \sum_{i\in S} x^i \le \sum_{i \in S} z^i \right\} \text{ for every } S \in 2^N \setminus \{\emptyset\},$$

let $U(\{\ell\}) = -\mathbb{R}^{\{\ell\}}_+$, and let $U(\emptyset) = \emptyset$. If $S \subseteq \widehat{N}$ such that $\ell \in S \neq \{\ell\}$, then let $T = S \setminus \{\ell\}$, let $a^S = |T|\chi^S_{\{\ell\}} - \chi^S_T$, let $L_S = \{ta^S \mid t \in \mathbb{R}\}$, and define $U(S) = \left(V(T) \times \{0^{\{\ell\}}\}\right) + L_S$. Clearly, $(\widehat{N},U)$ satisfies (11.3.1) – (11.3.3). Also, (11.3.4) and non-levelness are satisfied for $S \subseteq N$ and for $S = \{\ell\}$. In the remaining case, (11.3.4) and non-levelness may be proved as in (4) of the proof of Theorem 12.4.8.

Therefore, $(\widehat{N},U) \in \widehat{\Gamma}_{\mathcal{U}}$. Moreover, $\widehat{z} = (z, 0^\ell)$ is the unique member of $\mathcal{C}(\widehat{N},U)$. Hence, $\widehat{z} \in \sigma(\widehat{N},U)$ by NE. Also, $z = \widehat{z}^N$ and $U^M_{N,\widehat{z}} = V$. By RGP^M, $z \in \sigma(N,V)$.

q.e.d.

Note that if $V = V_v$ for some balanced TU game (N,v), then the game $(\widehat{N},U)$ constructed in the last part of the preceding proof, is associated with some balanced TU game. Hence, as remarked in (2) of Section 3.9, NE, IR, and RGP^M characterize the core on the set of balanced TU games.

Theorem 12.5.1 is due to Tadenuma (1992). The first inclusion of the uniqueness part of the present proof, however, differs from the corresponding part of Tadenuma's proof.

Some other properties of TU game solutions, relevant for the core, may be generalized to NTU game solutions. Let σ be a solution on $\widehat{\Gamma}$ and let $(N,V) \in \widehat{\Gamma}_{\mathcal{U}}$. For $i \in N$ denote

$$\begin{aligned} &b^i_{\min}(N,V) \\ &\quad = \min_{S \subseteq N\setminus\{i\}} \max \left\{ t^{\{i\}} \in \mathbb{R}^{\{i\}} \,\middle|\, (t^{\{i\}}, y^{N\setminus\{i\}}) \in V_0(S \cup \{i\}) \forall y \in V_0(S) \right\} \end{aligned}$$

where V_0 is defined by (12.3.2). (Here we use $\max \emptyset = -\infty$ as a convention.) By (11.3.1) – (11.3.4), the set

$$\{x \in V(N) \mid x^i \ge b^i_{\min}(N,V) \text{ for all } i \in N\}$$

is nonempty. Now, σ satisfies *reasonableness from below* (REBE) if, for every $(N, V) \in \widehat{\Gamma}$, $x \in \sigma(N, V)$ implies that $x^i \geq b^i_{\min}(N, V)$ for all $i \in N$ (see (2) of Definition 2.3.9). Also, σ is *covariant* (COV) if the following condition is satisfied: If (N, V), $(N, W) \in \widehat{\Gamma}$, $\alpha, b \in \mathbb{R}^N$, $\alpha \gg 0$, and if $W = \alpha * V + b$ (here we denote $\alpha * X = \{(\alpha^j x^j)_{j \in S} \mid x^S \in X\}$ for every $S \subseteq N$ and any $X \subseteq \mathbb{R}^S$), then $\sigma(N, W) = \alpha * \sigma(N, v) + b$. The generalization of *anonymity* (AN) to σ is straightforward.

Note that the core on $\widehat{\Gamma}$ satisfies REBE, because it satisfies IR. Clearly, it satisfies COV and AN. Unfortunately, it does not satisfy RCP in general (see Exercise 12.4.3). However, we shall show that it satisfies the following variant of RCP.

Definition 12.5.2. *The solution σ on $\widehat{\Gamma}$ satisfies the* **weak reconfirmation property** *(WRCP), if the following condition is satisfied for every $(N, V) \in \widehat{\Gamma}$, $S \in 2^N \setminus \{\emptyset\}$, and $x \in \sigma(N, V)$: If $(S, V_{S,x}) \in \widehat{\Gamma}$ and if $y^S \in \sigma(N, V_{S,x})$ satisfies*

$$\begin{aligned} &\textstyle\bigcup_{Q \subseteq N \setminus S} \{z^S \in \mathbb{R}^S \mid z^S \gg 0^S, (z^S + y^S, x^Q) \in V(S \cup Q)\} \\ \subseteq &\textstyle\bigcup_{Q \subseteq N \setminus S} \{z^S \in \mathbb{R}^S \mid z^S \gg 0^S, (z^S + x^S, x^Q) \in V(S \cup Q)\}, \end{aligned} \tag{12.5.1}$$

then $(y^S, x^{N \setminus S}) \in \sigma(N, V)$.

Hence, WRCP differs from RCP only inasmuch as it additionally requires that S only consider adjustments with the property that if S can improve upon this adjustment with the help of some Q, then it can improve upon x^S in the same way. Note that RCP implies WRCP. Further, note that, translated to TU games, WRCP and PO imply RCP.

Lemma 12.5.3. *The core on $\widehat{\Gamma}$ satisfies* WRCP.

Proof: Let $\sigma = \mathcal{C}$ and assume that (N, V), S, x, y^S satisfy the conditions of Definition 12.5.2. Let $z = (y^S, x^{N \setminus S})$. Assume, on the contrary, that $z \notin \mathcal{C}(N, V)$. Then there exist $\emptyset \neq T$ and $\widetilde{x}^T \in V(T)$ such that $\widetilde{x}^T \gg z^T$. Then $T \cap S \neq \emptyset$, because $x \in \mathcal{C}(N, V)$. If $S \not\subseteq T$, then $\widetilde{x}^{T \cap S} \in V_{S,x}(T \cap S)$ and $\widetilde{x}^{T \cap S} \gg y^{T \cap S}$, contradicting the fact that $y^S \in \mathcal{C}(S, V_{S,x})$. Finally, if $S \subseteq T$, then $T = S \cup Q$ for some $Q \subseteq N \setminus S$. By comprehensiveness, $(\widetilde{x}^S, x^Q) \in V(T)$. Also, $\widetilde{x}^S - y^S \gg 0^S$; thus, by (12.5.1), there exists $\widehat{Q} \subseteq N \setminus S$ such that $(\widetilde{x}^S, x^{\widehat{Q}}) \in V(S \cup \widehat{Q})$. By non-levelness there exists $\widehat{x}^{S \cup \widehat{Q}} \in V(S \cup \widehat{Q})$ such that $\widehat{x}^{S \cup \widehat{Q}} \gg x^{S \cup \widehat{Q}}$. Hence, a contradiction has been obtained. **q.e.d.**

Hwang and Sudhölter (2000) used the aforementioned properties to axiomatize the core on many remarkable subsets of $\widehat{\Gamma}_{\mathcal{U}}$. In order to characterize the core on, e.g., $\widehat{\Gamma}_{\mathcal{U}}$, the following "minimal nonemptiness requirement" of a solution σ

on a set $\widehat{\Gamma} \subseteq \widehat{\Gamma}_{\mathcal{U}}$ is useful: σ satisfies the *inessential two-person game property*, if $\sigma(N,V) \neq \emptyset$ for every *inessential* two-person game $(N,V) \in \widehat{\Gamma}$. (An NTU game (N,V) is inessential, if $0^S \in \partial V(S)$ for every coalition S.) We conclude with the following result which is proved in the aforementioned paper.

Theorem 12.5.4. *Assume that* $|\mathcal{U}| \geq 5$. *Then there exists a unique solution on* $\widehat{\Gamma}_{\mathcal{U}}$ *that satisfies the inessential two-person game property,* AN, COV, WRGP, WRCP, CRGP, *and* REBE, *and it is the core.*

Remark 12.5.5. Serrano and Volij (1998) show that the core on the set of all non-levelled NTU games may be characterized by nonemptiness on one-person games, IR, RGP, and a variant of CRGP. This variant requires that every Pareto optimal and individually rational payoff vector belong to the solution to a game, if every one of its restrictions to a nonempty proper subcoalition belongs to the solution to the corresponding reduced game.

Exercises

Exercise 12.5.1. Let $\widehat{\Gamma}_{\mathcal{U}}^{\mathcal{C}} \subseteq \widehat{\Gamma} \subseteq \widehat{\Gamma}_{\mathcal{U}}$. Show that the core on $\widehat{\Gamma}$ satisfies RGP^M.

Exercise 12.5.2. Let σ be a solution on a set $\widehat{\Gamma} \subseteq \widehat{\Gamma}_{\mathcal{U}}$ that satisfies IR and RGP^M. Show that σ satisfies PO.

12.6 Notes and Comments

Let (N,V) be an NTU game. If $V(\cdot) = V_\alpha(\cdot, G)$ or if $V(\cdot) = V_\beta(\cdot, G)$ for some cooperative game in strategic form $G = (N, (\Sigma(S))_{\emptyset \neq S \subseteq N}, (u^i)_{i\in N})$, then $V(N)$ is the *comprehensive hull* of

$$H = \{(u^i(\sigma))_{i\in N} \mid \sigma \in \Sigma(N)\},$$

that is, $V(N) = \{x \in \mathbb{R}^N \mid \exists\, y \in H \text{ such that } y \geq x\}$. Thus, H is the set of utility profiles that may be reached by cooperation.

Aumann (1961) showed the following variant of Theorem 12.1.5: Let (N,V) be a superadditive NTU game, let $H \subseteq \mathbb{R}^N$ such that $V(N)$ is the comprehensive hull of H. Also, let H_e, H_{ir}, and $H_{e,ir}$ denote the intersection of H and $V(N)_e$, $V(N)_{ir}$, and $V(N)_{e,ir}$, respectively. If H is a compact convex polyhedron, then

$$\mathcal{C}(N,V) = \mathcal{C}(N,V,H_e) = \mathcal{C}(N,V,H_{ir}) = \mathcal{C}(N,V,H_{e,ir}).$$

In fact, the crucial assumption that H is a convex compact polyhedron is satisfied in many important examples. E.g., if G is the cooperative game in strategic form which is associated with a finite game in strategic form, where $\Sigma(S)$, $S \in 2^N \setminus \{\emptyset\}$, is the set of correlated strategies (see Example 11.1.2), then H is a compact convex polyhedron.

13

The Shapley NTU Value and the Harsanyi Solution

This chapter is devoted to two NTU extensions of the TU Shapley value.

In Section 13.1 the Shapley NTU value is introduced. An existence theorem is stated and proved in Section 13.1. In Section 13.2 we show that the Shapley NTU value correspondence is characterized by nonemptiness, Pareto optimality, conditional additivity, scale covariance, unanimity, and independence of irrelevant alternatives.

The Harsanyi value and the Harsanyi solution are discussed in Section 13.3 and an existence proof is given. Section 13.4 is devoted to a characterization due to Hart (1985). We show that the Harsanyi solution is uniquely determined by suitable expansions of the foregoing axioms to payoff configuration solutions, that is, of nonemptiness, Pareto optimality, scale covariance, unanimity, and independence of irrelevant alternatives.

13.1 The Shapley Value of NTU Games

Let N be a finite nonempty set, let (N, V) be an NTU game, and let $\Delta_{++}(N) = \Delta_{++} = \{\lambda \in \mathbb{R}^N_{++} \mid \lambda(N) = 1\}$. For every $\lambda \in \Delta_{++}$ define the mapping $v_\lambda : 2^N \to \mathbb{R} \cup \{\infty\}$ by

$$v_\lambda(S) = \sup\{\lambda^S \cdot x^S \mid x^S \in V(S)\} \text{ for every } S \in 2^N \setminus \{\emptyset\} \qquad (13.1.1)$$

and $v_\lambda(\emptyset) = 0$. A vector $\lambda \in \Delta_{++}$ is *viable* if $v_\lambda(S) \in \mathbb{R}$ for all $S \subseteq N$. Denote by Δ^V_{++} the set of all viable $\lambda \in \Delta_{++}$. If $\lambda \in \Delta^V_{++}$, then v_λ is a TU game and we may compute the Shapley value $\phi(v_\lambda)$ of v_λ. (Throughout Sections 13.1 and 13.2 we fix the player set and identify a game with its coalition function.)

Definition 13.1.1. *Let (N, V) be an NTU game. A vector $x \in V(N)$ is a* **Shapley NTU value** *of (N, V) if there exists $\lambda \in \Delta^V_{++}$ such that $\lambda * x = \phi(v_\lambda)$. (Recall that $y * z = (y^i z^i)_{i \in N}$ for all $y, z \in \mathbb{R}^N$.)*

Now, let x be a Shapley NTU value of (N, V), let $\lambda \in \Delta^V_{++}$ such that $\lambda * x = \phi(v_\lambda)$, and let $i, j \in N$, $i \neq j$. If i "offers" $t^i > 0$ units of his utility to j at x, then j receives at most $\frac{\lambda^j}{\lambda^i} t^i$ additional utility units, that is, if $(x^i - t^i, x^j + t^j, x^{N\setminus\{i,j\}}) \in V(N)$, then $\frac{t^j}{t^i} \leq \frac{\lambda^i}{\lambda^j}$. Hence, the quotients of the coefficients of λ may be regarded as upper bounds for the transfer rates of utilities between the players. Let $(t^i_\ell)_{\ell \in \mathbb{N}}$ be a sequence such that $t^i_\ell > 0$ for all $\ell \in \mathbb{N}$ and $\lim_{\ell \to \infty} t^i_\ell = 0$. Then $t^j_\ell = \max\{t \in \mathbb{R} \mid (x^i - t^i_\ell, x^j + t, x^{N\setminus\{i,j\}}) \in V(N)\}$ exists for every $\ell \in \mathbb{N}$. We now recall that a convex subset A of $\mathbb{R}^N$ is *smooth* if at every $x \in \partial A$ there exists a unique supporting hyperplane to A. In the case that $V(N)$ is convex and smooth, we remark that $\lim_{\ell \to \infty} \frac{t^j_\ell}{t^i_\ell} = \frac{\lambda^i}{\lambda^j}$. Hence, the quotient λ^i/λ^j may be interpreted as the local transfer rate of utility from i to j at x.

Now, let $W = V_{v_\lambda}$ be the NTU game associated with the TU game v_λ. We consider the Shapley value of (N, v_λ) as its equitable solution. It is straightforward (Exercise 13.1.1) to prove that the unique Shapley NTU value of (N, W) coincides with $\phi(v_\lambda) = y = \lambda * x$. Thus, y remains equitable for (N, W). Clearly, the Shapley NTU value satisfies COV. Hence, x is the unique Shapley NTU value of and equitable in (N, U), where $U(S) = W(S) * (1/\lambda^i)_{i \in S}$ for every coalition S in N. Hence, a fortiori, x may be regarded as equitable in (N, V) by the above arguments. Also, it should be noted that the game (N, U) contains (N, V), that is, $V(S) \subseteq U(S)$ for all $S \subseteq N$. Hence we might use a variant of *independence of irrelevant alternatives* in the sense of Nash (1950) to support the definition of the Shapley NTU value.

The following example shows that the Nash solution to a normalized two-person bargaining problem coincides with its unique Shapley NTU value.

Example 13.1.2. Let $N = \{1, 2\}$, let $V(\{i\}) = -\mathbb{R}^{\{i\}}_+$, $i = 1, 2$, let $V(N)$ be convex and non-levelled such that $0 \in \mathbb{R}^N$ is an interior point of $V(N)$. Then $x \in \partial V(N)$ is a Shapley value of (N, V) iff there exists $\lambda \in \Delta^V_{++}$ such that

$$\lambda^1 y^1 + \lambda^2 y^2 \leq \lambda^1 x^1 + \lambda^2 x^2 \text{ for all } y \in V(N), \tag{13.1.2}$$

$$x^i = \frac{\lambda^1 x^1 + \lambda^2 x^2}{2\lambda^i} \text{ for all } i \in N. \tag{13.1.3}$$

From (13.1.3) we obtain

$$\frac{x^2}{x^1} = \frac{\lambda^1}{\lambda^2}. \tag{13.1.4}$$

Let $A = \{y \in \mathbb{R}^N \mid \lambda \cdot y = \lambda \cdot x\}$. Then the unique Shapley NTU value x is determined by Figure 13.1.1.

That is, the unique Shapley NTU value coincides with the Nash (1950) solution to the bargaining problem $(V(N), (0, 0))$.

We are now ready to state the main result of this section.

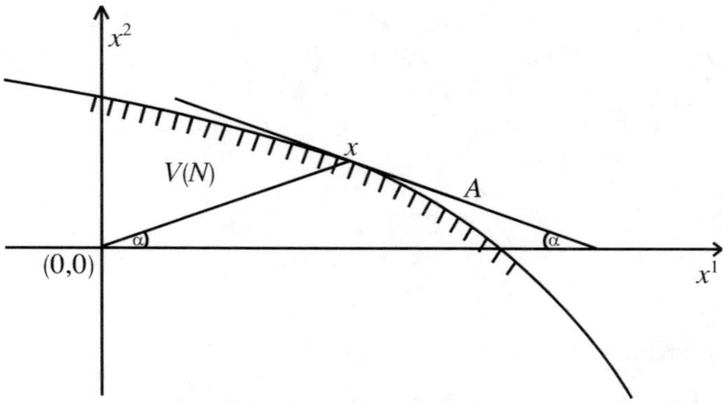

Fig. 13.1.1. The Shapley Value

Theorem 13.1.3. *Let (N, V) be an NTU game that has the following property: For every $S \in 2^N \setminus \{\emptyset\}$, there exist a compact set $C(S) \subseteq \mathbb{R}^S$ and a cone $K(S) \subseteq \mathbb{R}^S$ such that*

$$V(S) = C(S) + K(S), \tag{13.1.5}$$

$$K(N) \supseteq K(S) \times \{0^{N \setminus S}\}, \tag{13.1.6}$$

$$C(N) \text{ and } K(N) \text{ are convex, and} \tag{13.1.7}$$

$$V(N) \text{ is non-levelled.} \tag{13.1.8}$$

Then (N, V) has an NTU Shapley value.

We postpone the proof of Theorem 13.1.3 and shall first deduce some simple assertions and prove a useful lemma. Let (N, V) be an NTU game satisfying the conditions of Theorem 13.1.3.

We first observe that $\Delta_{++}^V \neq \emptyset$. Indeed, let $x \in \partial V(N)$. As $V(N)$ is convex, there exists $\widetilde{\lambda} \in \mathbb{R}^N$ such that $\widetilde{\lambda} \neq 0$ and $\widetilde{\lambda} \cdot x \geq \widetilde{\lambda} \cdot y$ for all $y \in V(N)$. As $K(N)$ is a cone, $\widetilde{\lambda} \cdot y \leq 0$ for all $y \in K(N)$. Hence, by (13.1.6), $\widetilde{\lambda}^S \cdot y^S \leq 0$ for all $y^S \in K(S)$ and all $S \in 2^N \setminus \{\emptyset\}$. Also, as $V(N)$ is comprehensive and non-levelled, $\widetilde{\lambda} \gg 0$. Thus, $\widetilde{\lambda}/\widetilde{\lambda}(N) \in \Delta_{++}^V$.

We conclude from the foregoing proof of $\Delta_{++}^V \neq \emptyset$ that

$$\Delta_{++}^V = \{\lambda \in \mathbb{R}^N \mid \lambda \cdot x \leq 0 \text{ for all } x \in K(N) \text{ and } \lambda(N) = 1\}. \tag{13.1.9}$$

By (13.1.9), Δ_{++}^V is closed and convex. Hence, there exists $\delta > 0$ such that $\lambda^i > \delta$ for all $\lambda \in \Delta_{++}^V$ and all $i \in N$.

Lemma 13.1.4. *The mapping $\Delta_{++}^V \to \mathbb{R}^{2^N}$ defined by $\lambda \mapsto v_\lambda$ is a continuous function.*

Proof: Let $\lambda(t) \in \Delta^V_{++}$, $t \in \mathbb{N}$, such that $\lim_{t\to\infty} \lambda(t) = \widehat{\lambda}$. Let $S \in 2^N \setminus \{\emptyset\}$. Then, for every $t \in \mathbb{N}$, there exists $x^S(t) \in C(S)$ such that $v_{\lambda(t)}(S) = \lambda^S(t) \cdot x^S(t)$. By compactness of $C(S)$ there exists a convergent subsequence $(x^S(t_k))_{k\in\mathbb{N}}$ of $(x^S(t))_{t\in\mathbb{N}}$. Let $\widehat{x}^S = \lim_{k\to\infty} x^S(t_k)$ and let $x^S \in V(S)$ such that $v_{\widehat{\lambda}}(S) = \widehat{\lambda}^S \cdot x^S$. Then

$$v_{\widehat{\lambda}}(S) = \widehat{\lambda}^S \cdot x^S = \lim_{k\to\infty} \lambda^S(t_k) \cdot x^S \leq \lim_{k\to\infty} \lambda^S(t_k) \cdot x_S(t_k) = \widehat{\lambda}^S \cdot \widehat{x}^S \leq v_{\widehat{\lambda}}(S).$$

Hence, $v_{\widehat{\lambda}}(S) = \widehat{\lambda}^S \cdot \widehat{x}^S = \lim_{k\to\infty} v_{\lambda(t_k)}(S)$. **q.e.d.**

We proceed with the proof of the theorem.

Proof of Theorem 13.1.3: Let $\lambda \in \Delta^V_{++}$. Define

$$x_\lambda = \left(\frac{\phi^i(v_\lambda)}{\lambda^i}\right)_{i\in N}.$$

By Lemma 13.1.4, x_λ is a continuous function of λ. Also, x_λ is not an interior point of $V(N)$. Let y_λ be the closest point to x_λ in $V(N)$. Then y_λ is a continuous function of λ. Define

$$\varphi(\lambda) = \{\eta \in \Delta^V_{++} \mid \eta \cdot y_\lambda \geq \eta \cdot x \text{ for all } x \in V(N)\}.$$

As $y_\lambda \in \partial V(N)$, $\varphi(\lambda) \neq \emptyset$. Also, φ is a convex-valued and upper hemicontinuous correspondence. Hence, by Kakutani's fixed-point theorem, there exists $\widehat{\lambda} \in \Delta^V_{++}$ such that $\widehat{\lambda} \in \varphi(\widehat{\lambda})$. We claim that $y_{\widehat{\lambda}}$ is a Shapley value of (N, V), that is, $y_{\widehat{\lambda}} = x_{\widehat{\lambda}}$. Assume, on the contrary, that $y_{\widehat{\lambda}} \neq x_{\widehat{\lambda}}$. Then $z = x_{\widehat{\lambda}} - y_{\widehat{\lambda}}$ is a normal to a supporting hyperplane to $V(N)$ at $y_{\widehat{\lambda}}$. Hence, $z \gg 0$, contradicting $\widehat{\lambda} \cdot x_{\widehat{\lambda}} = \sum_{i\in N} \phi^i(v_{\widehat{\lambda}}) = \widehat{\lambda} \cdot y_{\widehat{\lambda}}$. **q.e.d.**

We remark that there is a significant and interesting class of NTU games that do not have a Shapley NTU value. An NTU game (N, V) is a *hyperplane game* if there exist $\lambda_S \in \mathbb{R}^S_{++}$ and $r_S \in \mathbb{R}$ such that

$$V(S) = \{x \in \mathbb{R}^S \mid \lambda_S \cdot x \leq r_S\} \text{ for all } \emptyset \neq S \subseteq N. \tag{13.1.10}$$

Remark 13.1.5. Note that the hyperplane game defined by (13.1.10) does not have a Shapley NTU value, unless all hyperplanes are parallel, that is, λ_S is proportional to λ^S_N for all coalitions S in N. In Chapter 14 a further generalization of the Shapley value is presented that results in a unique solution to any hyperplane game.

Remark 13.1.6. Definition 13.1.1 slightly differs from that given by Shapley (1969), who defined v_λ for all $\lambda \in \{\lambda \in \mathbb{R}^N_+ \mid \lambda(N) = 1\}$. Moreover, he proved the existence of an NTU value of any game (N, V) that is compactly generated, that is, every $V(S)$ is the comprehensive hull of some compact subset of $\mathbb{R}^S$. Our approach is similar to that of Kern (1985) and the idea of the proof of Theorem 13.1.3 is due to Aumann (1985).

Exercises

Exercise 13.1.1. Let (N,v) be a TU game. Prove that the unique Shapley NTU value of (N,V_v) (see (11.3.5)) is $\phi(N,v)$.

Exercise 13.1.2. Let $N=\{1,2\}$, let $V(\{i\})=-\mathbb{R}_+^{\{i\}}$, $i=1,2$, and let $V(N)=\{x\in -\mathbb{R}_{++}^N \mid x^1\leq \frac{1}{x^2}\}$. Prove that every $x\in\partial V(N)$ is an NTU Shapley value of (N,V).

Exercise 13.1.3. Let $N=\{1,2\}$, let $V(\{i\})=-\mathbb{R}_+^{\{i\}}$, $i=1,2$, and let $V(N)=\{x\in -\mathbb{R}_{++}^N \mid x^1\leq -\frac{1}{(x^2)^2}\}$. Prove that (N,V) has no Shapley value.

Exercise 13.1.4. Let N be a finite nonempty set. If $K\subseteq\mathbb{R}^N$ is a cone, then let $K^*=\{y\in\mathbb{R}^N \mid y\cdot x\leq 0 \text{ for all } x\in K\}$. ($K^*$ is the *dual* cone of K.) Let (N,V) be an NTU game which is represented as in Theorem 13.1.3, i.e., for every coalition S, $V(S)=C(S)+K(S)$ for some compact set $C(S)$ and some cone $K(S)$. Assume that $(C(S),K(S))_{\emptyset\neq S\subseteq N}$ satisfies (13.1.5) and (13.1.8). Show that

(1) $\Delta_{++}^V=\bigcap_{S\in 2^N\setminus\{\emptyset\}}\left(K(S)\times\{0^{N\setminus S}\}\right)^*\cap\left\{x\in\mathbb{R}^N \middle| x(N)=1\right\}$;

(2) if (N,V) is *weakly superadditive*, that is, $\emptyset\neq S\subseteq N$ and $i\in N\setminus S$ imply $V(\{i\})\times V(S)\subseteq V(\{i\}\cup S)$, and if $V(N)$ is convex, then (N,V) has a Shapley NTU value.

13.2 A Characterization of the Shapley NTU Value

In this section we present Aumann's (1985) characterization of the Shapley NTU value. We consider the set of all NTU games (N,V) with a fixed set N of players, which satisfy the following properties:

(1) $V(N)$ is convex and non-levelled.

(2) $V(N)$ is smooth.

(3) For each $S\in 2^N\setminus\{\emptyset,N\}$ there exists $x_S\in\mathbb{R}^N$ such that

$$V(S)\times\{0^{N\setminus S}\}\subseteq V(N)+x_S.$$

Conditions (1) - (3) imply that for every $x\in\partial V(N)$ the pair (N,v_λ) is a TU game, where λ is a normal to a supporting hyperplane to $V(N)$ at x.

Denote by $\widehat{\Gamma}_N$ the set of all games which satisfy (1) – (3) and by $\widehat{\Gamma}_N^\Phi$ the subset of all games in $\widehat{\Gamma}_N$ that possess at least one Shapley NTU value. Moreover, denote by $\Phi(N,V)=\Phi(V)$ the set of all Shapley NTU values of a game $(N,V)\in\widehat{\Gamma}_N$. If $\widehat{\Gamma}\subseteq\widehat{\Gamma}_N^\Phi$, then the correspondence Φ on $\widehat{\Gamma}$ satisfies NE and PO (see Subsection 12.4.2), by definition. Furthermore, it satisfies the following axioms which are formulated for a general solution σ on a set $\widehat{\Gamma} \subseteq \widehat{\Gamma}_N$:

(1) If $V, W \in \widehat{\Gamma}$ such that $U = V + W \in \widehat{\Gamma}$, then

$$\sigma(U) \supseteq (\sigma(V) + \sigma(W)) \cap \partial U(N) \quad (\textit{conditional additivity } (\text{CADD})).$$

(2) If $T \in 2^N \setminus \{\emptyset\}$, if U_T is the NTU unanimity game on T, that is,

$$U_T(S) = \begin{cases} \{x \in \mathbb{R}^S \mid x(S) \leq 1\}, & \text{if } N \supseteq S \supseteq T, \\ \{x \in \mathbb{R}^S \mid x(S) \leq 0\}, & \text{if } \emptyset \neq S \not\supseteq T, \end{cases}$$

and if $U_T \in \widehat{\Gamma}$, then $\sigma(U_T) = \{\chi_T / |T|\}$ (*unanimity* (UNA)).

(3) If $\lambda \in \mathbb{R}^N_{++}$ and $V, \lambda * V \in \widehat{\Gamma}$, then $\sigma(\lambda * V) = \lambda * \sigma(V)$ (*scale covariance* (SCOV)).

(4) If $V, W \in \widehat{\Gamma}$, $V(N) \subseteq W(N)$, and $V(S) = W(S)$ for all $S \subsetneq N$, then $\sigma(V) \supseteq \sigma(W) \cap V(N)$ (*independence of irrelevant alternatives* (IIA)).

Note that SCOV is one part of COV as defined in Section 12.5. The proof of the remaining assertions of the next lemma is left as Exercise 13.2.1.

Lemma 13.2.1. *Let $\widehat{\Gamma} \subseteq \widehat{\Gamma}^{\Phi}_N$. The Shapley NTU value correspondence Φ on $\widehat{\Gamma}$ satisfies* NE, PO, CADD, UNA, SCOV, *and* IIA.

Notice that if (N, v) is a TU game, then $V_v \in \widehat{\Gamma}^{\Phi}_N$. Denote by

$$\widehat{\Gamma}^{TU}_N = \{V_v \mid (N, v) \text{ is a TU game}\}.$$

Using (11.3.5) we may identify $\widehat{\Gamma}^{TU}_N$ and the set Γ_N of all TU games with player set N. The next lemma applies to general solutions on every $\widehat{\Gamma}$, $\Gamma_N \subseteq \widehat{\Gamma} \subseteq \widehat{\Gamma}^{\Phi}_N$.

Lemma 13.2.2. *Let σ be a solution on a set $\widehat{\Gamma}$, $\widehat{\Gamma}^{TU}_N \subseteq \widehat{\Gamma} \subseteq \widehat{\Gamma}^{\Phi}_N$, that satisfies* NE, PO, CADD, UNA, *and* SCOV. *Then $\sigma(V_v) = \{\phi(v)\}$ for every TU game (N, v).*

Proof: By PO and CADD, σ is superadditive (see Definition 2.3.8) on $\widehat{\Gamma}^{TU}_N$. Let $(N, v) \in \Gamma^{\Phi}_N$ and $\alpha \in \mathbb{R}$. Denote $V^{\alpha} = V_{\alpha v}$. UNA and SUPA imply

$$\left\{\frac{\chi_N}{|N|}\right\} + \sigma(V^0) = \sigma(U_N) + \sigma(V^0) \subseteq \sigma(U_N + V^0) = \sigma(U_N) = \left\{\frac{\chi_N}{|N|}\right\}.$$

Hence, by NE, $\sigma(V^0) = \{0\}$. Also,

$$\sigma(V^1) + \sigma(V^{-1}) \subseteq \sigma(V^1 + V^{-1}) = \sigma(V^0) = \{0\}.$$

Hence, $\sigma(V^{-1}) = -\sigma(V^1)$ and, moreover, σ is single-valued and additive on $\widehat{\Gamma}_N^{TU}$. Finally, if $\alpha > 0$, then $V^\alpha = \alpha\chi_N * V^1$. Therefore SCOV implies $\sigma(V^\alpha) = \alpha\chi_N * \sigma(V^1) = \alpha\sigma(V^1)$. Thus, σ is linear and $\sigma(U_T) = \Phi(U_T)$ for all $T \in 2^N \setminus \{\emptyset\}$. **q.e.d.**

We proceed with the following result.

Lemma 13.2.3. *Let σ be a solution on $\widehat{\Gamma}_N^\Phi$ that satisfies* NE, PO, CADD, UNA, *and* SCOV. *Then $\sigma(V) \subseteq \Phi(V)$ for every $V \in \widehat{\Gamma}_N^\Phi$.*

Proof: Let $V \in \widehat{\Gamma}_N^\Phi$ and $y \in \sigma(V)$. By PO, $y \in \partial V(N)$. Hence there exists $\lambda \in \Delta_{++}(N)$ such that $\lambda \cdot y \geq \lambda \cdot x$ for all $x \in V(N)$. As V satisfies Property (3), $(N, v_\lambda) \in \Gamma_N$. Let $V^\lambda = V_{v_\lambda}$ and let $V_\lambda = \lambda * V$. Further, let V^0 be given by $V^0(S) = \{x \in \mathbb{R}^S \mid x(S) \leq 0\}$ for all $S \in 2^N \setminus \{\emptyset\}$ and $V^0(\emptyset) = \emptyset$. Then $V^0 \in \widehat{\Gamma}_N^\Phi$ and, by Lemma 13.2.2,

$$\begin{aligned}\lambda * y \in \sigma(V_\lambda) \cap \partial(V_\lambda + V^0) &= (\sigma(V_\lambda) + 0) \cap \partial(V_\lambda + V^0)\\ &= (\sigma(V_\lambda) + \sigma(V^0)) \cap \partial(V_\lambda + V^0)\\ &\subseteq \sigma(V_\lambda + V^0) = \sigma(V^\lambda) = \{\phi(v_\lambda)\}.\end{aligned}$$

Thus, $\lambda * y = \phi(v_\lambda)$, that is, $y \in \Phi(V)$. **q.e.d.**

Hence we obtain the following result.

Theorem 13.2.4. *The Shapley NTU value correspondence Φ is the maximum solution on $\widehat{\Gamma}_N^\Phi$ that satisfies* NE, PO, CADD, UNA, *and* SCOV.

We proceed with a complete characterization of Φ.

Theorem 13.2.5. *There exists a unique solution on $\widehat{\Gamma}_N^\Phi$ that satisfies* NE, PO, CADD, UNA, SCOV, *and* IIA, *and it is the Shapley NTU value Φ.*

Proof: We only have to show the uniqueness part. Let σ be a solution on $\widehat{\Gamma}_N^\Phi$ that satisfies NE, PO, CADD, UNA, SCOV, and IIA. By Lemma 13.2.3, it suffices to prove that Φ is a subsolution of σ. To this end let $V \in \widehat{\Gamma}_N^\Phi$, let $x \in \Phi(V)$, and let $\lambda \in \Delta_{++}^V$ such that $x = \phi(v_\lambda)$. Define the NTU game W by

$$W(S) = \begin{cases} \{y \in \mathbb{R}^N \mid y(N) \leq \lambda \cdot x\}, & \text{if } S = N,\\ \lambda * V(S), & \text{otherwise.}\end{cases}$$

Then $\lambda * x \in \Phi(W)$, so $W \in \widehat{\Gamma}_N^\Phi$. Hence $\sigma(W) \neq \emptyset$. Let V^0 correspond to the zero TU game and let V^λ correspond to v_λ. Then

$$\begin{aligned}\{\phi(v_\lambda)\} = \sigma(V^\lambda) &= \sigma(W + V^0)\\ &\supseteq (\sigma(W) + \sigma(V^0)) \cap \partial(W(N) + V^0(N))\\ &= (\sigma(W) + 0) \cap \partial W(N) = \sigma(W).\end{aligned}$$

Thus, $\lambda * x = \phi(v_\lambda) \in \sigma(W)$. By IIA, $\lambda * x \in \sigma(\lambda * V)$. By SCOV, $x \in \sigma(V)$. **q.e.d.**

We shall show by means of examples that each of the axioms NE, PO, CADD, UNA, and SCOV is independent of the remaining axioms employed in Theorem 13.2.5, provided that $|N| \geq 3$. (The logical independence of IIA is an open problem.)

Let $\widehat{\Gamma}^1 = \{\lambda * V_v \mid \lambda \in \Delta_{++}, v \in \Gamma_N\}$ and define $\widehat{\Gamma}^2 \subseteq \widehat{\Gamma}^{\Phi}_N$ by the following requirement: $V \in \widehat{\Gamma}^{\Phi}_N$ is a member of $\widehat{\Gamma}^2$ if there exists $W \in \widehat{\Gamma}^1$ such that $V(N) \subseteq W(N)$ and $V(S) = W(S)$ for all $S \subsetneqq N$. The solution σ^1 on $\widehat{\Gamma}^{\Phi}_N$ is defined by

$$\sigma^1(V) = \begin{cases} \Phi(V) \text{ , if } V \in \widehat{\Gamma}^2, \\ \emptyset \text{ , if } V \in \widehat{\Gamma}^{\Phi}_N \setminus \widehat{\Gamma}^2. \end{cases}$$

Then σ^1 satisfies PO, CADD, UNA, SCOV, and IIA, but it violates NE, because $|N| \geq 3$.

We define $\widehat{\Gamma}^3 \subseteq \widehat{\Gamma}^{\Phi}_N$ by the following requirement: Let $W = V_{-u_N}$, where u_N is the TU unanimity game on N. Then $V \in \widehat{\Gamma}^{\Phi}_N$ is a member of $\widehat{\Gamma}^3$ if there exists $\lambda \gg 0$ such that

(1) $V(S) = \lambda * W(S)$ for all $S \in 2^N \setminus \{N, \emptyset\}$,

(2) $V(N) \subseteq \lambda * W(N)$, and

(3) $\partial V(N) \cap \partial(\lambda * W)(N) \neq \emptyset$.

We remark that $\widehat{\Gamma}^3$ is closed under scale covariance, that is, $\alpha * V \in \widehat{\Gamma}^3$ for every $V \in \widehat{\Gamma}^3$ and every $\alpha \in \mathbb{R}^N_{++}$. Also, if $V \in \widehat{\Gamma}^3$ and $\lambda \in \mathbb{R}^N_{++}$ such that $\lambda * W(S) = V(S)$ for all $\emptyset \neq S \subsetneqq N$, then $\Delta^V_{++} = \{\lambda/\lambda(N)\}$ and, hence, $\Phi(V) = \{-\frac{\lambda * \chi_N}{\lambda(N)|N|}\}$. Let the solution σ^2 be defined by

$$\sigma^2(V) = \begin{cases} \{\alpha x \mid \alpha \geq 1, x \in \Phi(V)\} \text{ , if } V \in \widehat{\Gamma}^3, \\ \Phi(V) \text{ , if } V \in \widehat{\Gamma}^{\Phi}_N \setminus \widehat{\Gamma}^3. \end{cases}$$

Clearly, σ^2 satisfies NE, CADD, UNA, SCOV, and violates PO. Also, by construction, σ^2 satisfies IIA.

Let $\widehat{\Gamma}^4 = \{\lambda * U_T \mid \lambda \in \mathbb{R}^N_{++}, T \in 2^N \setminus \{\emptyset\}\}$ and let $\widehat{\Gamma}^5 \subseteq \widehat{\Gamma}^{\Phi}_N$ be defined by the following requirement: $V \in \widehat{\Gamma}^{\Phi}_N$ is a member of $\widehat{\Gamma}^5$ if there exists a game $W \in \widehat{\Gamma}^4$ such that $V(N) \subseteq W(N)$ and $V(S) = W(S)$ for all $S \subsetneqq N$. Define the solution σ^3 by

$$\sigma^3(V) = \begin{cases} \Phi(V) \text{ , if } V \in \widehat{\Gamma}^5, \\ \partial V(N) \text{ , if } V \in \widehat{\Gamma}^{\Phi}_N \setminus \widehat{\Gamma}^5. \end{cases}$$

Then σ^3 satisfies all axioms except CADD.

The easy proof that UNA is independent of the remaining axioms is left as Exercise 13.2.3. Kalai and Samet (1985) introduced *egalitarian values* and showed that the *symmetric* egalitarian value satisfies NE, PO, CADD, UNA, and IIA. However, it violates SCOV.

Aumann (1985, Section 10) contains the following remark and corollary.

Remark 13.2.6. Theorem 13.2.5 remains valid if the domain $\widehat{\Gamma}_N^{\Phi}$ is replaced by any domain $\widehat{\Gamma}$ of NTU games with player sets N that satisfies the following properties:

(1) $\widehat{\Gamma} \subseteq \widehat{\Gamma}_N^{\Phi}$, that is, $\Phi(V) \neq \emptyset$ for all $V \in \widehat{\Gamma}$.

(2) $\widehat{\Gamma}_N^{TU} \subseteq \widehat{\Gamma}$.

(3) If $V \in \widehat{\Gamma}$ and $\lambda \in \Delta_{++}(N)$, then $\lambda * V \in \widehat{\Gamma}$.

(4) If $V \in \widehat{\Gamma}$, if $\lambda \in \Delta_{++}^V$, and if (N, W) is the NTU game that differs from (N, V) only inasmuch as $W(N) = \{y \in \mathbb{R}^N \mid \lambda \cdot y \leq v_\lambda(N)\}$, then $W \in \widehat{\Gamma}$.

Indeed, if only games in $\widehat{\Gamma}$ have to be considered, then in the proofs of Lemma 13.2.3 and Theorem 13.2.5 just NTU games are constructed that belong to $\widehat{\Gamma}$, provided that $\widehat{\Gamma}$ satisfies the properties of the foregoing remark.

The following corollary is a direct consequence of Remark 13.2.6 and Theorem 13.2.5.

Corollary 13.2.7. *Let $\widehat{\Gamma}$ be the set of all NTU games with player sets N that satisfy the required property of Theorem 13.1.3. The Shapley NTU value Φ is the unique solution on $\widehat{\Gamma}$ that satisfies* NE, PO, CADD, UNA, SCOV, *and* IIA.

Exercises

Exercise 13.2.1. Prove Lemma 13.2.1 (see Aumann (1985)).

Exercise 13.2.2. Construct two-person NTU games V and W satisfying Properties (1) and (3) such that $\Phi(U) \not\supseteq (\Phi(V) + \Phi(W)) \cap \partial U(N)$, where $U = V + W$.

Exercise 13.2.3. Assume that $|N| \geq 3$. Prove that UNA is independent of the remaining axioms of Theorem 13.2.5.

13.3 The Harsanyi Solution

In this section we shall be interested in a further extension of the TU Shapley value. The Harsanyi NTU value is due to Harsanyi (1963).

We may extend the definition of a subgame (see Definition 3.2.2) to NTU games. Let (N, V) be an NTU game and let S be a coalition in N. The *subgame* (S, V^S) is defined by $V^S(R) = V(R)$ for all $R \subseteq S$ and it is sometimes denoted by (S, V). A *payoff configuration* for N is a collection $\boldsymbol{x} = (x_S)_{S \in 2^N \setminus \{\emptyset\}}$ such that $x_S \in \mathbb{R}^S$ for all $\emptyset \neq S \subseteq N$.[1] Let $\lambda \in \Delta_{++}(N)$. Let $(N, v_{\lambda,\boldsymbol{x}})$ be the TU game defined by

$$v_{\lambda,\boldsymbol{x}}(S) = \lambda^S \cdot x_S \text{ for all } \emptyset \neq S \subseteq N.$$

Definition 13.3.1. *Let (N, V) be an NTU game. A payoff configuration $\boldsymbol{x} = (x_S)_{S \in 2^N \setminus \{\emptyset\}}$ for N is a* **Harsanyi NTU solution** *of (N, V) if there exists $\lambda \in \Delta_{++}(N)$ such that*

$$x_S \in \partial V(S) \textit{ for all } \emptyset \neq S \subseteq N, \tag{13.3.1}$$

$$\lambda \cdot x_N = \max_{y \in V(N)} \lambda \cdot y, \tag{13.3.2}$$

$$\lambda^S * x_S = \phi(S, v_{\lambda,\boldsymbol{x}}) \textit{ for all } \emptyset \neq S \subseteq N. \tag{13.3.3}$$

Moreover, $x \in \mathbb{R}^N$ is a **Harsanyi NTU value** *of (N, V) if there exists a Harsanyi NTU solution $\boldsymbol{x}$ of (N, V) such that $x_N = x$.*

Remark 13.3.2. In order to compare Definitions 13.1.1 and 13.3.1 we may reformulate the definition of the Shapley NTU value as follows: A vector $x \in \mathbb{R}^N$ is a Shapley NTU value of (N, V) if there exist $\lambda \in \Delta_{++}(N)$ and a payoff configuration $\boldsymbol{x}$ of N with $x = x_N$ such that

$$x_S \in \partial V(S) \text{ for all } \emptyset \neq S \subseteq N, \tag{13.3.4}$$

$$\lambda^S \cdot x_S = \max_{y \in V(S)} \lambda^S \cdot y, \text{ for all } \emptyset \neq S \subseteq N, \tag{13.3.5}$$

$$\lambda * x_N = \phi(N, v_{\lambda,\boldsymbol{x}}). \tag{13.3.6}$$

Furthermore, a payoff configuration $\boldsymbol{x}$ is a *Shapley NTU solution* if it satisfies (13.3.4) - (13.3.6) for some viable λ. Note that the *efficiency* conditions, (13.3.1) and (13.3.4), respectively, coincide for both NTU solutions. The Harsanyi NTU solution satisfies *utilitarianism* only for the grand coalition (see (13.3.2)), whereas the Shapley NTU solution satisfies utilitarianism for every coalition (see (13.3.5)). Finally, the Shapley NTU solution satisfies *equity* only for the grand coalition (see (13.3.6)), whereas the Harsanyi NTU solution satisfies equity for every coalition (see (13.3.3)).

Clearly, the concepts of the Shapley NTU value and of the Harsanyi NTU value coincide for NTU games with at most two players. They may differ for three-person games. Indeed, the NTU game (N, V) defined in Exercise 13.3.2 has no Shapley NTU value, but a unique Harsanyi value.

[1] Formally, $\boldsymbol{x} \in \prod_{S \in 2^N \setminus \{\emptyset\}} \mathbb{R}^S$.

Furthermore, if (N, v) is a TU game, then $\phi(N, v)$ is the unique Harsanyi NTU value of (N, V_v). The straightforward proof is left as Exercise 13.3.1.

Let (N, V) be an NTU game. We denote by $\Delta_{++}^{V(N)}$ the subset of Δ_{++} whose members are viable for $V(N)$, that is,

$$\Delta_{++}^{V(N)} = \left\{ \lambda \in \Delta_{++}(N) \,\middle|\, \max_{y \in V(N)} \lambda \cdot y \text{ exists} \right\}.$$

Assume that $V(N)$ is non-levelled and that there exist $C, K \subseteq \mathbb{R}^N$ such that C is compact and convex, K is a convex cone, and $V(N) = C + K$.

Lemma 13.3.3. *For every $\lambda \in \Delta_{++}^{V(N)}$ there exists a unique payoff configuration $\boldsymbol{x}(\lambda) = \boldsymbol{x}$ that satisfies*

$$x_S \in \partial V(S) \text{ for all } \emptyset \neq S \subsetneqq N, \tag{13.3.7}$$

(13.3.2), and (13.3.3). Moreover, the mapping $\Delta_{++}^{V(N)} \to \mathbb{R}^{2^{|N|}}$, given by $\lambda \mapsto v_{\lambda, \boldsymbol{x}(\lambda)}$, is continuous.

Proof: Let $\lambda \in \Delta_{++}^{V(N)}$. We shall construct the coalition function $v^\lambda(S) = v_{\lambda, \boldsymbol{x}(\lambda)}(S)$ and $x_S = x_S(\lambda)$ recursively on $|S|$. If $|S| = 1$, then $S = \{i\}$ for some $i \in N$. Let $x_S = v^i$ and $v^\lambda(S) = \lambda^i x^i$. Thus, $v^\lambda(S)$ is continuous in λ. Assume now that $v^\lambda(S)$ is already constructed and it is continuous in λ for every $S \subseteq N$ such that $|S| < k$ for some $2 \leq k < |N|$. If $|S| = k$, then let $\mathcal{SYM}_S = \{\pi : S \to \{1, \ldots, k\} \mid \pi \text{ is bijective}\}$. Let $i \in S$ and $\pi \in \mathcal{SYM}_S$. If $\pi(i) < k$, then define

$$a_\pi^i(\lambda) = v^\lambda(\{j \in S \mid \pi(j) \leq \pi(i)\} - v^\lambda(\{j \in S \mid \pi(j) < \pi(i)\}.$$

If $\pi(i) = k$, then let $(S, v^{\lambda,t})$ be defined by $v^{\lambda,t}(T) = v^\lambda(T)$ for every $T \subsetneqq S$ and $v^{\lambda,t}(S) = t$. By (8.1.4),

$$\phi^i(S, v^{\lambda,t}) = \sum_{\pi \in \mathcal{SYM}_S : \pi(i) < k} \frac{a_\pi^i(\lambda)}{k!} + \frac{(k-1)!}{k!}(t - v^\lambda(S \setminus \{i\})).$$

Let

$$\alpha_\lambda^i = \sum_{\pi \in \mathcal{SYM}_S : \pi(i) < k} \frac{a_\pi^i(\lambda)}{\lambda^i k!} - \frac{v^\lambda(S \setminus \{i\})}{\lambda^i k}.$$

Then α_λ^i is continuous in λ. Let $t = t(\lambda) \in \mathbb{R}$ be maximal such that

$$\left(\alpha_\lambda^i + \frac{t}{\lambda^i k} \right)_{i \in S} \in V(S).$$

Then $t(\lambda)$ is continuous in λ. Let $v^\lambda(S) = v^{\lambda,t(\lambda)}(S)$ and $x_S^i = \alpha_\lambda^i + \frac{t(\lambda)}{\lambda^i k}$. Then $v^\lambda(S)$ is continuous in λ and $\lambda^S * x_S = \phi(S, v^\lambda)$.

Finally, let $v^\lambda(N) = \max_{y \in V(N)} \lambda \cdot y$, let $x_N \in \mathbb{R}^N$ be given by

$$x_N^i = \phi(N, v^\lambda)/\lambda^i,$$

and observe that $v^\lambda(N)$ is continuous in λ. **q.e.d.**

Note that a hyperplane game has always a unique Harsanyi value. This may be regarded as advantage of the Harsanyi value over the Shapley NTU value (see Remark 13.1.5). Indeed, Lemma 13.3.3 implies the following Corollary.

Corollary 13.3.4. *Let (N, V) be an NTU game such that $V(N)$ is a half-space, i.e., $V(N) = \{x \in \mathbb{R}^N \mid \lambda \cdot x \leq r\}$ for some $r \in \mathbb{R}$. Then $\boldsymbol{x}(\lambda)$ defined in Lemma 13.3.3 is the unique Harsanyi NTU solution of (N, V).*

Now we are ready to prove the main result of this section.

Theorem 13.3.5. *Let (N, V) be an NTU game. If $V(N)$ is non-levelled and if there exist a convex compact set $C \subseteq \mathbb{R}^N$ and a convex cone $K \subseteq \mathbb{R}^N$ such that $V(N) = C + K$, then there exists a Harsanyi NTU value of (N, V).*

Proof: By Exercise 13.1.4 (also see (13.1.9),

$$\Delta_{++}^{V(N)} = \{\lambda \in \mathbb{R}^N \mid \lambda(N) = 1,\ \lambda \cdot x \leq 0\ \forall x \in K\}.$$

Hence, $\Delta_{++}^{V(N)}$ is compact and convex. For $\lambda \in \Delta_{++}^{V(N)}$ define

$$x_\lambda = \left(\frac{\phi^i(N, v_{\lambda, \boldsymbol{x}(\lambda)})}{\lambda^i} \right)_{i \in N}.$$

Then, by Lemma 13.3.3, the proof can be completed as was the proof of Theorem 13.1.3. **q.e.d.**

Exercises

Exercise 13.3.1. Let (N, V_v) be the NTU game associated with a TU game (N, v). Show that $\phi(N, v)$ is the unique Harsanyi NTU value of (N, V_v).

Exercise 13.3.2. Let $N = \{1, 2, 3\}$ and let V be the coalition NTU function given by $V(\{2, 3\}) = \{x^{\{2,3\}} \in \mathbb{R}^{\{1,2\}} \mid 2x^2 + 3x^3 \leq 180\}$, $V(N) = \{x \in \mathbb{R}^N \mid x(N) \leq 120\}$, and $V(S) = \{x \in \mathbb{R}^S \mid x(S) \leq 0\}$ for all other coalitions. Show that $(16, 52, 52)$ is the unique Harsanyi value of (N, V).

It should be noted that Maschler and Owen (1989) used the NTU game of Exercise 13.3.2.

Exercise 13.3.3. Let $f : \mathbb{R} \to \mathbb{R}$ be defined by

$$f(x) = \begin{cases} 4x - x^2 & , \text{if } 0 \le x \le 1, \\ 4x & , \text{if } x < 0, \\ 2x + 1 & , \text{if } x > 1. \end{cases}$$

Let $N = \{1, 2\}$ and let V given by $V(N) = \{x \in \mathbb{R}^N \mid x^2 \le f(1 - x^1)\}$ and $V(S) = \{x \in \mathbb{R}^S \mid x(S) \le 0\}$ for all $\emptyset \ne S \subsetneq N$. Compute the unique Harsanyi value of (N, V).

13.4 A Characterization of the Harsanyi Solution

The Harsanyi solution may be characterized by simple axioms that are suitable modifications of NE, PO, CADD, UNA, SCOV, and IIA to *payoff configuration solutions* (see Theorem 13.2.5). A correspondence $\boldsymbol{\sigma}$ on a set $\widehat{\Gamma}$ of NTU games which assigns to each $(N, V) \in \widehat{\Gamma}$ a set of payoff configurations for N, called a *payoff configuration solution*, satisfies

(1) *nonemptiness* (NE) if $\boldsymbol{\sigma}(N, V) \ne \emptyset$ for all $(N, V) \in \widehat{\Gamma}$,

(2) *efficiency* (EFF) if, for all $(N, V) \in \widehat{\Gamma}$, $\boldsymbol{x} \in \boldsymbol{\sigma}(N, V)$ implies that $x_S \in \partial V(S)$ for all $S \in 2^N \setminus \{\emptyset\}$,

(3) *conditional additivity* (CADD) if the following condition is satisfied for all $(N, U), (N, V), (N, W) \in \widehat{\Gamma}$ and for all $\boldsymbol{x} \in \boldsymbol{\sigma}(N, V)$ and $\boldsymbol{y} \in \boldsymbol{\sigma}(N, W)$: If $U = V + W$ and $x_S + y_S \in \partial U(S)$ for every $\emptyset \ne S \subseteq N$, then $(x_S + y_S)_{S \in 2^N \setminus \{\emptyset\}} \in \boldsymbol{\sigma}(N, U)$,

(4) *unanimity* (UNA) if, for all $\emptyset \ne T \subseteq N$ with $(N, U_T) \in \widehat{\Gamma}$,

$$\boldsymbol{\sigma}(N, U_T) = \{\boldsymbol{z}\},$$

where $z_S = \chi_T^S / |T|$ if $T \subseteq S$ and $z_S = 0 \in \mathbb{R}^S$ if $T \setminus S \ne \emptyset$,

(5) *scale covariance* (SCOV) if, for all $(N, V) \in \widehat{\Gamma}$ and all $\lambda \in \mathbb{R}^N_{++}$ such that $(N, \lambda * V) \in \widehat{\Gamma}$,

$$\boldsymbol{\sigma}(N, \lambda * V) = \{(\lambda^S * x_S)_{S \in 2^N \setminus \{\emptyset\}} \mid \boldsymbol{x} \in \boldsymbol{\sigma}(N, V)\},$$

(6) *independence of irrelevant alternatives* (IIA) if the following condition is satisfied, for all $(N, V), (N, W) \in \widehat{\Gamma}$ and all $\boldsymbol{x} \in \boldsymbol{\sigma}(N, W)$: If $x_S \in V(S) \subseteq W(S)$ for all $\emptyset \ne S \subseteq N$, then $\boldsymbol{x} \in \boldsymbol{\sigma}(N, V)$.

Now, let N be a finite nonempty set of players and let $\widehat{\Gamma}_N^{\mathcal{H}}$ denote the set of all NTU games (N, V) such that $V(N)$ is non-levelled, convex, and smooth, and (N, V) has at least one Harsanyi solution. Moreover, let $\boldsymbol{\Phi}^{\mathcal{H}}$ denote the Harsanyi (payoff configuration) solution. The following lemma is analogous to Lemma 13.2.1.

Lemma 13.4.1. *The Harsanyi solution* $\boldsymbol{\Phi}^{\mathcal{H}}$ *on* $\widehat{\Gamma}_N^{\mathcal{H}}$ *satisfies* NE, EFF, CADD, UNA, SCOV, *and* IIA.

Proof: By the definition of $\widehat{\Gamma}_N^{\mathcal{H}}$, Definition 13.3.1, and Exercise 13.3.1, $\boldsymbol{\Phi}^{\mathcal{H}}$ satisfies NE, EFF, SCOV, and UNA. In order to show CADD, let $V, W, U \in \widehat{G}$, $\boldsymbol{x} \in \boldsymbol{\Phi}^{\mathcal{H}}(N, V), \boldsymbol{y} \in \boldsymbol{\Phi}^{\mathcal{H}}(N, W)$, $U = V + W$, and $x_S + y_S \in \partial U(S)$ for all $\emptyset \neq S \subseteq N$. Let $\boldsymbol{z} = (x_S + y_S)_{S \in 2^N \setminus \{\emptyset\}}$. As $z_N \in \partial U(N)$, there exists $\lambda \in \Delta_{++}(N)$ such that $\max_{y \in U(N)} \lambda \cdot y = \lambda \cdot z_N$. Hence

$$\max_{y \in V(N)} \lambda \cdot y = \lambda \cdot x_N; \tag{13.4.1}$$

$$\max_{y \in W(N)} \lambda \cdot y = \lambda \cdot y_N. \tag{13.4.2}$$

By the smoothness of $V(N)$ and $W(N)$, $\lambda \in \Delta_{++}(N)$ is uniquely determined by (13.4.1) respectively (13.4.2). Thus $\phi(S, v_{\lambda, \boldsymbol{x}}) = \lambda^S * x_S$ and $\phi(S, v_{\lambda, \boldsymbol{y}}) = \lambda^S * y_S$ for all $\emptyset \neq S \subseteq N$. Additivity of the Shapley TU value implies that $\phi(S, v_{\lambda, \boldsymbol{z}}) = \lambda^S * z_S$ so that $\boldsymbol{z} \in \boldsymbol{\Phi}^{\mathcal{H}}(N, U)$.

Finally, in order to show that the remaining axiom is valid, let $V, W, \boldsymbol{x}$ satisfy the conditions of IIA. Let $\lambda \in \Delta_{++}(N)$ such that $\lambda \cdot x_N = \max_{y \in W(N)} \lambda \cdot y = \lambda \cdot x_N$ and $\phi(S, v_{\lambda, \boldsymbol{x}}) = \lambda^S * x_S$ for all coalitions $S \subseteq N$. As $V(N) \subseteq W(N)$ and $x_N \in V(N)$, $\lambda \cdot x_N = \max_{y \in V(N)} \lambda \cdot y$. As $x_S \in \partial W(S) \cap V(S)$ and $V(S) \subseteq W(S)$ for any coalition $S \subseteq N$, $\boldsymbol{x} \in \boldsymbol{\Phi}^{\mathcal{H}}(N, V)$. **q.e.d.**

As N is fixed throughout this section, we may use V as an abbreviation for any NTU game (N, V). The following notation and lemmata are useful.

Notation 13.4.2. Let (N, V) be an NTU game, let $\boldsymbol{x}, \boldsymbol{y}$ be payoff configurations for N, $\alpha \in \mathbb{R}$, and $\lambda \in \mathbb{R}^N$. Denote

$$\boldsymbol{x} + \boldsymbol{y} = (x_S + y_S)_{S \in 2^N \setminus \{\emptyset\}}, \alpha \boldsymbol{x} = (\alpha x_S)_{S \in 2^N \setminus \{\emptyset\}}, \lambda * \boldsymbol{x} = (\lambda^S * x_S)_{S \in 2^N \setminus \{\emptyset\}},$$

and $\lambda \cdot \boldsymbol{x} = (\lambda^S \cdot x_S)_{S \in 2^N \setminus \{\emptyset\}}$. Also, denote

$$\partial V = \prod_{S \in 2^N \setminus \{\emptyset\}} \partial V(S) \subseteq \prod_{S \in 2^N \setminus \{\emptyset\}} \mathbb{R}^S.$$

Lemma 13.4.3. *Let* $\boldsymbol{\sigma}$ *be a payoff configuration solution on a set* $\widehat{\Gamma}$, $\widehat{\Gamma}_N^{TU} \subseteq \widehat{\Gamma} \subseteq \widehat{\Gamma}_N^{\mathcal{H}}$, *that satisfies* NE, EFF, CADD, UNA, *and* SCOV. *Then* $\boldsymbol{\sigma}(V_v) = \left\{(\phi(S, v))_{S \in 2^N \setminus \{\emptyset\}}\right\}$ *for every TU game* (N, v).

Proof: We proceed similarly as in the proof of Lemma 13.2.2. Let (N, v) be a TU game and, for $\alpha \in \mathbb{R}$, denote $V^\alpha = V_{\alpha v}$. Let $\{\boldsymbol{z}\} = \boldsymbol{\Phi}^{\mathcal{H}}(U_N)$. By UNA, $\boldsymbol{\sigma}(U_N) = \{\boldsymbol{z}\}$. By EFF and CADD,

$$\{\boldsymbol{z}\} + \boldsymbol{\sigma}(V^0) = \boldsymbol{\sigma}(U_N) + \boldsymbol{\sigma}(V^0) \subseteq \boldsymbol{\sigma}(U_N + V^0) = \boldsymbol{\sigma}(U_N) = \{\boldsymbol{z}\}.$$

Hence, by NE, $\boldsymbol{\sigma}(V^0) = \left\{(0_S)_{S\in 2^N\setminus\{\emptyset\}}\right\}$, where 0_S denotes $0 \in \mathbb{R}^S$. Also, $\boldsymbol{\sigma}(V^1) + \boldsymbol{\sigma}(V^{-1}) \subseteq \boldsymbol{\sigma}(V^1 + V^{-1}) = \boldsymbol{\sigma}(V^0)$, so that $\boldsymbol{\sigma}(V^1) = -\boldsymbol{\sigma}(V^{-1})$. Hence, $\boldsymbol{\sigma}$ is single-valued and additive on $\widehat{\Gamma}_N^{TU}$. Finally, if $\alpha > 0$, then $V^\alpha = \alpha\chi_N * V^1$. Therefore SCOV implies $\boldsymbol{\sigma}(V^\alpha) = \{\alpha\boldsymbol{x}\}$. Thus, $\boldsymbol{\sigma}$ is linear and $\boldsymbol{\sigma}(U_T) = \boldsymbol{\Phi}^{\mathcal{H}}(U_T)$ for all $T \in 2^N \setminus \{\emptyset\}$. **q.e.d.**

Now, we are ready to prove the following characterization result of $\boldsymbol{\Phi}^{\mathcal{H}}$.

Theorem 13.4.4. *There exists a unique payoff configuration solution on* $\widehat{\Gamma}_N^{\mathcal{H}}$ *that satisfies* NE, EFF, CADD, UNA, SCOV, *and* IIA, *and it is the Harsanyi solution* $\boldsymbol{\Phi}^{\mathcal{H}}$.

Proof: By Lemma 13.4.1 only uniqueness has to be shown. Let $\boldsymbol{\sigma}$ be a payoff configuration solution on $\widehat{\Gamma}_N^{\mathcal{H}}$ that satisfies NE, EFF, CADD, UNA, SCOV, and IIA. Let $V \in \widehat{\Gamma}_N^{\mathcal{H}}$. We first show that $\boldsymbol{\sigma}(V) \subseteq \boldsymbol{\Phi}^{\mathcal{H}}(V)$.

Let $\boldsymbol{x} \in \boldsymbol{\sigma}(V)$ and (N, V^0) again be defined by

$$V^0(S) = \{y \in \mathbb{R}^S \mid y(S) \leq 0\} \text{ for all } S \in 2^N \setminus \{\emptyset\}.$$

By Lemma 13.4.3, $\boldsymbol{\sigma}(V^0) = \boldsymbol{\Phi}^{\mathcal{H}}(V^0) = \mathbf{0}$, where $\mathbf{0} = (0_S)_{S\in 2^N\setminus\{\emptyset\}}$. Define (N, U^0) by $U^0(S) = -\mathbb{R}_+^S$ for $\emptyset \neq S \subsetneqq N$ and $U^0(N) = V^0(N)$. By IIA, $\mathbf{0} \in \boldsymbol{\sigma}(U^0)$. Let $\lambda \in \Delta_{++}(N)$ such that $\lambda \cdot x_N = \max_{y\in V(N)}(\lambda \cdot y)$ and define $\mu \in \mathbb{R}_{++}^N$ by $\mu^i = \frac{1}{\lambda^i}$ for $i \in N$. Define auxiliary NTU games (N, W_k), $k = 1, 2, 3$, by

$$W_1(T) = \begin{cases} V(T) \text{ for } T \in 2^N \setminus \{N, \emptyset\}, \\ \{y \in \mathbb{R}^N \mid \lambda \cdot y \leq \lambda \cdot x\} \text{ for } T = N; \end{cases} \tag{13.4.3}$$

$$W_2(T) = \begin{cases} x_T - \mathbb{R}_+^T \text{ for } T \in 2^N \setminus \{N, \emptyset\}, \\ W_1(N) \text{ for } T = N; \end{cases} \tag{13.4.4}$$

$$W_3(S) = \{y \in \mathbb{R}^S \mid \lambda^S \cdot y \leq \lambda^S \cdot x_S\} \text{ for } S \in 2^N \setminus \{\emptyset\}. \tag{13.4.5}$$

By Corollary 13.3.4, $W_k \in \widehat{\Gamma}_N^{\mathcal{H}}$. As $W_1 = V + \mu * U^0$, $\boldsymbol{x} + \mathbf{0} = \boldsymbol{x} \in \boldsymbol{\sigma}(W_1)$ by SCOV and CADD. By IIA, $\boldsymbol{x} \in \boldsymbol{\sigma}(W_2)$. As $W_3 = W_2 + \mu * V^0$, $\boldsymbol{x} + \mathbf{0} = \boldsymbol{x} \in \boldsymbol{\sigma}(W_3)$ again by SCOV and CADD. By Lemma 13.4.3 and SCOV, $\boldsymbol{\sigma}(W_3) = \boldsymbol{\Phi}^{\mathcal{H}}(W_3) = \{\boldsymbol{x}\}$, a singleton. By Definition 13.3.1, $\boldsymbol{x} \in \boldsymbol{\Phi}^{\mathcal{H}}(V)$.

In order to show the opposite inclusion let, now, $\boldsymbol{x} \in \boldsymbol{\Phi}^{\mathcal{H}}(V)$ and W_1 again be defined by (13.4.3). By IIA of $\boldsymbol{\Phi}^{\mathcal{H}}$ and Corollary 13.3.4, $\boldsymbol{x}$ is the unique element of $\boldsymbol{\Phi}^{\mathcal{H}}(W_1)$. By NE of $\boldsymbol{\sigma}$ and the preceding part of the proof, $\boldsymbol{x}$ is also the unique element of $\boldsymbol{\sigma}(W_1)$. As $V(N)$ is convex, IIA completes the proof. **q.e.d.**

Let $|N| \geq 3$. We now define payoff configuration solutions $\boldsymbol{\sigma}^i$, $i = 1, \ldots, 4, 6$, on $\widehat{\Gamma}_N^{\mathcal{H}}$ that satisfy all axioms of Theorem 13.4.4 except the i-th one. Let $V \in \widehat{\Gamma}_N^{\mathcal{H}}$, let $\boldsymbol{\Phi}$ denote the Shapley NTU solution, that is,

$$\boldsymbol{\Phi}(V) = \left\{ \boldsymbol{x} \in \prod_{\emptyset \neq S \subseteq N} \mathbb{R}^S \,\middle|\, \boldsymbol{x} \text{ satisfies (13.3.4) - (13.3.6) for some } \lambda \in \Delta_{++}^V \right\},$$

and define

$$\boldsymbol{\sigma}^1(V) = \boldsymbol{\Phi}^{\mathcal{H}}(V) \cap \boldsymbol{\Phi}(V); \tag{13.4.6}$$

$$\boldsymbol{\sigma}^2(V) = \begin{cases} \boldsymbol{\Phi}^{\mathcal{H}}(V), \text{ if } V(N) \text{ is a half-space or } \mathbf{0} \notin V, \\ \boldsymbol{\Phi}^{\mathcal{H}}(V) \cup \{\mathbf{0}\}, \text{ otherwise;} \end{cases} \tag{13.4.7}$$

$$\boldsymbol{\sigma}^3(V) = \begin{cases} \partial V \quad , \text{ if } V(N) \text{ is not a half-space,} \\ \boldsymbol{\Phi}^{\mathcal{H}}(V) \text{ , otherwise;} \end{cases} \tag{13.4.8}$$

$$\boldsymbol{\sigma}^4(V) = \partial V; \tag{13.4.9}$$

$$\boldsymbol{\sigma}^6(V) = \{\boldsymbol{x} \in \partial V \mid \boldsymbol{x} \leq \boldsymbol{y} \text{ for some } \boldsymbol{y} \in \boldsymbol{\Phi}^{\mathcal{H}}(N,V)\}, \tag{13.4.10}$$

where $\boldsymbol{x} \leq \boldsymbol{y}$ is defined by $x_S \leq y_S$ for all $\emptyset \neq S \subseteq N$.

Clearly, $\boldsymbol{\sigma}^i \neq \boldsymbol{\Phi}^{\mathcal{H}}$ for $i = 1, \ldots, 4$. It is straightforward to verify that $\boldsymbol{\sigma}^1$ satisfies EFF, CADD, UNA, SCOV, and IIA, so that it violates NE. Also, it is easy to show that $\boldsymbol{\sigma}^2$ satisfies NE, UNA, and SCOV. CADD follows from Exercise 13.4.2 and IIA is left to the reader (Exercise 13.4.3). It is straightforward to verify that $\boldsymbol{\sigma}^3$ satisfies NE, EFF, UNA, and SCOV, and that $\boldsymbol{\sigma}^4$ satisfies all axioms except UNA. By Exercise 13.4.3, $\boldsymbol{\sigma}^3$ satisfies IIA. The symmetric *egalitarian solution* (see Kalai and Samet (1985)) violates exclusively SCOV and, finally, $\boldsymbol{\sigma}^6$ violates exclusively IIA (the proof of CADD is left as Exercise 13.4.4).

The following statements are similar to Remark 13.2.6 and Corollary 13.2.7. Let $\widehat{\Gamma} \subseteq \widehat{\Gamma}_N^{\mathcal{H}}$ satisfy (2) and (3) of Remark 13.2.6, and the condition that differs from (4) of Remark 13.2.6 only inasmuch as Δ_{++}^V is replaced by $\Delta_{++}^{V(N)}$. It is easy to check that the proof of Theorem 13.4.4 remains valid if $\widehat{\Gamma}_N^{\mathcal{H}}$ is replaced by $\widehat{\Gamma}$. Hence, we have deduced the following corollary.

Corollary 13.4.5. *Let $\widehat{\Gamma}$ be the set of NTU games with player sets N that satisfy the conditions of Theorem 13.3.5. The Harsanyi solution $\boldsymbol{\Phi}^{\mathcal{H}}$ is the unique solution on $\widehat{\Gamma}$ that satisfies* NE, EFF, CADD, UNA, SCOV, *and* IIA.

Exercises

Exercise 13.4.1. Let $N = \{1, 2\}$. Show by means of an example that $\boldsymbol{\Phi}^{\mathcal{H}}$ does not satisfy CADD on the set of NTU games (N, V) with convex and non-levelled $V(N)$ that have a nonempty Harsanyi solution.

Exercise 13.4.2. Show that the Harsanyi solution satisfies the *zero-inessential games property* due to Hart (1985): If $\mathbf{0} \in \partial V$, then $\mathbf{0} \in \boldsymbol{\Phi}^{\mathcal{H}}(V)$.

Exercise 13.4.3. Prove that the solutions $\boldsymbol{\sigma}^2$ and $\boldsymbol{\sigma}^3$ (see (13.4.7) and (13.4.8)) satisfy IIA on $\widehat{\Gamma}_N^{\mathcal{H}}$.

Exercise 13.4.4. Verify that $\boldsymbol{\sigma}^6$ defined by (13.4.10) satisfies CADD.

13.5 Notes and Comments

The Shapley NTU value generalizes the Nash (1950) solution which has an axiomatic foundation and is defined only for bargaining problems (see Figure 13.1.1). The formal definition of a bargaining problem is provided in Section 14.3.1. It should be noted that there are further well-known solutions of bargaining problems that are axiomatized (see Kalai and Smorodinsky (1975) and Perles and Maschler (1981)). Moreover, Rosenmüller (2000) in Chapter VIII and Peters (1992) present comparisons of several axiomatizations. Some of the solutions may even be extended to n-person bargaining problems (see, e.g., Calvo and Gutiérrez (1994)). As bargaining problems are very special cooperative games, further considerations of axiomatic bargaining have to be waived.

An existence proof of the Harsanyi value of compactly generated NTU games is contained in Rosenmüller (1981). In his approach, however, the transfer vector λ, corresponding to a Harsanyi value, may not be positive. Only $\lambda \geq 0$ is required.

Theorem 13.4.4 and several modifications, some of them replacing NE by the zero-inessential game property, are due to Hart (1985).

The intersection of the Harsanyi solution and the Shapley solution, $\boldsymbol{\sigma}^1$ (see (13.4.6)), is called *Harsanyi-Shapley solution.* Chang and Hwang (2003) contains a characterization of the Harsanyi-Shapley solution.

14

The Consistent Shapley Value

This chapter is devoted to a further extension of the Shapley value, namely the *consistent* Shapley value. In a first step, in Section 14.1, the Shapley value is extended to hyperplane games in a way that generalizes the "random order procedure" defined by (8.1.4). We show that the consistent Shapley value is k-consistent. Moreover, we prove that the consistent Shapley value on hyperplane games is characterized by single-valuedness, Pareto optimality, covariance, and 2-consistency.

In Section 14.2 the consistent Shapley value is extended to p-smooth games. We show that this solution is nonempty when applied to any p-smooth game whose normal vectors have coordinates that are uniformly above some positive constant.

Finally, in Section 14.3 we prove that, on classes of uniformly p-smooth games, the consistent Shapley solution is the maximum solution that is nonempty, efficient, strongly monotonic, and satisfies the symmetry axiom. Moreover, the "maximality condition" may be replaced by independence of irrelevant alternatives, provided that the games under consideration are convex-valued.

14.1 For Hyperplane Games

We shall first define the consistent Shapley value for hyperplane games. Let (N, V) be a hyperplane game (see (13.1.10)). Recall that (N, V) does not have a Shapley NTU value, unless all hyperplanes are parallel (see Remark 13.1.5). However, we may extend Formula (8.1.4) as follows. Assume without loss of generality that $N = \{1, \ldots, n\}$. Let $\pi \in \mathcal{SYM}_N$, let $i \in N$, and let $a_\pi \in \mathbb{R}^N$ be defined as follows. If $\pi(i) = 1$, let $a^i_\pi(V) = v^i$. Assume now that the $a^j_\pi(V)$ are defined for all $j \in P^i_\pi = P$ (see (8.1.3)). Let

$$a^i_\pi(V) = \max\{x^i \in \mathbb{R} \mid (x^i, a^P_\pi) \in V(P \cup \{i\})\} \tag{14.1.1}$$

and define

$$\phi(N,V) = \sum_{\pi \in \mathcal{SYM}_N} \frac{a_\pi(V)}{n!}. \tag{14.1.2}$$

Then $\phi(N,V)$ is called the *consistent Shapley value* of (N,V).

We shall now provide a characterization of the consistent Shapley value for hyperplane games, which is similar to Theorem 8.3.6. This requires us to generalize some concepts of Section 8.3 to NTU games.

Let $\mathcal{U}$ be a set of players and let $\widehat{\Gamma}^h_{\mathcal{U}} = \widehat{\Gamma}^h$ be the set of all hyperplane games with players in $\mathcal{U}$. If σ is a single-valued solution on $\widehat{\Gamma}^h$, then we say that σ is *consistent* if $(S, V_{S,\sigma}) \in \widehat{\Gamma}^h$ and $\sigma(S, V_{S,\sigma}) = \sigma^S(N,V)$ for all $(N,V) \in \widehat{\Gamma}^h$ and all $S \in 2^N \setminus \{\emptyset\}$. Here, $V_{S,\sigma}$ is the coalition function of the σ-*reduced game* (compare with (8.3.1)) which is defined by

$$V_{S,\sigma}(T) = \{x^T \in \mathbb{R}^T \mid (x^T, \sigma^{N\setminus S}(T \cup (N \setminus S), V)) \in V(T \cup (N \setminus S))\}$$

for all $T \in 2^S \setminus \{\emptyset\}$ and $V_{S,\sigma}(\emptyset) = \emptyset$.

Proofs of the following simple remarks are left as Exercises 14.1.1 and 14.1.2.

Remark 14.1.1. A subgame of a hyperplane game is a hyperplane game. Moreover, if σ is a single-valued solution on $\widehat{\Gamma}^h$ and $(N,V) \in \widehat{\Gamma}^h$, then every σ-reduced game of (N,V) is a hyperplane game.

Remark 14.1.2. The solution ϕ satisfies COV and PO on every set of hyperplane games.

The consistent Shapley value does not satisfy consistency on hyperplane games in general (see Exercise 14.1.3), but it satisfies a related axiom.

The single-valued solution σ on $\widehat{\Gamma}^h$ is said to be k-*consistent* ($k \in \mathbb{N}$), if for every $(N,V) \in \widehat{\Gamma}^h$ such that $|N| \geq k$ and for all $i \in N$,

$$\sum_{T \subseteq N : T \ni i, |T| = k} \sigma^i(T, V_{T,\sigma}) = \binom{n-1}{k-1} \sigma^i(N,V). \tag{14.1.3}$$

This property has the following interpretation. If player $i \in N$ is faced with the game $(T, V_{T,\sigma})$ such that $T \ni i$ and $|T| = k$, then he may ask to adjust his payoff by $\sigma^i(T, V_{T,\sigma}) - \sigma^i(N,V)$. Then σ is k-consistent, if the adjustments to each player in all coalitions of size k cancel out.

Note that, if σ is consistent, then it is k-consistent for every $k \in \mathbb{N}$.

Theorem 14.1.3. *On $\widehat{\Gamma}^h_{\mathcal{U}}$ the mapping ϕ is k-consistent for every $k \in \mathbb{N}$.*

Proof: Let $(N,V) \in \widehat{\Gamma}^h$ defined by (13.1.10). If $\lambda_N = \chi_N$ and $i \in N$, then

$$\sum_{j \in N \setminus \{i\}} (\phi^i(N \setminus \{j\}, V) - \phi^j(N \setminus \{i\}, V)) + r_N = |N| \phi^i(N,V). \tag{14.1.4}$$

The straightforward proof of (14.1.4) is left to the reader (see Exercise 14.1.4).

We proceed by induction on k. Clearly, ϕ is 1-consistent. Assume now that ϕ is k-consistent for all $k < m$ and some $m \geq 2$. Now, let $k = m$ and let $(N, V) \in \widehat{\Gamma}^h_{\mathcal{U}}$ such that V is given by (13.1.10) and $|N| \geq k$. If $n = |N| = k$, then the proof is finished. Hence we may assume that $k < n$. By COV, $\lambda_N = \chi_N$ is assumed. Let $S \subseteq N$ such that $|S| = k$. Applying (14.1.4) to $(S, V_{S,\phi})$ yields

$$k\phi^i(S, V_{S,\phi}) = \sum_{j \in S \setminus \{i\}} \left(\phi^i(S \setminus \{j\}, V_{S,\phi}) - \phi^j(S \setminus \{i\}, V_{S,\phi})\right) + r_N - \sum_{\ell \in N \setminus S} \phi^\ell(N, V)$$

and, thus,

$$\begin{aligned} k \sum_{S \ni i, |S|=k} \phi^i(S, V_{S,\phi}) &= \sum_{S \ni i, |S|=k} \sum_{j \in S \setminus \{i\}} \left(\phi^i(S \setminus \{j\}, V_{S,\phi}) - \phi^j(S \setminus \{i\}, V_{S,\phi})\right) \\ &\quad + \binom{|N|-1}{k-1} r_N - \sum_{S \ni i, |S|=k} \sum_{\ell \in N \setminus S} \phi^\ell(N, V). \end{aligned}$$

Rearranging the order of summation and applying Exercise 14.1.5 to $\sigma = \phi$, $(V_{S,\phi})^{S \setminus \{\ell\}}$, $\ell = i, j$, and $T = S \setminus \{\ell\}$, yields

$$\begin{aligned} k \sum_{S \ni i, |S|=k} \phi^i(S, V_{S,\phi}) &= \sum_{j \in N \setminus \{i\}} \left(\sum_{T: j \notin T \ni i, |T|=k-1} \phi^i\left(T, \left(V^{N \setminus \{j\}}\right)_{T,\phi}\right) - \right. \\ &\quad \left. \sum_{T: j \in T \not\ni i, |T|=k-1} \phi^j\left(T, \left(V^{N \setminus \{i\}}\right)_{T,\phi}\right)\right) \\ &\quad + \binom{|N|-1}{k-1} r_N - \binom{|N|-2}{k-1} \sum_{\ell \in N \setminus \{i\}} \phi^\ell(N, V). \end{aligned}$$

By Pareto optimality, $r_N - \sum_{\ell \in N \setminus \{i\}} \phi^\ell(N, V) = \phi^i(N, V)$. Hence, applying the induction hypothesis yields

$$\begin{aligned} &k \textstyle\sum_{S \ni i, |S|=k} \phi^i(S, V_{S,\phi}) \\ &= \tbinom{|N|-2}{k-2} \textstyle\sum_{j \in N \setminus \{i\}} \left(\phi^i(N \setminus \{j\}, V) - \phi^j(N \setminus \{i\}, V)\right) \\ &\qquad + \tbinom{|N|-2}{k-2} r_N + \tbinom{|N|-2}{k-1} \phi^i(N, V). \end{aligned}$$

By (14.1.4),

$$k \sum_{S \ni i, |S|=k} \phi^i(S, V_{S,\phi}) = |N| \binom{|N|-2}{k-2} \phi^i(N,V) + \binom{|N|-2}{k-1} \phi^i(N,V)$$

and, thus, $\sum_{S \ni i, |S|=k} \phi^i(S, V_{S,\phi}) = \binom{|N|-1}{k-1} \phi^i(N,V)$. **q.e.d.**

Let (N, V) be an NTU game and let $k, \ell \in N$, $k \neq \ell$. Then k and ℓ are *substitutes* (with respect to (N, V)), if the following property is satisfied for all $x \in \mathbb{R}^N$ and every $S \subseteq N \setminus \{k, \ell\}$: If $y \in \mathbb{R}^N$ is defined by $y^{N \setminus \{k,\ell\}} = x^{N \setminus \{k,\ell\}}$, $y^k = x^\ell$, and $y^\ell = x^k$, then

$$x^{S \cup \{k\}} \in V(S \cup \{k\}) \Leftrightarrow y^{S \cup \{\ell\}} \in V(S \cup \{\ell\});$$
$$x^{S \cup \{k,\ell\}} \in V(S \cup \{k,\ell\}) \Leftrightarrow y^{S \cup \{k,\ell\}} \in V(S \cup \{k,\ell\}).$$

A solution σ on a set $\widehat{\Gamma}$ satisfies the *equal treatment property* (ETP) if $x^k = x^\ell$ for all $x \in \sigma(N, V)$ whenever k and ℓ are substitutes and $(N, V) \in \widehat{\Gamma}$. Now we are ready to state the characterization.

Theorem 14.1.4. *There is a unique single-valued solution on $\widehat{\Gamma}^h$ that satisfies* PO, ETP, COV, *and* 2*-consistency, and it is the consistent Shapley value.*

Proof: By Remark 14.1.2 and Theorem 14.1.3, ϕ satisfies PO, COV, and 2-consistency. Clearly, it satisfies ETP as well. Let σ be a single-valued solution that satisfies the desired properties, let $(N, V) \in \widehat{\Gamma}^h$, defined by (13.1.10), and let $n = |N|$. By COV we may assume that $\lambda_N = \chi_N$. We proceed as in the proof of Theorem 8.3.6. For $n \leq 2$, (N, V) is the NTU game associated with a TU game; hence $\sigma(N, V) = \phi(N, V)$ is deduced as in the aforementioned proof. If $n \geq 3$ and $i, j \in N$, $i \neq j$, and $S = \{i, j\}$, then, by the induction hypothesis,

$$\sigma^i(S, V_{S,\sigma}) - \phi^i(S, V_{S,\phi}) = \sigma^j(S, V_{S,\sigma}) - \phi^j(S, V_{S,\phi}), \tag{14.1.5}$$

which is deduced as (8.3.3). Also, by PO,

$$\begin{aligned} \sigma^i(S, V_{S,\sigma}) + \sigma^j(S, V_{S,\sigma}) &= \sigma^i(N, V) + \sigma^j(N, V), \\ \phi^i(S, V_{S,\phi}) + \phi^j(S, V_{S,\phi}) &= \phi^i(N, V) + \phi^j(N, V). \end{aligned} \tag{14.1.6}$$

Hence,

$$\begin{aligned} &\sigma^j(S, V_{S,\sigma}) - \phi^j(S, V_{S,\phi}) = \\ &\sigma^i(N,V) + \sigma^j(N,V) - \sigma^i(S, V_{S,\sigma}) - \phi^i(N,V) - \phi^j(N,V) + \phi^i(S, V_{S,\phi}). \end{aligned}$$

Thus, by (14.1.5),

$$2\left(\sigma^i(S, V_{S,\sigma}) - \phi^i(S, V_{S,\phi})\right) = \sigma^i(N,V) + \sigma^j(N,V) - \phi^i(N,V) - \phi^j(N,V).$$

By PO, $\sum_{j \in N} \sigma^j(N, V) = r_N = \sum_{j \in N} \phi^j(N, V)$, and therefore

$$2\sum_{j\in N\setminus\{i\}}(\sigma^i(\{i,j\},V_{\{i,j\},\sigma}) - \phi^i(\{i,j\},V_{\{i,j\},\phi})) \\ = (n-2)\left(\sigma^i(N,V) - \phi^i(N,V)\right).$$

Finally, by 2-consistency,

$$2(n-1)(\sigma^i(N,V) - \phi^i(N,V)) = (n-2)(\sigma^i(N,V) - \phi^i(N,V)).$$

We conclude that $\sigma(N,V) = \phi(N,V)$. **q.e.d.**

In Section 14.2 we shall show the existence of a suitable extension of the consistent Shapley value to a wider class of NTU games.

Exercises

Exercise 14.1.1. Prove Remark 14.1.1.

Exercise 14.1.2. Prove Remark 14.1.2.

Exercise 14.1.3. Let (N,V) be the game defined in Exercise 13.3.2. Show that $\phi(N,V) = (15,55,50)$ and use this example to show that ϕ is not consistent (see Maschler and Owen (1989)).

Exercise 14.1.4. Let $(N,V) \in \widehat{\Gamma}^h$ defined by (13.1.10) and let $i \in S \subseteq N$. Prove that

$$\sum_{j\in S\setminus\{i\}} \phi^i(S\setminus\{j\},V) + \frac{1}{\lambda_S^i}\left(r_S - \sum_{j\in S\setminus\{i\}} \lambda_S^j \phi^j(S\setminus\{i\},V)\right) = |S|\phi^i(S,V)$$

(see Lemma 1 of Maschler and Owen (1989)).

Exercise 14.1.5. Let (N,V) be an NTU game and let σ be a single-valued solution on the set of all subgames of (N,V) (that is, on the set $\{(T,V^T) \mid T \in 2^N \setminus \{\emptyset\}\}$). Show that, if $\ell \in S \subseteq N$ and $|S| \geq 2$, then

$$\left(V^{N\setminus\{\ell\}}\right)_{S\setminus\{\ell\},\sigma} = (V_{S,\sigma})^{S\setminus\{\ell\}}$$

(see Lemma 2 of Maschler and Owen (1989)).

14.2 For p-Smooth Games

We shall first extend the consistent Shapley value to a richer class of NTU games. Let (N,V) be an NTU game.

Definition 14.2.1. *Let $\emptyset \neq S \subseteq N$. Then $V(S)$ is* **positively smooth** *(p-smooth) if*

(1) at every $x \in \partial V(S)$ there exists a unique tangent hyperplane $H_x = H_{x,V(S)}$ to $V(S)$,

(2) the unique normal vector $\lambda_{x,V(S)} = \lambda_x \in \mathbb{R}^S_+$ (determined by the requirements $\lambda_x(S) = 1$ and $H_x = \{y \in \mathbb{R}^S \mid \lambda_x \cdot y = \lambda_x \cdot x\}$) satisfies $\lambda_x \gg 0$, and

(3) the mapping $\partial V(S) \to \Delta_{++}(S), x \mapsto \lambda_x$, is continuous.

The game (N, V) is **p-smooth** *if $V(S)$ is p-smooth for each $S \in 2^N \setminus \{\emptyset\}$.*

Let $V(S)$ be convex and smooth (see Section 13.1). Then $V(S)$ is p-smooth iff $V(S)$ is non-levelled.

Now we may extend the definition of the consistent Shapley value to p-smooth games.

Definition 14.2.2. *Let (N, V) be a p-smooth NTU game. A* **consistent Shapley solution** *of (N, V) is a payoff configuration $(x_S)_{\emptyset \neq S \subseteq N}$ with the following properties:*

(1) $x_S \in \partial V(S)$ for all coalitions S in N.

(2) If $\emptyset \neq S \subsetneqq N$, then $(x_R)_{\emptyset \neq R \subseteq S}$ is a consistent Shapley solution of the subgame (S, V).

(3) If (N, W) is the hyperplane game determined by the hyperplanes $H_{x_S,V(S)}$ (the tangent hyperplane through x_S at $\partial V(S)$), $\emptyset \neq S \subseteq N$, then $x_N = \phi(N, W)$.

A vector $x \in \mathbb{R}^N$ is a **consistent Shapley value** *of (N, V) if there exists a consistent Shapley solution $(x_S)_{S \in 2^N \setminus \{\emptyset\}}$ such that $x = x_N$.*

For any p-smooth game (N, V), let $\boldsymbol{\Phi}^{\mathcal{MO}}(N, V)$ denote the set of all consistent Shapley solutions to (N, V) and let $\Phi^{\mathcal{MO}}(N, V)$ denote the set of all consistent Shapley values to (N, V), that is, $\Phi^{\mathcal{MO}}(N, V) = \{x_N \mid \boldsymbol{x} \in \boldsymbol{\Phi}^{\mathcal{MO}}(N, V)\}$. In general, there are p-smooth games that do not have any consistent Shapley value (see Exercise 14.2.2). However, under a mild additional condition concerning the normal vectors, we now show the existence of a consistent Shapley value.

Theorem 14.2.3. *Let $0 < \delta < \frac{1}{|N|}$ and let (N, V) be a p-smooth NTU game such that*

$$\lambda^i_{x,V(S)} > \delta \ \forall x \in \partial V(S) \ \forall i \in S \ \forall S \in 2^N \setminus \{\emptyset\}. \tag{14.2.1}$$

Then $\boldsymbol{\Phi}^{\mathcal{MO}}(N, V) \neq \emptyset$.

Let (N, v) be an NTU game satisfying the condition of Theorem 14.2.3. Then (N, V) is non-levelled. Also, for every coalition S in N, the subgame (S, V) is a p-smooth game that satisfies (14.2.1).

Proof of Theorem 14.2.3: We may assume that $N = \{1, \ldots, n\}$. As $\boldsymbol{\Phi}^{\mathcal{MO}}$ is translation covariant (see Exercise 14.2.1), we may assume that

$$0^S \in V(S) \text{ for all } \emptyset \neq S \subseteq N. \tag{14.2.2}$$

If $n = 1$, then $\boldsymbol{\Phi}^{\mathcal{MO}}(N, V)$ is a singleton. Hence it suffices to show the following claim for $n \geq 2$: If $x_S \in \mathbb{R}^S, \emptyset \neq S \subsetneqq N$, such that $(x_T)_{\emptyset \neq T \subseteq S} \in \boldsymbol{\Phi}^{\mathcal{MO}}(S, V)$ for any $\emptyset \neq S \subsetneqq N$, then there exists $x_N \in \mathbb{R}^N$ such that $(x_S)_{\emptyset \neq S \subseteq N} \in \boldsymbol{\Phi}^{\mathcal{MO}}(N, V)$.

In order to prove our claim, let, for any $z \in \partial V(N)$, $\lambda = \lambda_{z,V(N)}$, $r_N(z) = \lambda_z \cdot z$, (N, V_z) be the hyperplane game defined by

$$\begin{aligned} V_z(S) &= \{y \in \mathbb{R}^S \mid \lambda_{x_S,V(S)} \cdot y \leq \lambda_{x_S,V(S)} \cdot x_S\} \forall \emptyset \neq S \subsetneqq N, \\ V_z(N) &= \{y \in \mathbb{R}^N \mid \lambda_z \cdot y \leq r_N(z)\}, \end{aligned}$$

and $\varphi(z) = \phi(N, V_z)$. By Definition 14.2.2, $x_S = \phi(S, V_z)$ for all $\emptyset \neq S \subsetneqq N$. So, for all $i \in N$, Exercise 14.1.4 applied to $S = N$ yields

$$n\varphi^i(z) = \sum_{j \in N \setminus \{i\}} x^i_{N \setminus \{j\}} + \frac{1}{\lambda^i_z} \left(r_N(z) - \sum_{j \in N \setminus \{i\}} \lambda^j_z x^j_{N \setminus \{i\}} \right). \tag{14.2.3}$$

The following constructions are used to show that $\varphi : \partial V(N) \to \mathbb{R}^N$ has a fixed point. If $\alpha > \max_{i,j \in N, i \neq j} |x^i_{N \setminus \{j\}}|$, then

$$\left| \sum_{j \in N \setminus \{i\}} x^i_{N \setminus \{j\}} \right| < (n-1)\alpha < (n-1)\frac{\alpha}{\delta},$$

$$\frac{1}{\lambda^i_z} \left| \sum_{j \in N \setminus \{i\}} \lambda^j_z x^j_{N \setminus \{i\}} \right| < \frac{\alpha}{\lambda^i_z} \leq \frac{\alpha}{\delta}.$$

We may conclude from (14.2.3) that

$$\left| \varphi^i(z) - \frac{r_N(z)}{n\lambda^i_z} \right| < \frac{\alpha}{\delta} \text{ for all } i \in N. \tag{14.2.4}$$

Let $\beta = \frac{\alpha(\delta+1)}{\delta^2}$, let $b = (\underbrace{\beta, \ldots, \beta}_{n})$, let $Z = \partial V(N) \cap ((-b) + \mathbb{R}^N_+)$, and define $\widetilde{\varphi} : Z \to \mathbb{R}^N$ by $\widetilde{\varphi}^i(z) = \max\{\varphi^i(z), -\beta\}$ for all $z \in Z$ and $i \in N$. It suffices to show that

$$\exists z \in Z : \widetilde{\varphi}(z) = z, \tag{14.2.5}$$

$$z \in Z, \widetilde{\varphi}(z) = z \Rightarrow \widetilde{\varphi}(z) = \varphi(z). \tag{14.2.6}$$

In order to show (14.2.5), we first prove that

$$z \in Z \Rightarrow \widetilde{\varphi}(z) \neq -b. \tag{14.2.7}$$

Indeed, let $i \in N$ such that $\lambda_z^i \geq \lambda_z^j$ for all $j \in N$. As $0 \in V(N)$, there exists $k \in N$ such that $z^k \geq 0$. Hence,

$$\frac{r_N(z)}{\lambda_z^i} \geq \sum_{j \in N: z^j < 0} \frac{\lambda_z^j}{\lambda_z^i} z^j \geq -(n-1)\beta.$$

As $\delta + 1 \geq 1 \geq n\delta$, we have $\frac{\alpha}{\delta} \leq \frac{\beta}{n}$ so that (14.2.4) implies

$$\varphi^i(z) > -\frac{\alpha}{\delta} - \frac{n-1}{n}\beta \geq -\beta.$$

Now, let $X = \{x \in (-b) + \mathbb{R}_+^N \mid x(N) = 0\}$ and let

$$f : ((-b) + \mathbb{R}_+^N) \setminus \{-b\} \to X$$

be the projection centered at $-b$, that is, $f(y)$ is the unique element of X that belongs to the line through $-b$ and y for every $y \geq -b, y \neq -b$. The restriction of f to Z, denoted by g, is bijective and g^{-1} is continuous. Also, as $V(N)$ is positively smooth, φ and, hence, $\widetilde{\varphi}$ is continuous. By (14.2.7), the composition $h = f \circ \widetilde{\varphi} \circ g^{-1} : X \to X$ is well-defined and, hence, continuous. As X is a compact convex set (a simplex), h has a fixed point x. Clearly, $g^{-1}(x)$ is a fixed point of $\widetilde{\varphi}$.

In order to show (14.2.6) let $z \in Z$ be a fixed point of $\widetilde{\varphi}$. It suffices to show that $z \gg -b$. Assume, on the contrary, that there exists $i \in N$ with $z^i = -\beta$. As $\frac{\alpha}{\delta} = \frac{\delta}{\delta+1}\beta$, (14.2.4) implies that

$$-\beta \geq \varphi^i(z) > -\frac{\alpha}{\delta} + \frac{r_N(z)}{n\lambda_z^i} \geq -\beta\frac{\delta}{\delta+1} + \frac{r_N(z)}{n\lambda_z^i}.$$

We conclude that $\frac{r_N(z)}{n\lambda_z^i} < -\beta\frac{1}{\delta+1}$, hence

$$\frac{r_N(z)}{n} < -\beta\frac{\delta}{\delta+1}. \tag{14.2.8}$$

Using the remaining inequality in (14.2.4) yields

$$\varphi^j(z) < \beta\frac{\delta}{\delta+1} - \beta\frac{\delta}{(\delta+1)\lambda_z^j} < 0 \;\forall j \in N. \tag{14.2.9}$$

As $0 \in V(N)$ there exists $k \in N$ with $z^k \geq 0$. By (14.2.9), $\varphi^k(z) < 0$, so $\widetilde{\varphi}^k(z) < 0 \leq z^k$ and the desired contradiction has been obtained. **q.e.d.**

Exercises

Exercise 14.2.1. Show that on the set of p-smooth games with players in U, $\boldsymbol{\Phi}^{\mathcal{MO}}$ is *translation covariant*, that is, if (N, V), $N \subseteq U$, is a p-smooth game and $x \in \mathbb{R}^N$, then $\boldsymbol{\Phi}^{\mathcal{MO}}(N, V + x) = \boldsymbol{\Phi}^{\mathcal{MO}}(N, V) + \boldsymbol{x}$, where $\boldsymbol{x}$ is the payoff configuration $(x^S)_{\emptyset \neq S \subseteq N}$.

Exercise 14.2.2. Let (N, V) be the NTU game defined in Exercise 13.1.3. Show that (N, V) does not have any consistent Shapley value.

Exercise 14.2.3. Let (N, V) be a p-smooth two-person NTU game such that $V(N)$ is convex. Show that x is a consistent Shapley value of (N, V) iff x is a Harsanyi value of (N, V).

14.3 Axiomatizations

We shall present a characterization of the consistent Shapley value that is a generalization of Theorem 8.2.6. Let N, $|N| \geq 2$, be a finite set. An NTU game (N, V) is *uniformly* p-smooth if it is p-smooth and if there exists δ with $0 < \delta < \frac{1}{|N|}$ such that (14.2.1) is satisfied. Let $\widehat{\Gamma}_N^{\text{ups}}$ denote the set of all uniformly p-smooth NTU games (N, V).

By Exercise 14.3.1, the consistent Shapley value does not satisfy ETP. However, it satisfies anonymity. In this section we employ the weaker *symmetry* axiom (see Remark 2.3.5). A *symmetry* of (N, v) is a permutation π of N such that $\pi V(S) = V(\pi S)$ for all $\emptyset \neq S \subseteq N$. A payoff configuration solution $\boldsymbol{\sigma}$ on a set $\widehat{\Gamma}$ of NTU games is *symmetric* (satisfies SYM) if, for any $(N, V) \in \widehat{\Gamma}$ and any symmetry π of (N, v), $\pi\boldsymbol{\sigma}(N, V) = \boldsymbol{\sigma}(N, V)$. Here we use the following notation for any permutation π of N and any payoff configuration $\boldsymbol{x} = (x_S)_{\emptyset \neq S \subseteq N}$: $\pi(\boldsymbol{x}) = (y_S)_{\emptyset \neq S \subseteq N}$ is defined by $y_S = \pi\left(x_{\pi^{-1}(S)}\right)$ for all $\emptyset \neq S \subseteq N$.

Remark 14.3.1. For any payoff configuration solution $\boldsymbol{\sigma}$ on a set $\widehat{\Gamma}$ of NTU games the *corresponding solution* σ is defined by

$$\sigma(N, V) = \{x_N \mid (x_S)_{\emptyset \neq S \subseteq N} \in \boldsymbol{\sigma}(N, V)\} \text{ for all } (N, V) \in \widehat{\Gamma}.$$

Note that, if $\boldsymbol{\sigma}$ has SYM, then σ has SYM (in the sense that $\pi\sigma(N, V) = \sigma(N, V)$ for all $(N, V) \in \widehat{\Gamma}$ and all symmetries π of (N, V)).

In order to generalize Definition 8.2.5 we first generalize "incremental contribution" (see Section 2.3).

Let $(N, V) \in \widehat{\Gamma}$, $i \in N$ and $\boldsymbol{x} = (x_S)_{\emptyset \neq S \subseteq N}$ such that $x_S \in \partial V(S)$ for all $S \in 2^N \setminus \{\emptyset\}$. Let (N, W) denote the hyperplane game determined by

$H_{x_S,V(S)}$, $\emptyset \neq S \subseteq N$ (see Definition 14.2.1). The *incremental contribution* of i to a coalition $S \subseteq N$ with $i \in S$ at $\boldsymbol{x}$, $D^i(\boldsymbol{x}, S, V)$, is defined by

$$D^i(\boldsymbol{x}, S, V) = \max \left\{ t^{\{i\}} \in \mathbb{R}^{\{i\}} \,\middle|\, \left(t^{\{i\}}, x_{S\setminus\{i\}}\right) \in W(S) \right\}, \tag{14.3.1}$$

where $x_\emptyset$ is the unique element of $\mathbb{R}^\emptyset$. By definition of $W(S)$, (14.3.1) is equivalent to

$$t = D^i(\boldsymbol{x}, S, V) \Leftrightarrow \lambda_{x_S,V(S)} \cdot \left(t, x_{S\setminus\{i\}}\right) = \lambda_{x_S,V(S)} \cdot x_S. \tag{14.3.2}$$

Note that for any TU game (N, v) (see (11.3.5) for the definition of V_v)

$$\boldsymbol{x} \in \partial V_v \Rightarrow D^i(\boldsymbol{x}, S, V_v) = v(S) - v(S \setminus \{i\}) \text{ for all } S \subseteq N, i \in S. \tag{14.3.3}$$

We are now ready to define "strong monotonicity" of a payoff configuration solution.

Definition 14.3.2. *A payoff configuration solution $\boldsymbol{\sigma}$ on a set $\widehat{\Gamma}$ of NTU games is* **strongly monotonic** *if the following condition is satisfied for all $(N, U), (N, V) \in \widehat{\Gamma}$, $\boldsymbol{x} \in \boldsymbol{\sigma}(N, U) \cap \partial U, \boldsymbol{y} \in \boldsymbol{\sigma}(N, V) \cap \partial V$, and $i \in N$:*

$$D^i(\boldsymbol{x}, S, U) \leq D^i(\boldsymbol{y}, S, V) \;\forall S \subseteq N, S \ni i \Rightarrow x^i_S \leq y^i_S \;\forall S \subseteq N, S \ni i.$$

Remark 14.3.3. Let $\boldsymbol{\sigma}$ be a strongly monotonic efficient payoff configuration solution on $\widehat{\Gamma}$ and let Γ be the set of TU games (N, v) such that $(N, V_v) \in \widehat{\Gamma}$. The solution σ' on Γ defined by $\sigma'(N, v) = \sigma(N, V_v)$ for all $(N, v) \in \Gamma$, where σ is the solution that corresponds to $\boldsymbol{\sigma}$, is strongly monotonic in the sense that, if $(N, u), (N, v) \in \Gamma$ satisfy the conditions of Definition 8.2.5 for some $i \in N$ and if $x \in \sigma'(N, u)$ and $y \in \sigma'(N, v)$, then $x^i \geq y^i$. Indeed, the incremental contributions of i are solely determined by the coalition functions u and v (see (14.3.3)).

The following reformulation of Definition 14.2.2 is useful. Let (N, v) be a p-smooth NTU game, let $\boldsymbol{x} \in \partial V$ (see Notation 13.4.2) and let $(N, V_{\boldsymbol{x}})$ be the hyperplane game defined by the tangent hyperplanes $H_{x_S,V(S)}$, $\emptyset \neq S \subseteq N$. Then (see Maschler and Owen (1992))

$$\boldsymbol{x} \in \boldsymbol{\Phi}^{\mathcal{MO}}(N, V) \Leftrightarrow x_S = \phi(S, V_{\boldsymbol{x}}) \;\forall \emptyset \neq S \subseteq N. \tag{14.3.4}$$

Let $\boldsymbol{x} \in \boldsymbol{\Phi}^{\mathcal{MO}}(N, V)$. For each $i \in N$ and $S \subseteq N$ with $S \ni i$ we may express x^i_S as expectation of incremental contributions of i as follows. Let $\boldsymbol{R}_{S,i} = \boldsymbol{R}$ be the random coalition $P^i_\pi \cup \{i\} = \{j \in S \mid \pi(j) \leq \pi(i)\}$ obtained from a random order π of S chosen uniformly in $\mathcal{SYM}_S$ (where $\mathcal{SYM}_S = \{\pi : S \to \{1, \ldots, |S|\} \mid \pi \text{ is bijective}\}$), that is, any $\pi \in \mathcal{SYM}_S$ occurs with probability $\frac{1}{|S|!}$. Thus, for any $T \subseteq S$ with $T \ni i$,

$$Pr(\boldsymbol{R} = T) = \frac{(|T| - 1)!(|S| - |T|)!}{|S|!}.$$

Let "$\mathcal{E}$" denote "expectation". By (14.1.1), (14.1.2), and (14.3.2), for any $\boldsymbol{x} \in \partial V$, the foregoing criterion (14.3.4) is equivalent to

$$\boldsymbol{x} \in \boldsymbol{\Phi}^{\mathcal{MO}}(N,V) \Leftrightarrow x^i_S = \mathcal{E}\Big(D^i\left(\boldsymbol{x}, \boldsymbol{R}_{S,i}, V\right)\Big) \; \forall i \in S \subseteq N. \tag{14.3.5}$$

Note that (14.3.5) is due to Hart (2005).

The following two statements will be used in the characterization of the consistent Shapley value solution.

Lemma 14.3.4. *On* $\widehat{\Gamma} \subseteq \widehat{\Gamma}^{\text{ups}}_N$, $\boldsymbol{\Phi}^{\mathcal{MO}}$ *satisfies* NE, EFF, SYM, IIA, *and* strong monotonicity.

Proof: By Theorem 14.2.3 and by Definition 14.2.2, $\boldsymbol{\Phi}^{\mathcal{MO}}$ satisfies NE and EFF. It is straightforward to verify that $\boldsymbol{\Phi}^{\mathcal{MO}}$ satisfies the suitable generalization of AN, hence SYM. In order to show IIA, let $(N,V),(N,W) \in \widehat{\Gamma}$, $\boldsymbol{x} \in \boldsymbol{\Phi}^{\mathcal{MO}}(N,W)$ and assume that $x_S \in V(S) \subseteq W(S)$ for all $\emptyset \neq S \subseteq N$. By EFF, $\boldsymbol{x} \in \partial V$. Consequently, $H_{x_S,V(S)} = H_{x_S,W(S)}$ for all $S \in 2^N \setminus \{\emptyset\}$ and IIA follows from (14.3.4). Moreover, (14.3.5) implies strong monotonicity.
q.e.d.

Lemma 14.3.5. *Let* $\widehat{\Gamma} \subseteq \widehat{\Gamma}^{\text{ups}}_N$ *and* $\Gamma = \{(N,v) \mid (N,V_v) \in \widehat{\Gamma}\}$. *If* $\boldsymbol{\sigma}$ *on* $\widehat{\Gamma}$ *satisfies* NE, EFF, SYM *and* strong monotonicity, *then* σ *on* Γ, *given by* $\sigma(N,v) = \{x_N \mid \boldsymbol{x} \in \boldsymbol{\sigma}(N,V_v)\}$ *for all* $(N,v) \in \Gamma$, *satisfies* SIVA, ETP *and* strong monotonicity.

Proof: Let $(N,u),(N,v) \in \Gamma$, $i \in N$, and $x \in \sigma(N,u), y,z \in \sigma(N,v)$. Then there exist $\boldsymbol{x} \in \boldsymbol{\sigma}(N,V_u)$ and $\boldsymbol{y},\boldsymbol{z} \in \boldsymbol{\sigma}(N,V_v)$. By EFF and strong monotonicity, (14.3.3) yields $y^j_S = z^j_S$ for all $S \subseteq N$ that contain j for all $j \in N$. Hence, NE implies SIVA. Moreover, SIVA together with SYM implies ETP.

Now, if $v(S \cup \{i\}) - v(S) \leq u(S \cup \{i\}) - u(S)$ for all $S \subseteq N \setminus \{i\}$, then, by EFF and (14.3.3), strong monotonicity implies $x^i \geq y^i$ and the proof is complete.
q.e.d

Theorem 14.3.6. *Let* $\widehat{\Gamma} \subseteq \widehat{\Gamma}^{\text{ups}}_N$ *such that* $(N,V_v) \in \widehat{\Gamma}$ *for any TU game* (N,v). *The consistent Shapley solution on* $\widehat{\Gamma}$ *is the maximum solution that satisfies* NE, EFF, SYM, *and* strong monotonicity.

Proof: By Lemma 14.3.4, $\boldsymbol{\Phi}^{MO}$ has the desired properties. Let $\boldsymbol{\sigma}$ be a payoff configuration solution that satisfies the foregoing axioms and let $(N,V) \in \widehat{\Gamma}$. It remains to show that $\boldsymbol{\sigma}(N,V) \subseteq \boldsymbol{\Phi}^{MO}(N,V)$. Let $\boldsymbol{x} \in \boldsymbol{\sigma}(N,V)$. By Lemma 14.3.5 and Theorem 8.2.6, if $V = V_v$ for some TU game (N,v), then $\boldsymbol{x} = (\phi(S,v))_{S \in 2^N \setminus \{\emptyset\}}$. In general, let for $i \in N$ the TU game (N,w^i) be defined by

$$w^i(S) = \begin{cases} D^i(\boldsymbol{x}, S, V) & \text{, if } i \in S \subseteq N, \\ 0 & \text{, if } S \subseteq N \setminus \{i\}. \end{cases}$$

By EFF and strong monotonicity, (14.3.3) implies that $x^i_S = \phi^i(S, w^i)$ for all $S \subseteq N$ with $i \in S$. By (14.3.5) we may conclude that $\boldsymbol{x} \in \boldsymbol{\Phi}^{MO}(N, V)$. **q.e.d.**

In Subsection 14.3.2 we show that each of the axioms of Theorem 14.3.6 is logically independent of the remaining axioms, provided $\widehat{\Gamma}$ is rich enough.

Subsection 14.3.1 is devoted to show that independence of irrelevant alternatives (IIA) may replace the "maximum condition" in Theorem 14.3.6 if only NTU games (N, V) are considered such that $V(S)$ is convex for each coalition S.

14.3.1 The Role of IIA

Let $\widehat{\Gamma}_N^{\text{ups,c}}$ be the set of al NTU games $(N, V) \in \widehat{\Gamma}_N^{\text{ups}}$ such that, for all $S \in 2^N \setminus \{\emptyset\}$, $V(S)$ is convex.

Theorem 14.3.7. *Let $\widehat{\Gamma} \subseteq \widehat{\Gamma}_N^{\text{ups,c}}$ such that any hyperplane game with player set N belongs to $\widehat{\Gamma}$. The consistent Shapley solution on $\widehat{\Gamma}$ is the unique payoff configuration solution that satisfies* NE, EFF, SYM, IIA *and* strong monotonicity.

Proof: Let $\boldsymbol{\sigma}$ satisfy the desired properties, let $(N, V) \in \widehat{\Gamma}$, and $\boldsymbol{x} \in \boldsymbol{\Phi}^{\mathcal{MO}}$ (N, V). By Theorem 14.3.6 it suffices to show that $\boldsymbol{x} \in \boldsymbol{\sigma}(N, V)$. Let $(N, V_{\boldsymbol{x}})$ be the hyperplane game determined by $H_{x_S, V(S)}$, $\emptyset \neq S \subseteq N$. By convexity of $V(S)$, $H_{x_S, V(S)}$ is a supporting hyperplane of $V(S)$. Thus, $V(S) \subseteq V_{\boldsymbol{x}}(S)$ for all $S \in 2^N \setminus \{\emptyset\}$. As $\boldsymbol{\sigma}(N, V_{\boldsymbol{x}}) \subseteq \boldsymbol{\Phi}^{\mathcal{MO}}(N, V_{\boldsymbol{x}})$, NE and (14.3.4) imply that $\boldsymbol{x} \in \boldsymbol{\sigma}(N, V_{\boldsymbol{x}})$. By IIA, $\boldsymbol{x} \in \boldsymbol{\sigma}(N, V)$. **q.e.d**

We now show that the statement of Theorem 14.3.7 does not remain valid for $\widehat{\Gamma} = \widehat{\Gamma}_N^{\text{ups}}$ (i.e., if all uniformly p-smooth games have to be considered).

Let $S \subseteq N$ with $|S| = 2$. Throughout this subsection let, for any game $(S, V) \in \widehat{\Gamma}_S^{\text{ups}}$, $\beta_V \in \mathbb{R}^S$ be the disagreement point of (S, V), that is $\beta_V^i = \max V(\{i\})$ for $i \in S$. We say that (S, V) is a *bargaining problem* (BP) if $\beta_V \in V(S) \setminus \partial V(S)$. For any BP (S, V), the maximal *Nash product*,

$$\gamma_V = \max \left\{ \prod_{i \in S} (x^i - \beta_V^i)_+ \,\middle|\, x \in V(S) \right\}, \tag{14.3.6}$$

exists. We may now define the solution σ for any game $(S, V) \in \widehat{\Gamma}_N^{\text{ups}}$ by

$$\sigma(S, V) = \left\{ x \in V(S) \,\middle|\, \prod_{i \in S} (x^i - \beta_V^i)_+ = \gamma_V \right\}, \text{ if } (S, V) \text{ is a BP}, \tag{14.3.7}$$

and by $\sigma(S, V) = \Phi^{\mathcal{MO}}(S, V)$, if (S, V) is not a BP. Note that, if (S, V) is a BP and $x \in \sigma(S, V)$, then $\lambda_{x_S, V(S)} = \lambda_{x_S, W(S)}$, where

$$W(S) = \left\{ z \in \mathbb{R}^S \,\middle|\, \prod_{i \in S} (z^i - \beta_V^i)_+ \leq \gamma_V \right\}.$$

By Exercise 14.3.2, $\sigma(S,V) \subseteq \Phi^{\mathcal{MO}}(S,V)$. Now, let $\boldsymbol{\sigma}$ be that payoff configuration solution that is defined by

$$\boldsymbol{\sigma}(N,V) = \left\{ \boldsymbol{x} \in \boldsymbol{\Phi}^{\mathcal{MO}}(N,V) \mid x_S \in \sigma(S,V) \; \forall S \subseteq N, |S| = 2 \right\}$$

for all $(N,V) \in \widehat{\Gamma}_N^{\text{ups}}$. As σ is a nonempty subsolution of the consistent Shapley value correspondence on uniformly p-smooth 2-person games, a careful inspection of the proof of Theorem 14.2.3 shows that $\boldsymbol{\sigma}$ satisfies NE. As any subsolution of the consistent Shapley solution is strongly monotonic, $\boldsymbol{\sigma}$ satisfies the axiom as well. The straightforward proofs of SYM and IIA are skipped. Finally, Exercise 14.3.4 shows that $\sigma \neq \Phi^{\mathcal{MO}}$.

14.3.2 Logical Independence

The following examples of payoff configuration solutions that are defined on $\widehat{\Gamma}_N^{\text{ups}}$ and that may, thus, be restricted to $\widehat{\Gamma}_N^{\text{ups,c}}$, show that each axiom of Theorem 14.3.6 or of Theorem 14.3.7, respectively, is logically independent of the remaining axioms, provided that $\widehat{\Gamma} = \widehat{\Gamma}_N^{\text{ups}}$ or $\widehat{\Gamma} = \widehat{\Gamma}_N^{\text{ups,c}}$, respectively.

Example 14.3.8. Choose a TU game (N,v) whose symmetry group is trivial (e.g., choose $x \in \mathbb{R}^N$ such that $x^i \neq x^j$ for all $i,j \in N, i \neq j$, and define $v(S) = x(S)$ for all $S \subseteq N$.) Also, let $\boldsymbol{y} \in \partial V_v$ such that $y_N \neq \phi(N,v)$. Now, define $\boldsymbol{\sigma}^1(N,V_v) = \{\boldsymbol{y}\}$ and $\boldsymbol{\sigma}^1(N,V) = \emptyset$ for all $(N,V) \in \widehat{\Gamma} \backslash \{(N,V_v)\}$. Then $\boldsymbol{\sigma}^1$ satisfies EFF and SYM and it is not contained in the consistent Shapley solution. Thus, NE is needed in Theorem 14.3.6 in order to guarantee that a solution is contained in the consistent Shapley solution.

Example 14.3.9. Define

$$\boldsymbol{\sigma}^2(N,V) = \begin{cases} \boldsymbol{\Phi}^{\mathcal{MO}}(N,V) & \text{, if } \mathbf{0} \notin V, \\ \{\mathbf{0}\} & \text{, if } \mathbf{0} \in V. \end{cases}$$

Clearly, $\boldsymbol{\sigma}^2$ satisfies NE and SYM, violates EFF, and is not a subsolution of the consistent Shapley solution. IIA is easy to verify. Moreover, note that if, for a game $(N,V) \in \widehat{\Gamma}$, $\mathbf{0} \in \partial V$, then $\mathbf{0} \in \boldsymbol{\Phi}^{\mathcal{MO}}(N,V)$. We conclude that $\boldsymbol{\sigma}^2$ is strongly monotonic.

Example 14.3.10. We may assume without loss of generality that $N = \{1,\ldots,n\}$. Now, let $(N,V) \in \widehat{\Gamma}$, let π be a permutation of N and let $a_\pi = a_\pi(V)$ be defined by 14.1.1. Note that the maximum indeed exists, because (N,V) is **uniformly** p-smooth. Also, for $\emptyset \neq S \subseteq N$ and $i \in S$, with $P = P_\pi^i \cap S$, we recursively define

$$a^i_{\pi,S} = \max\{x^i \in \mathbb{R} \mid (x^i, a^P_{\pi,S}) \in V(P \cup \{i\})\},$$

where $a^{\emptyset}_{\pi,S}$ is the unique element of $\mathbb{R}^{\emptyset}$. Define

$$\boldsymbol{\sigma}^3(N,V) = \left\{(a_{\pi,S})_{S \in 2^N \setminus \{\emptyset\}}\right\}.$$

It is straightforward to verify that $\boldsymbol{\sigma}^3$ satisfies NE, IIA, strong monotonicity. Also, $\boldsymbol{\sigma}^3$ violates SYM and is not contained in $\boldsymbol{\Phi}^{\mathcal{MO}}$.

Example 14.3.11. We shall now modify the definition of the payoff configuration solution $\boldsymbol{\sigma}$ of Subsection 14.3.1. We first extend (14.3.6) to any game $(S,V) \in \widehat{\Gamma}^{\text{ups}}_S$, $|S| = 2$, that is not a bargaining problems, that is, the disagreement point β_V does not belong to $V(S) \setminus \partial V(S)$, by defining

$$\gamma_V = \max\left\{ \prod_{i \in S} (\beta^i_V - x^i)_+ \,\middle|\, x \in \partial V(S) \right\}.$$

We then may also extend (14.3.7) by defining, for any $(S,V) \in \widehat{\Gamma}^{\text{ups}}_S$ with $|S| = 2$,

$$\sigma^4(S,V) = \sigma(S,V), \text{ if } (S,V) \text{ is a BP},$$
$$\sigma^4(S,V) = \left\{ x \in \partial V(S) \,\middle|\, \prod_{i \in S} (\beta^i_V - x^i)_+ = \gamma_V \right\}, \text{ otherwise.}$$

The proof that $\sigma^4(S,V) \subseteq \Phi^{\mathcal{MO}}(S,V)$ even if $\beta_V \notin V(S) \setminus \partial V(S)$ is similar to the proof that $\sigma(S,V) \subseteq \Phi^{\mathcal{MO}}(S,V)$ of Subsection 14.3.1 for bargaining problems (see Exercises 14.2.3 and 13.1.2). Now we define, for any $(N,V) \in \widehat{\Gamma}^{\text{ups}}_N$,

$$\boldsymbol{\sigma}^4(N,V) = \left\{ \boldsymbol{x} \in \boldsymbol{\Phi}^{\mathcal{MO}}(N,V) \mid x_S \in \sigma^4(S,V)\ \forall S \subseteq N, |S| = 2 \right\}.$$

It is straightforward to verify that $\boldsymbol{\sigma}^4$ satisfies NE, EFF, SYM, and does not coincide with $\boldsymbol{\Phi}^{\mathcal{MO}}$. Hence, it violates IIA and it simultaneously shows that the "maximum" condition may not be deleted in Theorem 14.3.6.

The *empty* payoff configuration solution may be used to show that NE is logically independent of the remaining axioms in Theorem 14.3.7. Finally, the *equal split* configuration solution that assigns to each subgame (S,V) of (N,V) the unique element $y_S \in \partial V(S)$ that satisfies $y^i_S = y^j_S$ for all $i,j \in S$, shows that strong monotonicity is logically independent of the remaining axioms in both theorems.

Remark 14.3.12. It should be noted that Theorem 14.3.7 and Examples 14.3.8 and 14.3.10 are due to Hart (2005), who also used the equal split solution. Moreover, Hart (2005) contains the variant of Theorem 14.3.6 for convex-valued games.

Exercises

Exercise 14.3.1. Show that $\boldsymbol{\Phi}^{\mathcal{MO}}$ does not satisfy ETP on $\widehat{\Gamma}_N^{\text{ups,c}}$. (Exercises 14.2.3 and 13.1.2 are useful.)

Exercise 14.3.2. Let $N = \{1,2\}$ and (N, V) be the zero-normalized bargaining problem determined by $V(N) = \{x \in \mathbb{R}^N \mid x_+^1 x_+^2 \leq 1\}$. Show that $\Phi^{\mathcal{MO}}(N, V) = \partial V(N)$.

Exercise 14.3.3. Let $(N, V) \in \widehat{\Gamma}_N^{\text{ups,c}}$ and $\boldsymbol{x} \in \partial V$. (Use the convention that $x_\emptyset$ is the unique member of $\mathbb{R}^\emptyset$.) For $i \in S \subseteq N$ let $D^i(\boldsymbol{x}, S, V)$ be defined by (14.3.1). Show that (see Hart (2005))

$$D^i(\boldsymbol{x}, S, V) = \sup_{\varepsilon>0} \max \left\{t \in \mathbb{R} \,\middle|\, (1-\varepsilon)x_S + \varepsilon\left(t, x_{S\setminus\{i\}}\right) \in V(S)\right\}.$$

Exercise 14.3.4. Let σ be defined by (14.3.7). Find a bargaining problem (S, V) such that $\sigma(S, V) \subsetneqq \Phi(S, V)$.

14.4 Notes and Comments

Theorems 14.1.3 and 14.1.4 are due to Maschler and Owen (1989).

We remark that Theorem 14.1.4 remains valid for TU games, that is, the Shapley value is the unique single-valued solution on $\Gamma_{\mathcal{U}}$, the set of all TU games with player sets contained in $\mathcal{U}$, that satisfies PO, COV, ETP, and 2-consistency. Though 2-consistency is less intuitive than consistency, it is implied by consistency. Also, it should be noted that there does not exist any solution on the set $\widehat{\Gamma}_{\mathcal{U}}^h$ of hyperplane games that satisfies PO, COV, ETP, and consistency (see Exercise 14.1.3).

ETP is used in the proof of Theorem 14.1.4 only for two-person TU games. Note that COV is only used for two-person hyperplane games. Moreover, if 2-consistency is replaced by k-consistency, $k = 1, 2$, then PO may be deduced (see Lemma 8.3.8).

Remark 14.4.1. The extension of the consistent Shapley value to p-smooth NTU games and Theorem 14.2.3 is due to Maschler and Owen (1992). We had to deviate from their proof of the foregoing theorem because of some inaccuracy. Basically, they state that the number $r_N(z)$ in (14.2.3) is always nonnegative, which may not be true, unless $V(N)$ is convex.

Note that as in Section 8.2 (see Remark 8.2.7) also in Section 14.3 the *strong monotonicity* axiom may be replaced by a weaker condition called "marginality" (see Hart (2005)).

Finally, it should be remarked that another axiomatization of the consistent Shapley solution may be found in de Clippel, Peters, and Zank (2004).

15

On the Classical Bargaining Set and the Mas-Colell Bargaining Set for NTU Games

In this chapter the classical bargaining set and the Mas-Colell bargaining set, which have been discussed in Chapter 4, are generalized to NTU games. Simple majority voting games are used to investigate some of the properties and to compare these solution concepts.

In Section 15.1 the basic definitions of the bargaining set of an NTU game with coalition structure, of the Mas-Colell bargaining set of an NTU game, and of simple majority voting games are presented. Moreover, the voting paradox with three voters on three alternatives is discussed and it is shown that a fourth alternative may lead to an empty bargaining set, whereas the Mas-Colell bargaining set remains nonempty.

In Section 15.2 it is shown by means of examples that the Mas-Colell bargaining set may be empty for suitable simple majority voting games with any even number of players not less than four and any number $m \geq 6$ of alternatives. Also, existence is shown for simple majority voting games on less than six alternatives.

Section 15.3 is devoted to show that the Mas-Colell bargaining set of an NTU game may be empty even if the game is superadditive and non-levelled.

Existence results of the bargaining sets on majority voting games with many voters, whose preferences are drawn in a specified way, are deduced in Section 15.4. In a simple probabilistic model, we show that both bargaining sets are nonempty with probability tending to one as the number of voters tends to infinity. Moreover, we show that the Mas-Colell bargaining set is nonempty for any suitable k-fold replication with k sufficiently large, whereas the classical bargaining set may be empty for any k.

Finally, several comments are given in Section 15.5.

15.1 Preliminaries

This section is partitioned into three subsections. In the first subsection the classical bargaining set is generalized to NTU games with coalition structures and ordinal and cardinal solutions are introduced. In the second subsection the Mas-Colell bargaining set of an NTU game is defined and simple majority voting games are introduced.

15.1.1 The Bargaining Set $\mathcal{M}$

In this subsection we shall generalize the definition of the TU bargaining set to NTU games. Let (N, V) be an NTU game. If $S \in 2^N \setminus \{\emptyset\}$ and $A \subseteq \mathbb{R}^S$ then we denote by A_e the set of weakly Pareto optimal elements of A (see (12.1.1)), that is,

$$A_e = \{x^S \in A \mid \text{there exists no } y^S \in A \text{ such that } y^S \gg x^S\}.$$

We now generalize parts of Sections 3.8 and 4.1 to NTU games. We shall say that $(N, V, \mathcal{R})$ is an NTU game *with coalition structure* if $\mathcal{R}$ is a coalition structure for N (i.e., a partition of N). Let $\mathcal{R}$ be a coalition structure for N. We denote $X = X(N, V, \mathcal{R}) = \prod_{R \in \mathcal{R}} V(R)_e$ and $I = I(N, V, \mathcal{R}) = \{x \in X \mid x^i \geq v^i \text{ for all } i \in N\}$. Hence, X is the set of *weakly efficient* payoff vectors and I is the set of *individually rational* elements of X.

Let $x \in X$ and let $k, \ell \in R$, $k \neq \ell$, for some $R \in \mathcal{R}$. An *objection* of k against ℓ at x is a pair (P, y) such that

$$P \in \mathcal{T}_{k\ell}(N) \text{ (i.e. } k \in P \not\ni \ell\text{)}, \; y \in V(P), \text{ and } y \gg x^P.$$

Note that the definition of an objection coincides with the corresponding definition for TU games (see Exercise 4.1.6).

Let (P, y) be an objection of k against ℓ at x. A *counterobjection* to (P, y) is a pair (Q, z) such that

$$Q \in \mathcal{T}_{\ell k}(N), \; z \in V(Q), \; z^{Q \setminus P} \geq x^{Q \setminus P}, \text{ and } z^{P \cap Q} \geq y^{P \cap Q}$$

(see Definition 4.1.3). An objection (P, y) is *justified* if there is no counterobjection to (P, y). A vector $x \in X$ is *stable* if there is no justified objection at x (i.e., if each objection at x has a counterobjection). The *prebargaining set* of $(N, V, \mathcal{R})$, $\mathcal{PM}(N, V, \mathcal{R})$, is the set of stable payoff vectors in $X(N, V, \mathcal{R})$. Furthermore,

$$\mathcal{M}(N, V, \mathcal{R}) = \mathcal{PM}(N, V, \mathcal{R}) \cap I$$

is the *bargaining set* of $(N, V, \mathcal{R})$.

Remark 15.1.1. Let (N, v) be a TU game and (N, V_v) the associated NTU game. Then

$$\mathcal{M}(N, v, \mathcal{R}) = \mathcal{M}(N, V_v, \mathcal{R}) \text{ and } \mathcal{PM}(N, v, \mathcal{R}) = \mathcal{PM}(N, V_v, \mathcal{R}).$$

In view of Theorem 4.2.12, $\mathcal{M}(N, v, \mathcal{R}) \neq \emptyset$, provided that $I(N, v, \mathcal{R}) \neq \emptyset$. Moreover, by Exercise 4.2.3, $\mathcal{PM}(N, v, \mathcal{R}) \neq \emptyset$ for every TU game with coalition structure $(N, v, \mathcal{R})$. The next example shows that the results of Part I do not hold in the NTU case.

Example 15.1.2 (Peleg (1963)). Let (N, V) be specified by

$$\begin{aligned} N &= \{1, 2, 3, 4\}, \\ V(\{1, 2\}) &= \{x \in \mathbb{R}^{\{1,2\}} \mid x(S) \leq 1\}, \\ V(\{1, 3, 4\}) &= \{x \in \mathbb{R}^{\{1,3,4\}} \mid x^1 \leq 2, x^3 \leq 3, x^4 \leq 4\}, \\ V(\{2, 3, 4\}) &= \{x \in \mathbb{R}^{\{2,3,4\}} \mid x^2 \leq 2, x^3 \leq 4, x^4 \leq 3\}, \end{aligned}$$

and $V(S) = -\mathbb{R}^S_+$ for all other coalitions. Hence, for every coalition S, $(v^i)_{i \in S} \in V(S)$. Thus, $I(N, V, \mathcal{R}) \neq \emptyset$, for every coalition structure $\mathcal{R}$ for N. Let $\mathcal{R} = \{\{1, 2\}, \{3\}, \{4\}\}$. Then we claim that $\mathcal{PM}(N, V, \mathcal{R}) = \emptyset$. Indeed, if $x \in X(N, V, \mathcal{R})$, then $x(\{1, 2\}) = 1$ and $x^3 = x^4 = 0$. However, if $x^1 > 0$, then 2 has a justified objection against 1 via the coalition $\{2, 3, 4\}$. Analogously, if $x^2 > 0$, then 1 has a justified objection against 2 via $\{1, 3, 4\}$.

We shall now discuss an interesting property of the bargaining set regarded as a solution on a set $\widehat{\Delta}$ of NTU games with coalition structures. A *solution* on $\widehat{\Delta}$ is a function σ which associates with each game $(N, V, \mathcal{R}) \in \Gamma$ a subset $\sigma(N, V, \mathcal{R})$ of

$$X^*(N, V, \mathcal{R}) = \{x \in \mathbb{R}^N \mid x^R \in V(R) \text{ for all } R \in \mathcal{R}\}.$$

Let $N \neq \emptyset$ be a finite set of players. A *monotone transformation* for N is a mapping $f = \prod_{i \in N} f^i$ such that $f^i : \mathbb{R}^{\{i\}} \to \mathbb{R}^{\{i\}}$ is an increasing and bijective function. If $(N, V, \mathcal{R})$ is an NTU game with coalition structure and if f is a monotone transformation for N, then let

$$f * V = (f * V(S))_{S \in 2^N \setminus \{\emptyset\}}$$

be given by $f * V(S) = \{f * x^S \mid x^S \in V(S)\}$, where $f * x^S = (f^i(x^i))_{i \in S}$ for every $x^S \in \mathbb{R}^S$.

Remark 15.1.3. If $(N, V, \mathcal{R})$ is an NTU game with coalition structure and if f is a monotone transformation for N, then $(N, f * V, \mathcal{R})$ is a game with coalition structure.

The simple proof of this remark is left as Exercise 15.1.1.

Let σ be a solution on a set $\widehat{\Delta}$ of NTU games with coalition structures. The solution σ is *ordinal* if for every $(N,V,\mathcal{R}) \in \widehat{\Delta}$ and every monotone transformation f for N such that $(N, f * V, \mathcal{R}) \in \widehat{\Delta}$, $\sigma(N, f * V, \mathcal{R}) = f * \sigma(N,V,\mathcal{R})$. The straightforward proof of the following remark is left to the reader (see Exercise 15.1.2).

Remark 15.1.4. The prebargaining set $\mathcal{PM}$ is an ordinal solution on any set of NTU games with coalition structures.

It should be noted that every ordinal solution is covariant. A solution σ on $\widehat{\Delta}$ satisfies *covariance* (COV) if, for all $(N,V,\mathcal{R}) \in \widehat{\Delta}$ and $\alpha, b \in \mathbb{R}^N$ such that $\alpha \gg 0$, the following condition is satisfied for $W = \alpha * V + b$: If $(N,W,\mathcal{R}) \in \widehat{\Delta}$, then $\sigma(N,W,\mathcal{R}) = \alpha * \sigma(N,V,\mathcal{R}) + b$.

We say that a solution σ is a *cardinal solution* if it satisfies COV.

15.1.2 The Mas-Colell Bargaining Set $\mathcal{MB}$ and Majority Voting Games

As in Subsection 4.4.2 we just consider NTU games with the trivial coalition structure. Thus, we assume throughout that (N,V) is an NTU game and that $x \in X(N,V) = X(N,V,\{N\})$. Note that $X(N,V)$ is the set of *preimputations* of (N,V). Also, $I(N,V)$ is the set of *imputations* of (N,V). In order to state the definitions of the classical bargaining set and of the Mas-Colell bargaining set in a simple and parallel way, the following definition is useful.

Definition 15.1.5. *Let (N,V) be an NTU game and $x \in \mathbb{R}^N$. A pair (P,y) is an* **objection** *at x if $\emptyset \neq P \subseteq N$, y is Pareto optimal in the subgame (P,V), and $y > x^P$. An objection (P,y) is* **strong** *if $y \gg x^P$. The pair (Q,z) is a* **weak counterobjection** *to the objection (P,y) if $Q \subseteq N$, $Q \neq \emptyset, P$, if $z \in V(Q)$, and if $z \geq (y^{P\cap Q}, x^{Q\setminus P})$. A weak counterobjection (Q,z) is a* **counterobjection** *to the objection (P,y) if $z > (y^{P\cap Q}, x^{Q\setminus P})$.*

Remark 15.1.6. If (N,V) is an NTU game and $x \in X(N,V)$, then $x \in \mathcal{PM}(N,V)$ iff for any strong objection (P,y), any $k \in P$, and $\ell \in N \setminus P$ there is a weak counterobjection (Q,z) with $k \notin Q \ni \ell$.

We now generalize Definition 4.4.11 to NTU games.

Definition 15.1.7. *Let (N,V) be an NTU game. The* **Mas-Colell prebargaining set** *of (N,V), $\mathcal{PMB}(N,V)$, is the set of all $x \in X(N,V)$ such that any objection at x has a counterobjection. Moreover, the* **Mas-Colell bargaining set** *of (N,V) is the set*

$$\mathcal{MB}(N,V) = \mathcal{PMB}(N,V) \cap I(N,V).$$

In order to define majority voting games, let N, $n = |N| \geq 2$, be the finite set of voters (the players), and let $A = \{a_1, \ldots, a_m\}$, $m \geq 2$, be a set of m alternatives.

We shall now assume that each $i \in N$ has a linear order R^i on A. Thus, for every $i \in N$, R^i is a complete, transitive, and antisymmetric binary relation on A. Moreover, let u^i, $i \in N$, be a utility function that represents R^i. We shall always assume that

$$\min_{\alpha \in A} u^i(\alpha) = 0 \text{ for all } i \in N. \tag{15.1.1}$$

Let $u^N = (u^i)_{i \in N}$ be a utility profile that satisfies (15.1.1). We consider the strategic game $g = \big(N, (A)_{i \in N}, (v^i)_{i \in N}\big)$ in which every player votes for some alternative in A whose payoff functions $v^i, i \in N$, are defined as follows. If a strict majority of voters agrees on $\alpha \in A$, then the outcome is α, and every voter i gets utility $u^i(\alpha)$. Otherwise, if no majority forms, a deadlock results and every voter gets utility 0, so that, for any strategy profile $(\alpha^i)_{i \in N} \in A^N$,

$$v^k\left((\alpha^i)_{i \in N}\right) = \begin{cases} u^k(\alpha) & \text{, if } \exists \alpha \in A \text{ such that } |\{i \in N \mid \alpha^i = \alpha\}| > \frac{n}{2}, \\ 0 & \text{, otherwise.} \end{cases}$$

Now, let $G(g) = G$ denote the corresponding cooperative game in strategic form (see Example 11.1.3 for the definition of $G(g)$). We are ready to define the *simple majority voting game* (see Aumann (1967)) (N, V_{u^N}) by $V_{u^N} = V_\alpha(\cdot, G)$ (see (11.2.1))[1], that is, for any $\emptyset \neq S \subseteq N$,

$$V_{u^N}(S) = \{x \in \mathbb{R}^S \mid S \text{ is } \alpha\text{-effective for } x\}.$$

Let S be a coalition and $i \in S$. If $|S| \leq \frac{n}{2}$ and if all members of $N \setminus S$ select i's worst alternative, then S cannot guarantee any positive payoff to player i, because it is not possible to reach a majority for any but i's worst alternative. If $|S| > \frac{n}{2}$ and if each member of S selects the same alternative $\alpha \in A$, then S guarantees $u^i(\alpha)$ to any $i \in S$. Thus, for any $\emptyset \neq S \subseteq N$,

$$V_{u^N}(S) = \{x \in \mathbb{R}^S \mid x \leq 0\} \text{ if } |S| \leq \frac{n}{2}; \tag{15.1.2}$$

$$V_{u^N}(S) = \{x \in \mathbb{R}^S \mid \exists \alpha \in A \text{ with } x \leq u^S(\alpha)\} \text{ if } |S| > \frac{n}{2}, \tag{15.1.3}$$

where $u^S(\alpha) = (u^i(\alpha))_{i \in S}$. By Remark 11.2.2, (N, V_{u^N}) is a superadditive NTU game. Also, (N, V_{u^N}) is *zero-normalized*, that is, $V_{u^N}(\{i\}) = -\mathbb{R}_+^{\{i\}}$ for all $i \in N$ (compare with Definition 2.1.13).

Remark 15.1.8. A careful inspection of (11.2.2) shows (see Exercise 15.1.6) that β-effectiveness[2] again leads to the same NTU game (N, V_{u^N}) in our case, that is,

$$V_{u^N}(S) = V_\beta(S, G) = \{x^S \in \mathbb{R}^S \mid S \text{ is } \beta\text{-effective for } x^S\} \forall \emptyset \neq S \subseteq N.$$

---

[1] Here the symbol α does not refer to an element of A.

[2] Here the symbol β does not refer to an element of A.

Notation 15.1.9. In the sequel let $L = L(A)$ denote the set of linear orders on A. For $R \in L$ and for $k \in \{1, \ldots, m\}$, let $t_k(R)$ denote the k-th alternative in the order R. If $R^N \in L^N$ and $\alpha, \beta \in A$, $\alpha \neq \beta$, then α *dominates* β (abbreviated $\alpha \succ_{R^N} \beta$) if $|\{i \in N \mid \alpha \; R^i \; \beta\}| > \frac{n}{2}$. We shall say that an alternative $\alpha \in A$ is a *weak Condorcet winner* (with respect to R^N) if $\beta \not\succ_{R^N} \alpha$ for all $\beta \in A$. The alternative α is a *Condorcet winner*, if $\alpha \succ_{R^N} \beta$ for all $\beta \in A \setminus \{\alpha\}$. Also, if $R^N \in L^N$, then denote

$$\mathcal{U}^{R^N} = \{(u^i)_{i \in N} \mid u^i \text{ is a representation of } R^i \text{ with (15.1.1) } \forall i \in N\}.$$

Remark 15.1.10. Any two utility representations of the same preference profile lead to simple majority voting games that are derived from each other by an ordinal transformation (see Exercise 15.1.7).

15.1.3 The 3 × 3 Voting Paradox

Throughout this subsection let A be the set of alternatives and $N = \{1, \ldots, n\}$ be a coalition, let $R^N \in L(A)^N$, $u^N \in \mathcal{U}^{R^N}$, and $V = V_{u^N}$. We present two examples with three alternatives a, b, c and one modification with 4 alternatives. The first example shows that the core of a simple majority voting game may not be contained in the Mas-Colell bargaining set of the game.

Remark 15.1.11. (1) Note that $\mathcal{C}(N, V) \subseteq \mathcal{M}(N, V)$. Indeed, $x \in \mathcal{C}(N, V)$ is weakly Pareto optimal, individually rational, and it has no strong objection.

(2) The Mas-Colell bargaining set need not contain the core (see the following example). However, the Mas-Colell bargaining set of a simple majority voting game must contain a core element if the core is nonempty. The foregoing statement is immediately implied by Exercise 15.1.8.

(3) We conclude that, if a simple majority voting game has a nonempty core, then its bargaining set and its Mas-Colell bargaining set are nonempty as well.

Example 15.1.12. Let $n = 4$ and let R^N be given by Table 15.1.1. Then

Table 15.1.1. Preference Profile of a 4-Person Voting Problem

R^1	R^2	R^3	R^4
a	a	c	c
b	b	b	b
c	c	a	a

$x = (\min\{u^i(b), u^i(a)\})_{i \in N} \in \mathcal{C}(N, V)$, because there is no strong objection at x. However, $x \notin \mathcal{MB}(N,V)$ because $(N, u^N(a))$ is a justified objection in the sense of the Mas-Colell bargaining set at x.

The second example is the Voting Paradox of three voters and three alternatives.

Example 15.1.13. Let $n = 3$, and let $R^N \in L^N$ be given by Table 15.1.2.

Table 15.1.2. Preference Profile of the 3×3 Voting Paradox

R^1	R^2	R^3
a	c	b
b	a	c
c	b	a

We claim that $\mathcal{M}(N,V) = \{0\}$. Indeed, it is straightforward to verify that $0 \in \mathcal{M}(N,V)$. In order to show the opposite inclusion let $x \in \mathcal{M}(N,V)$. Then there exists $\alpha \in A$ such that $x \leq u^N(\alpha)$. Without loss of generality we may assume that $\alpha = a$. Assume, on the contrary, that $x > 0$. If $x^1 > 0$, then $(\{2,3\}, u^{\{2,3\}}(c))$ is a justified objection of 3 against 1 at x in the sense of the bargaining set. If $x^1 = 0$ and, hence, $x^2 > 0$, then $(\{1,3\}, u^{\{1,3\}}(b))$ is a justified objection of 1 against 2.

In order to compute $\mathcal{MB}(N,V)$, let $x = (u^1(b), u^2(a), 0)$ and claim that $x \in \mathcal{MB}(N,V)$. Indeed, let (P,y) be an objection at x. Then $|P| \geq 2$. As y is Pareto optimal in $V(P)$, $y \in \{u^P(\alpha) \mid \alpha \in A\}$. If $y = u^P(a)$, then (P,y) is countered by $(\{2,3\}, u^{\{2,3\}}(c))$. If $y = u^P(b)$, then $y > x^P$ implies that $P = \{1,3\}$. In this case (P,y) is countered by $(\{1,2\}, u^{\{1,2\}}(a))$. Finally, if $y = u^P(c)$, then $y > x^P$ implies that $P = \{2,3\}$ and that (P,y) is countered by $(\{1,3\}, u^{\{1,3\}}(b))$.

In order to show that every $\hat{x} \in \mathbb{R}^N$ satisfying $0 \leq \hat{x} \leq x$ is an element of $\mathcal{MB}(N,V)$, it should be noted that each objection at $\hat{x}$ is also an objection at x if $\hat{x}^1 > 0$ and $\hat{x}^2 > 0$. If $\hat{x}^1 = 0$ and $\hat{x}^2 > 0$, then the additional objections are of the form $(P, u^P(c))$ for some $P \subseteq N$ and these objections can be countered by $(\{1,3\}, u^{\{1,3\}}(b))$. Similarly, if $\hat{x}^1 > 0$ and $\hat{x}^2 = 0$, then the additional objections can be countered by $(\{1,2\}, u^{\{1,2\}}(a))$. Finally, if $\hat{x} = 0$, then each additional objection can be countered by one of the foregoing pairs $(\{1,3\}, u^{\{1,3\}}(b))$ or $(\{1,2\}, u^{\{1,2\}}(a))$.

Similarly, for $y = (u^1(b), 0, u^3(c))$ and $z = (0, u^2(a), u^3(c))$ we have that every $\hat{y} \in \mathbb{R}^N$ satisfying $0 \leq \hat{y} \leq y$ and every $\hat{z} \in \mathbb{R}^N$ satisfying $0 \leq \hat{z} \leq z$ is in $\mathcal{MB}(N,V)$.

We shall show now that there are no other elements in $\mathcal{MB}(N,V)$. Indeed, any remaining individually rational $\widetilde{x} \in V(N)$ must have a coordinate that is higher than the utility of that voter's second best alternative. Say, without loss of generality, that $\widetilde{x}^1 > u^1(b)$. Then $(\{2,3\}, u^{\{2,3\}}(c))$ is a justified objection in the sense of the Mas-Colell bargaining set at $\widetilde{x}$. We conclude that $\mathcal{MB}(N,V)$ is the intersection of $\mathbb{R}^N_+$ and the comprehensive hull of $\{x,y,z\}$.

Discussion: The singleton $\mathcal{M}(N,V)$ tells us that in order to achieve (coalitional) stability the players have to give up any profit above their individually protected levels of utility. There is no hint how an alternative of A will be chosen. The message of $\mathcal{MB}(N,V)$ is much more detailed. For example, the element $x = (u^1(b), u^2(a), 0)$ tells us that the alternative a may be chosen provided player 1 disposes of $u^1(a) - u^1(b)$ utiles. Thus, we also see here that lower utility levels guarantee stability. Actually, x implies that there is an agreement between 1 and 2, the alternative a is chosen as a result of the agreement, and the utility of 1 is reduced (because of the agreement) from $u^1(a)$ to $u^1(b)$. Note that cooperative game theory does not specify the details of agreements that support stable payoff vectors.

For the following modification of the 3×3 voting paradox with 4 alternatives the Mas-Colell bargaining set is still "similar" (see Exercise 15.1.9), but the classical prebargaining set is empty.

Example 15.1.14. Let $A = \{a,b,c,d\}$, let $N = \{1,2,3\}$, and let R^N be given by Table 15.1.3.

Table 15.1.3. Preference Profile of a 4-Alternative Voting Problem

R^1	R^2	R^3
a	c	b
b	a	c
d	d	d
c	b	a

We claim that $\mathcal{PM}(N,V) = \emptyset$. Let x be an imputation of (N,V). In order to show that $x \notin \mathcal{M}(N,V)$ we may assume without loss of generality that $x^1 \geq u^1(d)$. We distinguish the following possibilities:

(1) $x \leq u^N(a)$ or $x \leq u^N(d)$. Then $\left(\{2,3\}, u^{\{2,3\}}(c)\right)$ is a justified objection (in the sense of the bargaining set) of 3 against 1.

(2) $x \leq u^N(b)$. If $x^3 < u^3(c)$, then we may use the same justified strong objection as in the first possibility. If $x^3 \geq u^3(c)$, then $\left(\{1,2\}, u^{\{1,2\}}(a)\right)$ is a justified objection of 2 against 3.

Note that the Mas-Colell bargaining set of the game defined in Example 15.1.14 is nonempty (see Theorem 15.2.2).

Exercises

Exercise 15.1.1. Show that if (N, V) is an NTU game and f is a monotone transformation for N, then $f * V$ is an NTU coalition function.

Exercise 15.1.2. Show that both $\mathcal{PM}$ and $\mathcal{M}$ are ordinal solutions on any set of NTU games with coalition structures.

Exercise 15.1.3. Let $(N, V, \mathcal{R})$ be an NTU game with coalition structure. Let the *core* of $(N, V, \mathcal{R})$ be the set $\mathcal{C}(N, V, \mathcal{R})$ of elements of $X^*(N, V, \mathcal{R})$ that are not dominated by any vector in $\mathbb{R}^N$. Show that

$$\mathcal{C}(N, V, \mathcal{R}) \subseteq \mathcal{M}(N, V, \mathcal{R}).$$

Let $(N, V, \mathcal{R})$ be an NTU game with coalition structure, let $\emptyset \neq S \subseteq N$, and let $x \in X^*(N, V, \mathcal{R})$. Then the *reduced NTU game with coalition structure* with respect to S and x, $(S, V^{\mathcal{R}}_{S,x}, \mathcal{R}_{|S})$ (see the notation after Definition 3.8.7 for the definition of $\mathcal{R}_{|S}$), is defined as follows. Let $T \subseteq S$. If $T \notin \mathcal{R}_{|S}$, then $V^{\mathcal{R}}_{S,x}(T) = V_{S,x}(T)$. If $T = S \cap R \in \mathcal{R}_{|S}$, then

$$V^{\mathcal{R}}_{S,x}(T) = \{y^T \in \mathbb{R}^T \mid (y^T, x^{R \setminus T}) \in V(R)\}.$$

Now RGP may be generalized to solutions on $\widehat{\Delta}$ by replacing Δ by $\widehat{\Delta}$ and v by V in Definition 3.8.9, wherever these symbols occur.

Exercise 15.1.4. Let $\mathcal{U}$ be a set of players and let $\widehat{\Delta}$ be the set of all non-levelled NTU games with coalition structures whose player set is contained in $\mathcal{U}$. Show that $\mathcal{PM}$ satisfies RGP on $\widehat{\Delta}$.

Exercise 15.1.5. Show that, on any set of NTU games, $\mathcal{MB}$ and $\mathcal{PMB}$ are ordinal solutions.

Exercise 15.1.6. Prove Remark 15.1.6.

Let A be a set of alternatives and let $L = L(A)$.

Exercise 15.1.7. Let $(u_1^i)_{i \in N}, (u_2^i)_{i \in N} \in \mathcal{U}^{R^N}$. Show that there exists a monotone transformation f for N such that $f * V_{u_1^N} = V_{u_2^N}$.

Exercise 15.1.8. Let $R^N \in L(A)^N$, $u^N \in \mathcal{U}^{R^N}$, and $V = V_{u^N}$. Show the following statements.

(1) $\mathcal{C}(N, V) \neq \emptyset$ if and only if A has a weak Condorcet winner with respect to R^N.

(2) If α is a weak Condorcet winner, then $u^N(\alpha) \in \mathcal{C}(N,V) \cap \mathcal{MB}(N,V)$.

Exercise 15.1.9. Let (N,V) be the simple majority voting game defined in Example 15.1.14. Compute $\mathcal{MB}(N,V)$.

15.2 Voting Games with an Empty Mas-Colell Bargaining Set

The question of nonemptiness of the Mas-Colell bargaining set for superadditive NTU games is already mentioned in Section 6 of Mas-Colell (1989) and in Section 1 of Holzman (2000). As simple majority voting games are superadditive, Example 15.2.1 provides a negative answer to the foregoing question.

Throughout this section let A be the set of alternatives and N be a coalition (we assume without loss of generality that $N = \{1,\ldots,n\}$), let $R^N \in L(A)^N$, and let $u^N \in \mathcal{U}^{R^N}$. First we show by means of an example that $\mathcal{PMB}(N,V_{u^N})$ may be empty if $|A| = 6$.

Example 15.2.1. Let $n = 4$, $A = \{a_1,\ldots,a_4,b,c\}$, let $R^N \in L^N$ be given by Table 15.2.4, and denote $V = V_{u^N}$.

Table 15.2.4. Preference Profile leading to an empty $\mathcal{PMB}$

R^1	R^2	R^3	R^4
a_1	a_4	a_3	a_2
a_2	a_1	a_4	a_3
c	c	c	b
b	b	b	a_4
a_3	a_2	a_1	c
a_4	a_3	a_2	a_1

Claim: $\mathcal{PMB}(N,V) = \emptyset$.

Proof: Assume that there exists $x \in \mathcal{PMB}(N,V)$. Let $\alpha \in A$ such that $x \leq u^N(\alpha)$. Let

$$S_1 = \{1,2,3\}, S_2 = \{1,2,4\}, S_3 = \{1,3,4\}, S_4 = \{2,3,4\}.$$

We distinguish the following possibilities:

(1) $x \leq u^N(a_1)$. In this case $(S_4, u^{S_4}(a_4))$ is an objection at x. As there must be a counter objection to this, we conclude that $(S_3, u^{S_3}(a_3))$ is a counter

objection, and therefore also an objection at x. Hence, $x^1 \le u^1(a_3)$. To this objection, too, there must be a counter objection. We conclude that $(S_2, u^{S_2}(a_2))$ is a counter objection. Hence, $x^2 \le u^2(a_2)$ and therefore $x \ll u^N(b)$ and the desired contradiction has been obtained in this case.

(2) The possibilities $x \le u^N(\alpha)$ for $\alpha \in \{a_2, a_3, a_4\}$ may be treated similarly.

(3) $x \le u^N(b)$. Then $(S_1, u^{S_1}(c))$ is an objection at x. There are several possibilities for a counter objection to this. Each of them involves player 4 and one of the alternatives a_1, a_4, or c. We conclude that, in any case, $x^4 \le u^4(a_4)$. Hence, $(S_4, u^{S_4}(a_4))$ is an objection at x. Now we conclude that $(S_3, u^{S_3}(a_3))$ must be a counter objection and, hence, an objection at x. We continue by concluding that $(S_2, u^{S_2}(a_2))$ must be an objection and that, hence, $(S_1, u^{S_1}(a_1))$ is a counter objection. Therefore, $x \ll u^N(b)$ and the desired contradiction has been obtained.

(4) $x \le u^N(c)$. We consecutively deduce that

$$(S_4, u^{S_4}(a_4)), \ldots, (S_1, u^{S_1}(a_1))$$

are objections. The desired contradiction again is obtained by the observation that $x \ll u^N(b)$. **q.e.d.**

Example 15.2.1 may be generalized to any number $m \ge 6$ of alternatives. Also, it may be generalized to any even number $n \ge 4$ of voters: if $R_i = R^i$ for $i = 1, \ldots, 4$, if

$$R_5 = (a_2, a_1, c, b, a_3, a_4), R_6 = (a_4, a_3, c, b, a_1, a_2),$$

if $n = 4 + 2k$ for some $k \in \mathbb{N}$, if $\widetilde{R}^N \in L^N$ such that

$$|\{j \in N \mid \widetilde{R}^j = R_i\}| = \begin{cases} k \text{ , if } i = 5, 6, \\ 1 \text{ , if } i = 1, 2, 3, 4, \end{cases}$$

and if $\widetilde{V} = V_{u^N}$ for some $u^N \in \mathcal{U}^{\widetilde{R}^N}$, then $\mathcal{MB}(N, \widetilde{V}) = \emptyset$.

We now prove that Example 15.2.1 is minimal in the sense that a simple majority voting game with less than 6 alternatives has a nonempty Mas-Colell bargaining set.

Theorem 15.2.2. *If $|A| \le 5$, then $\mathcal{MB}(N, V_{u^N}) \ne \emptyset$.*

Let $V = V_{u^N}$. The following lemma is useful. Denote $m = |A|$ and recall that $t_j(R^i)$ is the j-best alternative of i for $j = 1, \ldots, m$ and $i \in N$ (see Notation 15.1.9).

Lemma 15.2.3. *Assume that there is no weak Condorcet winner. If $x \in \mathbb{R}^N_+$ satisfies $x^i \le u^i\big(t_{m-1}(R^i)\big)$ for all $i \in N$ and if x is weakly Pareto optimal in $V(N)$, then $x \in \mathcal{MB}(N, V)$.*

Proof: If (S, y) is an objection at x, then $|S| > n/2$ and there exists $\alpha \in A$ such that $u^S(\alpha) = y$. Choose $\beta \in A$ such that $\beta \succ \alpha$. Then there exists $T \subseteq N$, $|T| > n/2$ such that $u^T(\beta) \gg u^T(\alpha)$. Thus, $(T, u^T(\beta))$ is a counter objection. **q.e.d.**

Proof of Theorem 15.2.2: By Exercise 15.1.8 we may assume that A has no weak Condorcet winner. If $|A| \leq 3$, then $|A| = 3$. We claim that $0 \in \mathcal{MB}(N, V)$ in this case. Weak Pareto optimality of 0 is shown as soon we have proved that for any $\alpha \in A$ there exists $i \in N$ such that $t_3(R^i) = \alpha$. Assume, on the contrary, that $\alpha \in \{t_1(R^i), t_2(R^i)\}$ for all $i \in N$. As α is not a weak Condorcet winner, there exists $\beta \in A$ with $\beta \succ \alpha$ so that $|\{i \in N \mid t_1(R^i) = \beta\}| > \frac{n}{2}$. Thus, β is a Condorcet winner which was excluded. Moreover, again by the absence of weak Condorcet winners, any objection at 0 has a counterobjection.

We now prove the theorem for $m = 4$. For each $\alpha \in A$,

$$\text{there exists } i \in N \text{ such that } \alpha \in \{t_3(R^i), t_4(R^i)\}. \tag{15.2.1}$$

Indeed, if for some $\alpha \in A$, $\alpha \in \{t_1(R^i), t_2(R^i)\}$ for all $i \in N$, then $\beta \succ \alpha$ implies that β is a Condorcet winner which was excluded. For $\alpha \in A$, define $x_\alpha = \left(\min\{u^i(\alpha), u^i(t_3(R^i))\}\right)_{i \in N}$. By Lemma 15.2.3, $x_\alpha \in \mathcal{MB}(N, V)$, if x_α is weakly Pareto optimal. Hence, in order to complete the proof for $m = 4$, it suffices to show that there exists $\alpha \in A$ such that x_α is weakly Pareto optimal. Two possibilities may occur: If there exists $\alpha \in A$ such that $\alpha \neq t_4(R^i)$ for all $i \in N$, then, by (15.2.1), x_α is weakly Pareto optimal. Otherwise, any x_α is weakly Pareto optimal.

Now, let $m = 5$, let $A = \{a_1, \ldots, a_5\}$, and assume that $\mathcal{MB}(N, V) = \emptyset$. Then, for each $\alpha \in A$

(1) there exists $\beta \in A$ such that $\beta \succ \alpha$;

(2) $u^N(\alpha)$ is Pareto optimal (because $\mathcal{MB}$ is nonempty when we restrict our attention to the game corresponding to the restriction of u^N to $A \setminus \{\alpha\}$).

For $\alpha \in A$ denote $\ell(\alpha) = \max\{k \in \{1, \ldots, 5\} \mid \exists i \in N : t_k(R^i) = \alpha\}$. Let $\ell_{min} = \min_{\alpha \in A} \ell(\alpha)$. We distinguish cases:

(i) $\ell_{min} \geq 4$: Then there exists a weakly Pareto optimal $x \in V(N)$ such that $x^i \leq u^i(t_4(R^i))$ for all $i \in N$ which is impossible by Lemma 15.2.3.

(ii) $\ell_{min} \leq 2$: Let $\alpha, \beta \in A$ such that $\ell(\alpha) = \ell_{min}$ and $\beta \succ \alpha$. Then β is a Condorcet winner, which is impossible by (1).

(iii) $\ell_{min} = 3$: Let $B = \{\beta \in A \mid \ell(\beta) = 3\}$. If $|B| = 3$, then any $\alpha \in A \setminus B$ violates (2). If $|B| = 2$, let us say $B = \{\alpha, \beta\}$, then we may assume without loss of generality that $\alpha \not\succ \beta$. Let $\gamma \in A$ such that $\gamma \succ \beta$. Then none of the remaining $\delta \in A \setminus (\{\gamma\} \cup B)$ dominates any of the elements α, β, γ. By (1) we conclude that $\gamma \succ \beta \succ \alpha \succ \gamma$. Then $(\min\{u^i(\alpha), u^i(\beta)\})_{i \in N} \in \mathcal{MB}(N, V)$.

Now we turn to the case $|B| = 1$, let us say $B = \{a_3\}$. Let $\widehat{S} = \{i \in N \mid t_3(R^i) = a_3\}$. For any $k \in \widehat{S}$ there exists $x_k \in \mathbb{R}^N$ such that x_k is weakly Pareto optimal, $x_k^k = u^k(a_3)$, and $x_k^i \le u^i(t_4(R^i))$ for all $i \in N \setminus \{k\}$. As $x_k \notin \mathcal{MB}(N,V)$, there exists a justified objection $(S, u^S(\alpha))$ for some $S \subseteq N$, $|S| > n/2$, and some $\alpha \in A$. Let $\beta \in A$ such that $\beta \succ \alpha$. Then there exists $T \subseteq N$, $|T| > n/2$, such that $u^{S\cap T}(\beta) \gg u^{S\cap T}(\alpha)$ and $u^{T\setminus S}(\beta) \ge (u^i(t_4(R^i)))_{i\in T\setminus S}$. As $(T, u^T(\beta))$ is not a counter objection, we conclude that $k \in T$, $t_4(R^k) = \beta$, and $t_5(R^k) = \alpha$. We conclude that for any $k \in \widehat{S}$ the alternative $t_5(R^k)$ is only dominated by $t_4(R^k)$. If n is odd, we may now easily finish the proof by the observation that α dominates all other alternatives except β, and therefore $(\min\{u^i(\alpha), u^i(\beta)\})_{i\in N} \in \mathcal{MB}(N,V)$. Hence we may assume from now on that n is an even number. As $a_3 \not\succ \alpha$, $\{i \in N \mid u^i(\alpha) > u^i(a_3)\} \cap \{i \in N \mid u^i(\beta) > u^i(\alpha)\} \neq \emptyset$. Thus, there exists $j \in \widehat{S}$ such that $t_1(R^j) = \beta$ and $t_2(R^j) = \alpha$. So far we have for any $k \in \widehat{S}$, where $\alpha = t_5(R^k), \beta = t_4(R^k)$:

$$\alpha \text{ is only dominated by } \beta; \tag{15.2.2}$$

$$\text{There exists } j \in \widehat{S} \text{ such that } t_1(R^j) = \beta, t_2(R^j) = \alpha; \tag{15.2.3}$$

$$|\{i \in N \mid u^i(\alpha) > u^i(a_3)\}| \ge \tfrac{n}{2}. \tag{15.2.4}$$

Now, let $k, j \in \widehat{S}$ have the foregoing properties, let us say $k = 1$ and $j = 2$. We also may assume that $t_4(R^1) = a_4$, $t_5(R^1) = a_5$, $t_4(R^2) = a_1$, $t_5(R^2) = a_2$ (hence $R^2 = (a_4, a_5, a_3, a_1, a_2)$). So, for any $k \in \widehat{S}$, we have

$$\{t_4(R^k), t_5(R^k)\} = \{a_4, a_5\} \Rightarrow t_4(R^k) = a_4 \tag{15.2.5}$$

$$t_5(R^k) = a_5 \Rightarrow t_4(R^k) = a_4 \tag{15.2.6}$$

$$\{t_4(R^k), t_5(R^k)\} = \{a_1, a_2\} \Rightarrow t_4(R^k) = a_1 \tag{15.2.7}$$

$$t_5(R^k) = a_2 \Rightarrow t_4(R^k) = a_1 \tag{15.2.8}$$

We now show that there exists $k \in \widehat{S}$ such that $t_5(R^k) \notin \{a_5, a_2\}$. Assume the contrary. Then $\{i \in N \mid u^i(a_5) > u^i(a_3)\} \cap \{i \in N \mid u^i(a_2) > u^i(a_3)\} = \emptyset$ and, by (15.2.4), $a_5 \not\succ a_3$ and $a_2 \not\succ a_3$. Hence, by (1), $a_1 \succ a_3$ or $a_4 \succ a_3$. However, note that by our assumption $u^i(a_1) > u^i(a_3)$ implies $u^i(a_1) > u^i(a_5)$ for all $i \in N$. Thus, if $a_1 \succ a_3$, then $a_1 \succ a_5$ which contradicts (15.2.2). Similarly, $a_4 \succ a_3$ can be excluded.

Hence, we may assume without loss of generality, that there exists $k \in \widehat{S}$ such that $t_5(R^k) = a_1$. We now claim that there exists $j \in \widehat{S}$ such that $t_5(R^j) = a_4$. By (15.2.2) and the fact that $a_1 \succ a_2$, $t_4(R^k) \in \{a_4, a_5\}$. If $t_4(R^k) = a_4$, then by (15.2.3) there exists $j \in \widehat{S}$ such that $\{t_4(R^j), t_5(R^j)\} = \{a_2, a_5\}$. By (15.2.6), $a_5 \neq t_5(R^j)$, and by (15.2.8), $a_2 \neq t_5(R^j)$. Hence this possibility is ruled out. We conclude that $t_4(R^k) = a_5$. By (15.2.3) there exists $j \in \widehat{S}$ such that $\{t_4(R^j), t_5(R^j)\} = \{a_2, a_4\}$. By (15.2.8), $t_5(R^j) = a_4$. So our claim has been shown.

So far we have deduced there exist $k_j \in \widehat{S}$, $j = 1,2,4,5$, such that $t_5(R^{k_j}) = a_j$. By (15.2.4), $|\{i \in N \mid u^i(a_j) > u^i(a_3)\}| \geq \frac{n}{2}$ for all $j = 1,2,4,5$. We conclude that $a_3 = t_3(R^i)$ for all $i \in N$ and $|\{i \in N \mid u^i(a_j) > u^i(a_3)\}| = \frac{n}{2}$ for all $j = 1,2,4,5$. Therefore a_3 is not dominated by any alternative, which contradicts (1). **q.e.d.**

Exercises

Let A be a set of alternatives, let $R^N \in L(A)^N$, and let $u^N \in \mathcal{U}^{R^N}$ and $V = V_{u^N}$.

Exercise 15.2.1. Prove that if $x \in \mathcal{PMB}(N,V)$, then $x_+ \in \mathcal{MB}(N,V)$.

Exercise 15.2.2. Assume that $|N|$ is odd and let $x, y \in X(N,V)$ such that $x \leq y$. Prove that if $y \in \mathcal{PMB}(N,V)$, then $x \in \mathcal{PMB}(N,V)$.

15.3 Non-levelled NTU Games with an Empty Mas-Colell Prebargaining Set

Section 15.2 shows that the Mas-Colell bargaining set of a superadditive NTU game may be empty. However, a related problem is mentioned in Vohra (1991) who proved nonemptiness for any weakly superadditive non-levelled NTU game that satisfies one additional condition. An NTU game (N,V) is *weakly superadditive* if for every $i \in N$ and every $\emptyset \neq S \subseteq N \setminus \{i\}$, $V(S) \times V(\{i\}) \subseteq V(S \cup \{i\})$. In particular Vohra raised the question whether the additional condition is necessary for the nonemptiness of $\mathcal{MB}(N,V)$. It is the purpose of this section to show that there exists a superadditive non-levelled game whose Mas-Colell prebargaining set is empty. Note that simple majority voting games are not non-levelled.

Instead of providing an explicit example of an NTU game with the desired properties we shall prove the existence of such an example. The proof proceeds in steps. First we define an extension, $\mathcal{PMB}^*$, of the Mas-Colell prebargaining set. In Subsection 15.3.1 we present an example of a 4-person simple majority voting game on 10 alternatives whose extended bargaining set is empty. Finally, in Subsection 15.3.2 we show that in any neighborhood of a compactly generated superadditive game there exists a superadditive non-levelled game. Finally we use a restricted variant of upper hemicontinuity of $\mathcal{PMB}^*$ to conclude the existence of a superadditive non-levelled NTU game whose Mas-Colell bargaining set is empty.

Now our auxiliary solution, $\mathcal{PMB}^*$, is defined.

Definition 15.3.1. *Let (N, V) be an NTU game and $x \in X(N, V)$. A* **strongly justified strong objection** *at x is a strong objection at x that has no weak counterobjection. The* **extended prebargaining set** *of (N, V), $\mathcal{P}MB^*(N, V)$, is the set of all preimputations x that have no strongly justified strong objections at x. Moreover, $\mathcal{M}B^*(N, V) = \mathcal{P}MB^*(N, V) \cap I(N, V)$ is the* **extended bargaining set** *of (N, V).*

Remark 15.3.2. Note that for any NTU game (N, V), $\mathcal{P}MB(N, V) \subseteq \mathcal{P}MB^*(N, V)$. Moreover, the foregoing inclusion may be strict. Indeed, the extended bargaining set of the Example 15.2.1 is nonempty. The simple proof of this fact is left as Exercise 15.3.2.

15.3.1 The Example

This subsection is devoted to an example of a simple majority voting game with 10 alternatives and 4 voters whose extended prebargaining set is shown to be empty.

The preferences are given by Table 15.3.5.

Table 15.3.5. Preference Profile on 10 Alternatives

R^1	R^2	R^3	R^4
a_1	a_4	a_3	a_2
a_2	a_1	a_4	a_3
a_2^*	a_1^*	a_4^*	a_3^*
a_1^*	c	a_3^*	a_2^*
c	a_4^*	c	b
b	b	b	a_4^*
a_3^*	a_2^*	a_1^*	a_4
a_3	a_2	a_1	c
a_4^*	a_3^*	a_2^*	a_1^*
a_4	a_3	a_2	a_1

The corresponding domination relation, $\succ = \succ_{R^N}$, is depicted in Table 15.3.6.

Table 15.3.6. Domination Relation

$a_1 \succ a_2$	$a_2 \succ a_3$	$a_3 \succ a_4$	$a_4 \succ a_1$
$a_1 \succ a_2^*$	$a_2 \succ a_3^*$	$a_3 \succ a_4^*$	$a_4 \succ a_1^*$
	$a_4 \succ c$	$c \succ b$	

Theorem 15.3.3. *Let R^N be defined by Table 15.3.5 and let $V = V_{u^N}$. Then $\mathcal{PMB}^*(N,V) = \emptyset$.*

Proof: Assume, on the contrary, that there exists a preimputation x in the set $\mathcal{MB}^*(N,V)$. Let $S_1 = \{1,2,3\}$, $S_2 = \{1,2,4\}$, $S_3 = \{1,3,4\}$, and $S_4 = \{2,3,4\}$. Frequently used strong objections that use some of the foregoing coalitions may be constructed with the help of Table 15.3.7 which is deduced from Table 15.3.5 (see also Table 15.3.6). By (15.1.3), $x \leq u^N(\alpha)$ for some

Table 15.3.7. Constructions of Strong Objections

$u^{S_1}(a_1) \gg u^{S_1}(a_2)$	$u^{S_1}(a_1) \gg u^{S_1}(a_2^*)$	
$u^{S_2}(a_2) \gg u^{S_2}(a_3)$	$u^{S_2}(a_2) \gg u^{S_2}(a_3^*)$	$u^{S_4}(a_4) \gg u^{S_4}(c)$
$u^{S_3}(a_3) \gg u^{S_3}(a_4)$	$u^{S_3}(a_3) \gg u^{S_3}(a_4^*)$	$u^{S_1}(c) \gg u^{S_1}(b)$
$u^{S_4}(a_4) \gg u^{S_4}(a_1)$	$u^{S_4}(a_4) \gg u^{S_4}(a_1^*)$	

$\alpha \in A$. As A has 10 elements, we proceed by distinguishing the arising 10 possibilities. First we shall consider the following case:

$$x \leq u^N(a_1). \tag{15.3.1}$$

As $a_4 \succ a_1$, $(S_4, u^{S_4}(a_4))$ is a strong objection at x (see Table 15.3.7). Note that a_4 is the first, that is, the most preferred, alternative of player 2, the second alternative of player 3, and note that a_3 is the first alternative of player 3. So, if $\alpha \in A \setminus \{a_3, a_4\}$, then

$$|S_4 \cap \{i \in N \mid a_4 \; R^i \; \alpha\}| \geq 2.$$

Thus the foregoing objection can be weakly countered only by (S_3, y) for some $y \leq u^{S_3}(a_3)$, or by (T, z) for some $|T| \geq 3$ such that $1 \in T$ and some $z \leq u^T(a_4)$, or by $(\{1\}, 0)$ (if $x^1 = 0$). Hence $x^1 \leq u^1(a_3)$. From Table 15.3.7 we conclude that $(S_3, u^{S_3}(a_3^*))$ is a strong objection at x. Let $\alpha \in A$. If

$$|S_3 \cap \{i \in N \mid a_3^* \; R^i \; \alpha\}| < 2,$$

then $\alpha \in \{a_2, a_3^*, a_3\}$. Thus, if (P, y) is a weak counter objection to $(S_3, u^{S_3}(a_3^*))$, then $2 \in P$ and $y^2 \leq u^2(a_2)$. As $x \in \mathcal{MB}(N,V)$ is assumed, there exists a weak counter objection (P, y) to the foregoing strong objection. We conclude that $x^2 \leq y^2 \leq u^2(a_2)$. Thus, $x \ll u^N(b)$ and the desired contradiction has been obtained.

The following 3 cases may be treated similarly to (15.3.1):

$$x \leq u^N(\alpha) \text{ for some } \alpha \in \{a_2, a_3, a_4\}. \tag{15.3.2}$$

Indeed, if $i \in \{2,3,4\}$ and $x \leq u^N(a_i)$, then Table 15.3.7 shows that

$$(S_{i-1}, u^{S_{i-1}}(a_{i-1}))$$

is a strong objection at x. A careful inspection of the tables allows to specify one further strong objection, namely $(S_{i-2},u^{S_{i-2}}(a^*_{i-2}))$ if $i \neq 2$ and $(S_4,u^{S_4}(a^*_4))$ if $i = 2$, and again the existence of a weak counter objections implies that $x \ll u^N(b)$.

The next case is the following case.

$$x \leq u^N(a^*_1). \tag{15.3.3}$$

As $(S_4, u^{S_4}(a_4))$ is a strong objection at x (see Table 15.3.7), a careful inspection of Table 15.3.5 shows that we may proceed as in (15.3.1).

The following 3 cases may be treated similarly to (15.3.3):

$$x \leq u^N(\alpha) \text{ for some } \alpha \in \{a^*_2, a^*_3, a^*_4\}. \tag{15.3.4}$$

Now we shall consider the 9th possibility:

$$x \leq u^N(b). \tag{15.3.5}$$

In this case Table 15.3.7 shows that $(S_1, u^{S_1}(c))$ is a strong objection at x. If (P, y) is a weak counter objection to the foregoing strong objection, then an inspection of Table 15.3.5 shows that (P, y) satisfies at least one of the following properties:

$$\begin{aligned} y &\leq u^P(c) \text{ and } 4 \in P; \\ y &\leq u^P(a_1) \text{ and } P = S_2; \\ y &\leq u^P(a^*_1) \text{ and } P = S_2; \\ y &\leq u^P(a_4) \text{ and } P = S_4. \end{aligned}$$

Therefore $x^4 \leq u^4(a_4)$. We conclude that $(S_4, u^{S_4}(a^*_4))$ is a strong objection at x. Then

$$\{\alpha \in A \mid |S_4 \cap \{i \in N \mid a^*_4 \; R^i \; \alpha\}| < 2\} = \{a_3, a_4, a^*_4\}.$$

Hence $x^1 \leq u^1(a_3)$. Thus, $(S_3, u^{S_3}(a^*_3))$ is a strong objection at x. The observation that

$$\{\alpha \in A \mid |S_3 \cap \{i \in N \mid a^*_3 \; R^i \; \alpha\}| < 2\} = \{a_2, a_3, a^*_3\}$$

shows that $x^2 \leq u^2(a_2)$ and, thus, $(S_2, u^{S_2}(a^*_2))$ is a strong objection at x. We compute

$$\{\alpha \in A \mid |S_2 \cap \{i \in N \mid a^*_2 \; R^i \; \alpha\}| < 2\} = \{a_1, a_2, a^*_2\}.$$

Thus, if (P, y) is a weak counter objection to $(S_2, u^{S_2}(a^*_2))$, then $3 \in P$ and $y^3 \leq u^3(a_1)$. We conclude that $x^3 \leq u^3(a_1)$. Therefore, again, $x \ll u^N(b)$.

Finally, we have to consider the following case.

$$x \leq u^N(c). \tag{15.3.6}$$

Then $(S_4, u^{S_4}(a_4))$ is a strong objection at x (see Table 15.3.7). If (P, y) is a weak counter objection to $(S_4, u^{S_4}(a_4))$, then (1) $P = S_3$ and $y \leq u^P(a_3)$ or (2) $1 \in P$ and $y \leq u^P(a_4)$. Hence, $x^1 \leq u^1(a_3)$ and $(S_3, u^{S_3}(a_3^*))$ is a strong objection at x. We may now continue as in (15.3.5) and deduce that $x \ll u^N(b)$.

By (15.1.3), the domain defined by (15.3.1) – (15.3.6) is equal to V(N). Hence, we have derived a contradiction to the required weak Pareto optimality in all possible 10 cases. **q.e.d.**

15.3.2 Non-levelled Games

Let N be a finite nonempty set and let (N, V) be an NTU game. For any $\lambda \in \Delta_{++}^V$ (see Section 13.1 for the definition of "viable") let (N, v_λ) be the TU game defined in (13.1.1), and let $\varepsilon > 0$. For $\emptyset \neq S \subseteq N$ let $f_{\lambda^S}^\varepsilon : \mathbb{R}^S \to \mathbb{R}^S$ be defined by

$$f_{\lambda^S}^\varepsilon(x) = x + \varepsilon \left(1 - \frac{1}{1 + (v_\lambda(S) - \lambda^S \cdot x)_+}\right) \chi_S \text{ for all } x \in \mathbb{R}^S. \tag{15.3.7}$$

Note that

$$\|f_{\lambda^S}^\varepsilon(x) - x\|_\infty \leq \varepsilon \text{ for all } x \in \mathbb{R}^S; \tag{15.3.8}$$

$$x, \widetilde{x} \in \mathbb{R}^S, \lambda^S \cdot x = \lambda^S \cdot \widetilde{x} \Rightarrow \lambda^S \cdot f_{\lambda^S}^\varepsilon(x) = \lambda^S \cdot f_{\lambda^S}^\varepsilon(\widetilde{x}). \tag{15.3.9}$$

The following lemma is useful.

Lemma 15.3.4. *Let $x, \widetilde{x} \in \mathbb{R}^S$, $i \in S$, and assume that $\varepsilon \leq \frac{1}{\lambda(S)}$.*

(1) If $\lambda^S \cdot x \leq \lambda^S \cdot \widetilde{x}$ and $x^i > \widetilde{x}^i$, then $\left(f_{\lambda^S}^\varepsilon(x)\right)^i > \left(f_{\lambda^S}^\varepsilon(\widetilde{x})\right)^i$.

(2) If $\lambda^S \cdot x < v_\lambda(S)$, then $\lambda^S \cdot f_{\lambda^S}^\varepsilon(x) < v_\lambda(S)$.

(3) $\lambda^S \cdot x \leq \lambda^S \cdot \widetilde{x}$ if and only if $\lambda^S \cdot f_{\lambda^S}^\varepsilon(x) \leq \lambda^S \cdot f_{\lambda^S}^\varepsilon(\widetilde{x})$.

(4) The mapping $f_{\lambda^S}^\varepsilon$ is bijective.

Proof: (1) is an immediate consequence of (15.3.7). In order to verify the remaining statements, let $f = f_{\lambda^S}^\varepsilon$ and define, for $t \in \mathbb{R}$, $g(t) = \lambda^S \cdot f(t\chi_S)$. In order to prove (2), by (15.3.9) it suffices to show that, for all t such that $t\lambda(S) < v_\lambda(S)$, $g(t) < v_\lambda(S)$. As $\varepsilon \leq \frac{1}{\lambda(S)}$ and $1 + v_\lambda(S) - t\lambda(S) > 1$,

$$g(t) = \lambda(S)\left(t + \varepsilon \frac{v_\lambda(S) - t\lambda(S)}{1 + v_\lambda(S) - t\lambda(S)}\right) < v_\lambda(S).$$

In order to verify (3), it suffices to show that g is monotonic, that is, for $t, \widetilde{t} \in \mathbb{R}$,

$$t < \widetilde{t} \Rightarrow g(t) < g(\widetilde{t}). \tag{15.3.10}$$

If $t\lambda(S) \geq v_\lambda(S)$, then $\widetilde{t}\lambda(S) > v_\lambda(S)$ and, by (15.3.7), $f(t\chi_S) = t\chi_S$ and $f(\widetilde{t}\chi_S) = \widetilde{t}\chi_S$, so that (15.3.10) is valid in this case.

In view of (2) it suffices to show that $g'(t) > 0$ for all t satisfying $t\lambda(S) < v_\lambda(S)$. The observation that

$$g'(t) = \lambda(S)\left(1 - \varepsilon\frac{\lambda(S)}{(1 + v_\lambda(S) - t\lambda(S))^2}\right) > \lambda(S)(1 - \varepsilon\lambda(S))$$

completes the proof of (3).

Note that by (15.3.7) and (15.3.8)

$$\{f(x + t\chi_S) \mid t \in \mathbb{R}\} = \{x + t\chi_S \mid t \in \mathbb{R}\} \text{ for all } x \in \mathbb{R}^N.$$

Hence, by Statement (3), f is bijective. **q.e.d.**

For any $\emptyset \neq S \subseteq N$, let $V^\varepsilon_\lambda(S) = f^\varepsilon_{\lambda^S}(V(S))$. Also, put $V^\varepsilon_\lambda(\emptyset) = \emptyset$.

Lemma 15.3.5. *If $\varepsilon \leq 1$, then $(N, V^\varepsilon_\lambda)$ is an NTU game.*

Proof: Let $W = V^\varepsilon_\lambda$, let S be a coalition, and let $f = f^\varepsilon_{\lambda^S}$. Then $W(S) \neq \emptyset$, because $W(S) = f(V(S))$. By continuity of f and (15.3.8), the image of a closed set under f is closed. Hence $W(S)$ is closed. As $V(S) \cap (x + \mathbb{R}^S)$ is bounded for any $x \in \mathbb{R}^S$, (15.3.8) implies that $W(S) \cap (x + \mathbb{R}^S)$ is bounded.

In order to show that $W(S)$ is comprehensive, let $\widetilde{y} \in W(S)$ and $y < \widetilde{y}$. Then there is $\widetilde{x} \in V(S)$ such that $f(\widetilde{x}) = \widetilde{y}$. By the surjectivity (see (4) of Lemma 15.3.4) of f, there exists $x \in \mathbb{R}^S$ such that $f(x) = y$. By (1) and (3) of Lemma 15.3.4, $x \leq \widetilde{x}$. As $V(S)$ is comprehensive, $x \in V(S)$. **q.e.d.**

Lemma 15.3.6. *Let $\varepsilon \leq 1$.*

(1) If (N, V) is zero-normalized, then $(N, V^\varepsilon_\lambda)$ is zero-normalized.

(2) If (N, V) is superadditive, then $(N, V^\varepsilon_\lambda)$ is superadditive.

(3) $(N, V^\varepsilon_\lambda)$ is non-levelled.

Proof: Let $W = V^\varepsilon_\lambda$.

(1) Assume that (N,V) is zero-normalized and let $i \in N$. Then $v_\lambda(\{i\}) = 0$ and (2) of Lemma 15.3.4 shows (1).

(2) Assume that (N, V) is superadditive, let $S, T \subseteq N$ be disjoint coalitions, and let $y \in \mathbb{R}^{S \cup T}$ such that $y^S \in W(S)$ and $y^T \in W(T)$. Let $x \in \mathbb{R}^{S \cup T}$ be determined by the requirements that $f^\varepsilon_{\lambda^S}(x^S) = y^S$ and $f^\varepsilon_{\lambda^T}(x^T) = y^T$. As (N, V) is superadditive, we conclude that $x \in V(S \cup T)$ and that

$$v_\lambda(S \cup T) \geq \max\{\lambda^S \cdot x^S + v_\lambda(T), \lambda^T \cdot x^T + v_\lambda(S)\}.$$

Therefore, $f^\varepsilon_{\lambda^{S\cup T}}(x) \geq y$ and comprehensiveness of $W(S \cup T)$ completes the proof of (2).

(3) Let $\emptyset \neq S \subseteq N$, $f = f^\varepsilon_{\lambda^S}$, $\widetilde{y} \in W(S)$ and $y < \widetilde{y}$. It remains to show that there exists $z \in W(S)$ such that $z \gg y$. Let $\widetilde{x} \in V(S)$ such that $f(\widetilde{x}) = \widetilde{y}$ and let $i \in N$ be such that $y^i < \widetilde{y}^i$. For $\delta > 0$ define $\widetilde{x}_\delta = \widetilde{x} - \delta\chi_{\{i\}}$. By continuity of f there exists $\delta > 0$ such that $(f(\widetilde{x}_\delta))^i > y^i$. As $\lambda \cdot \widetilde{x}_\delta < \lambda \cdot \widetilde{x}$, $f(\widetilde{x}_\delta) \gg y$. Comprehensiveness of $V(S)$ guarantees that $\widetilde{x}_\delta \in V(S)$, hence $z = f(\widetilde{x}_\delta)$ has the desired properties. **q.e.d.**

Remark 15.3.7. Let (N, V) is a simple majority voting game. Then $\Delta^V_{++} = \Delta_{++}$, because (N, V) is compactly generated.

Lemma 15.3.8. *Let (N, V) is a simple majority voting game, $\lambda \in \Delta_{++}$, and let $(\varepsilon_k)_{k\in\mathbb{N}}$ be a real sequence such that $1 \geq \varepsilon_k > 0$ and $\lim_{k\to\infty} \varepsilon_k = 0$. If $\mathcal{PMB}^*(N, V^{\varepsilon_k}_\lambda) \neq \emptyset$ for all $k \in \mathbb{N}$, then $\mathcal{PMB}^*(N, V) \neq \emptyset$.*

Proof: Let $V_k = V^{\varepsilon_k}_\lambda$ and $x_k \in \mathcal{PMB}(N, V_k)$. Then $((x_k)_+)_{k\in\mathbb{N}}$ has a convergent subsequence, let us say, it is convergent and the limit is x. As $x_k - \varepsilon_k\chi_N \in V(N)$, we conclude that $(x_k - \varepsilon_k\chi_N)_+ \in V(N)$, so $x \in V(N)$. As $V(N) \subseteq V^\varepsilon_\lambda(N)$ for $\varepsilon > 0$, x is weakly Pareto optimal. In order to show that $x \in \mathcal{MB}(N, V)$ let (P, y) be a strong objection at x. As $x^P \geq 0$, $y \gg 0$. Also, if there exists $i \in N\backslash P$ such that $x^i = 0$, then $(\{i\}, 0)$ is a weak counterobjection to (P, y). So, we assume that $P \supseteq \{i \in N \mid x^i = 0\}$. As $V_k(P) \supseteq V(P)$, there exists y_k such that y_k is Pareto optimal in (P, V_k) and $y_k \geq y$. Therefore, for almost all k, say for all $k \in \mathbb{N}$, (P, y) is a strong objection at x_k with respect to V_k. As $x_k \in \mathcal{PMB}(N, V_k)$, there exists a weak counterobjection (Q_k, z_k) to (P, y_k). Then there exists $Q \subseteq N$ such that $Q = Q_k$ for almost all, say for all, $k \in \mathbb{N}$. Also, $((z_k)_+)_{k\in\mathbb{N}}$ has a convergent subsequence, say, it is convergent and the limit is z. We conclude that $z \in V(Q)$ and, hence, (Q, z) is a weak counterobjection to (P, y). **q.e.d.**

Now we are ready to prove the nonexistence result.

Theorem 15.3.9. *There exists a superadditive non-levelled NTU game (N, U) such that $\mathcal{PMB}(N, U) = \emptyset$.*

Proof: Let (N, V) be a simple majority voting game defined in Section 15.3.1. Hence $\mathcal{PMB}^*(N, V) = \emptyset$. Let $\lambda \in \Delta_{++}$. By Lemma 15.3.8 there exists $0 < \delta \leq 1$ such that $\mathcal{PMB}^*(N, V^\varepsilon_\lambda) = \emptyset$ for all $\varepsilon \leq \delta$. By Lemma 15.3.6 and Remark 15.3.2, $(N, V^\varepsilon_\lambda)$ has the desired properties. **q.e.d.**

Exercises

Let A be a set of alternatives, let $R^N \in L(A)^N$, and let $u^N \in \mathcal{U}^{R^N}$ and $V = V_{u^N}$.

Exercise 15.3.1. Prove the following statements.

(1) If $x \in \mathcal{PMB}^*(N,V)$, then $x_+ \in \mathcal{MB}^*(N,V)$.

(2) Let $|N|$ be odd and let $x, y \in X(N,V), x \leq y$. If $y \in \mathcal{PMB}^*(N,V)$, then $x \in \mathcal{PMB}^*(N,V)$.

Exercise 15.3.2. Let (N,V) be the simple majority voting game defined in Example 15.2.1. Find an element of $\mathcal{MB}^*(N,V)$.

15.4 Existence Results for Many Voters

We present here two models, in which special assumptions about the distribution of preferences in the population of voters lead to existence results when there are many voters.

The first model is probabilistic. Let A be a fixed set of m alternatives, and let $L = L(A)$. We assume that each $R \in L$ appears with positive probability $p_R > 0$ in the population of potential voters, where $\sum_{R \in L} p_R = 1$. Now let $(\mathcal{R}^i)_{i \in \mathbb{N}}$ be a sequence of independent and identically distributed random variables such that $Pr(\mathcal{R}^i = R) = p_R$ for all $i \in \mathbb{N}$, $R \in L$. Let $\mathcal{R}^N = (\mathcal{R}^1, \ldots, \mathcal{R}^n)$ be the corresponding random profile of preferences for n voters, and let $(N, V(\mathcal{R}^N))$ be the random simple majority voting game that is associated via some utility representation $u^{N,R^N} = (u^{i,R^i})_{i \in N}$ for each realization R^N of $\mathcal{R}^N$.

We are going to prove that in this model the limiting probability, as $n \to \infty$, that the bargaining set and the Mas-Colell bargaining set are nonempty, equals 1. We note that the analogous statement does not hold true for the core. In the case of the core, the limiting probability in question is that of the existence of a weak Condorcet winner. This has been studied quite a lot in the literature (see, e.g., Sen (1970) and Gehrlein (2002)). In the simplest set-up, where $p_R = \frac{1}{m!}$ for every $R \in L(A)$, it is known that the limiting probability that there exists a weak Condorcet winner is strictly less than 1 for every $m \geq 3$, and it tends to 0 as $m \to \infty$. In the more general set-up that we consider here, it is even possible to choose $p_R > 0$ so that this limiting probability will equal 0 (see Example 15.4.4 below).

Define, for $j = 1, \ldots, m$,

$$\varepsilon_j(p) = \varepsilon_j = \min_{\alpha \in A} \sum_{R \in L : \alpha = t_j(R)} p_R.$$

As $p_R > 0$ for all $R \in L$, $\varepsilon_j > 0$. Note that for any $\gamma < 1$,

$$\lim_{n \to \infty} Pr\left(\mathcal{R}^N \in \left\{ R^N \in L^N \,\middle|\, \min_{\alpha \in A} |\{i \in N \mid \alpha = t_j(R^i)\}| \geq \gamma \varepsilon_j n \right\}\right) = 1. \tag{15.4.1}$$

Theorem 15.4.1.

$$\lim_{n\to\infty} Pr\left(\mathcal{M}\left(N, V\left(\mathcal{R}^N\right)\right) \neq \emptyset\right) = \lim_{n\to\infty} Pr\left(\mathcal{MB}\left(N, V\left(\mathcal{R}^N\right)\right) \neq \emptyset\right) = 1.$$

Proof: Call $R^N \in L^N$ *good* if for all $\alpha \in A$ there exists $i \in N$ such that $\alpha = t_m(R^i)$. If R^N is good, then $0 \in \mathcal{M}(N, V)$, where $V = V_{u^{N,R^N}}$. Regarding $\mathcal{MB}(N, V)$ when R^N is good, we distinguish two cases. If there is a weak Condorcet winner α, then $u^{N,R^N}(\alpha) \in \mathcal{MB}(N, V)$. If no such α exists, then $0 \in \mathcal{MB}(N, V)$. Thus we see that in order to prove both parts of the theorem, it suffices to show that $\mathcal{R}^N$ is good with probability tending to 1 as n tends to infinity. This fact is implied by (15.4.1) applied to $j = m$. **q.e.d.**

As shown in the next theorem, the probability that a positive fraction of voters may receive maximal utility in the Mas-Colell bargaining set tends to 1 if n tends to infinity.

Theorem 15.4.2. *If $\varepsilon^* < \varepsilon_1(p)$, then*

$$\lim_{n\to\infty} Pr\left(\left|\left\{i \in N \middle| \exists\ x \in \mathcal{MB}(N,V(\mathcal{R}^N)) : x^i = u^{i,R^i}(t_1(R^i))\right\}\right| \geq \varepsilon^* n\right) = 1.$$

The following lemma is used in the proof of Theorem 15.4.2.

Lemma 15.4.3. *Let (N, V) be a zero-normalized superadditive NTU game, $i \in N$, and $x \in \mathbb{R}^N_+$ such that x is weakly Pareto optimal in $V(N)$ and $x^j = 0$ for all $j \in N \setminus \{i\}$. If $\mathcal{C}(N, V) = \emptyset$ and $\mathcal{C}(N \setminus \{i\}, V) = \emptyset$, then $x \in \mathcal{MB}(N, V)$.*

Proof: Let (P, y) be an objection such that P is maximal. By superadditivity $N\setminus\{i\} \subseteq P$. If $i \in P$, then there exists a counter objection, because $\mathcal{C}(N, V) = \emptyset$. If $P = N\setminus\{i\}$, then there exists a counter objection, because $\mathcal{C}(N\setminus\{i\}, V) = \emptyset$. **q.e.d.**

Proof of Theorem 15.4.2: By (15.4.1), by (2) of Remark 2.2, and by Lemma 15.4.3 it suffices to prove that

$$\lim_{n\to\infty} Pr\Big(\mathcal{C}\left(N,V(\mathcal{R}^N)\right) = \emptyset \text{ and } \exists i \in N : \mathcal{C}\left(N\setminus\{i\},V(\mathcal{R}^N)\right) \neq \emptyset\Big) = 0. \quad (15.4.2)$$

Let R^N be a realization of $\mathcal{R}^N$ and let $u^N = u^{N,R^N}$, $V = V_{u^N}$. Define a binary relation $\rhd_{R^N} = \rhd$ on A as follows. For $\alpha, \beta \in A, \alpha \neq \beta$, define

$$\alpha \rhd \beta \Leftrightarrow |\{i \in N \mid \alpha\ R^i\ \beta\}| = 1 + [n/2].$$

Claim: If $\mathcal{C}(N, V) = \emptyset$ and $\mathcal{C}(N \setminus \{i\}, V) \neq \emptyset$ for some $i \in N$, then there exist $\alpha, \beta \in A$, $\alpha \neq \beta$, such that $\alpha \rhd \beta$. Indeed, if $\mathcal{C}(N\setminus\{i\}, V) \neq \emptyset$, then there exists $\beta \in A$ such that $u^{N\setminus\{i\}}(\beta) \in \mathcal{C}(N \setminus \{i\}, V)$. As $u^N(\beta) \notin \mathcal{C}(N, V)$, there exists $\alpha \in A\setminus\{\beta\}$ such that $|\{j \in N \mid \alpha\ R^j\ \beta\}| > \frac{n}{2}$. Yet, $|\{j \in N\setminus\{i\} \mid \alpha\ R^j\ \beta\}| \leq \frac{n}{2}$ and therefore $\alpha \rhd \beta$.

Define $q^n_{\alpha,\beta} = Pr\left(\mathcal{R}^N \in \{R^N \in L^N \mid \alpha \rhd_{R^N} \beta\}\right)$.

In order to prove (15.4.2), it suffices to prove that $\lim_{n\to\infty} q^n_{\alpha,\beta} = 0$. Let $q = \sum_{R\in L:\alpha R\beta} p_R$. With $r = r(n) = 1+[n/2]$ we obtain $q^n_{\alpha,\beta} = \binom{n}{r} q^r(1-q)^{n-r}$. We distinguish two cases. If n is even, then $r = \frac{n}{2}+1$ and

$$q^n_{\alpha,\beta} = \binom{2r-2}{r} q^r(1-q)^{r-2} \le \binom{2r}{r} q^2\left(q(1-q)\right)^{r-2} \le \frac{(2r)!}{r!r!2^{2r-4}}.$$

Using $k! \approx \sqrt{2\pi k}\left(\frac{k}{e}\right)^k$ yields

$$\frac{(2r)!}{r!r!2^{2r-4}} \approx \frac{16}{\sqrt{\pi r}} \to_{r\to\infty} 0.$$

Similarly, we may approximate $q^n_{\alpha,\beta}$ if n is odd. **q.e.d.**

The next example shows that $\mathcal{MB}$ cannot be replaced by $\mathcal{M}$ in Theorem 15.4.2.

Example 15.4.4. Let $A = \{a, b, c\}$ and let p satisfy $p_{R^i} > \frac{1}{4}$ for $i = 1, 2, 3$, where the R^i are defined by Table 3.1. Then

$$\lim_{n\to\infty} Pr\left(\mathcal{R}^N \in \{R^N \mid a \succ_{R^N} b \succ_{R^N} c \succ_{R^N} a\}\right) = 1,$$

hence, the probability that the core is empty tends to 1 if n tends to infinity. Also, if (N, V) is a realization of $(N, V(\mathcal{R}^N))$ such that $\mathcal{C}(N, V) = \emptyset$, then $x \in \mathcal{M}(N, V)$ implies $x^i < u^{i,R^i}(t_1(R^i))$ for all $i \in N$. Indeed, this statement can be proved as follows: Let (N, V_{u^N}), $u^N \in \mathcal{U}^{R^N}$ for some $R^N \in L(A)^N$, be a majority voting game whose core is empty. Let $i \in N$ and let $x \in V(N) \cap \mathbb{R}^N_+$ satisfy $x^i = u^i(\alpha)$, where $\alpha = t_1(R^i)$. Hence, $x \le u^N(\alpha)$ and there exists $\beta \in A$ such that $\beta \succ_{R^N} \alpha$. Let $P = \{j \in N \mid \beta\ R^j\ \alpha\}$ and let $y = u^P(\beta)$. Then (P, y) is a justified strong objection in the sense of $\mathcal{M}$ of any voter $j \in P$ against i so that $x \notin \mathcal{M}(N, V)$.

The second model involves replication. Let A be a fixed set of m alternatives, and let $L = L(A)$. Let $N = \{1, \ldots, n\}$, let $R^N \in L^N$, and let $u^N \in \mathcal{U}^{R^N}$. In order to replicate the simple majority voting game (N, V_{u^N}), let $k \in \mathbb{N}$ and denote

$$kN = \{(j, i) \mid i \in N,\ j = 1, \ldots, k\}.$$

Furthermore, let $R^{(j,i)} = R^i$ and $u^{(j,i)} = u^i$ for all $i \in N$ and $j = 1, \ldots, k$. Then $(kN, V_{u^{kN}})$ is the *k-fold replication* of (N, V_{u^N}).

Theorem 15.4.5.

$$\textit{If } k \ge \left\{\begin{array}{l} n+2\ ,\ \textit{if } n \textit{ is odd,} \\ \frac{n}{2}+2\ ,\ \textit{if } n \textit{ is even,} \end{array}\right\} \textit{ then } \mathcal{MB}(kN, V_{u^{kN}}) \ne \emptyset.$$

Proof: If α is a weak Condorcet winner with respect to R^N, then $u^{kN}(\alpha) \in \mathcal{MB}(kN, V_{u^{kN}})$. Hence we may assume that for every $\alpha \in A$ there exists $\beta(\alpha) \in A$ such that $\beta(\alpha) \succ_{R^N} \alpha$. Let $\widetilde{x} \in \mathbb{R}^N_+$ be any weakly Pareto optimal element in $V_{u^N}(N)$. We define $x \in \mathbb{R}^{kN}$ by $x^{(1,i)} = \widetilde{x}^i$ and $x^{(j,i)} = 0$ for all $i \in N$ and $j = 2, \dots, k$ and claim that $x \in \mathcal{MB}(kN, V_{u^{kN}})$. Let (P, y) be an objection at x. Then there exists $\alpha \in A$ such that $y = u^P(\alpha)$. Let $\beta = \beta(\alpha)$ and let $T = \{i \in N \mid \beta \; R^i \; \alpha\}$. Then

$$|T| \geq \begin{cases} \frac{n+1}{2} & \text{, if } n \text{ is odd,} \\ \frac{n}{2} + 1 & \text{, if } n \text{ is even.} \end{cases} \tag{15.4.3}$$

Let $Q = \{(j,i) \mid i \in T, j = 2, \dots, k\}$ and define $z \in \mathbb{R}^Q$ by $z^{(j,i)} = u^i(\beta)$ for all $i \in T$ and $j = 2, \dots, k$. Then $|Q| = (k-1)|T|$ and $z > (y^{P \cap Q}, x^{Q \setminus P})$. By (15.4.3), $|Q| \geq \frac{kn+1}{2}$. So, (Q, z) is a counter objection to (P, y). **q.e.d.**

By means of an example it may be shown that replication may not guarantee non-emptiness of the classical bargaining set. Indeed, consider the game of Section 15.3.1 with $n = 4$ and $m = 10$ alternatives.

By Theorem 15.3.3 the extended prebargaining set is empty. The proof proceeds by contradiction and it may be modified in order to show that $\mathcal{M}(kN, V_{u^{kN}}) = \emptyset$ for every k. Assume that $x \in \mathcal{M}(kN, V_{u^{kN}})$. Then $x^{kN} \leq u^{kN}(\alpha)$ for some $\alpha \in A$. As in the proof of Theorem 15.3.3 we may distinguish 10 cases, because $|A| = 10$. The justified objections may be replaced by their k-fold replications; e.g., with $S_4 = \{2, 3, 4\}$, $(S_4, u^{S_4}(a_4))$ is a strong objection at any imputation $x \leq u^N(a_1)$. The k-fold replication of this objection is a strong objection at $x^{kN} \leq u^{kN}(a_1)$ of any copy $(j, 4)$ of player 4 against any copy $(\ell, 1)$ of player 1 and the existence of a weak counter objection of $(\ell, 1)$ against $(j, 4)$ implies that $x^{(\ell,1)} \leq u^1(a_3)$. We may continue along the lines of the proof of Theorem 15.3.3 and show that $x^{kN} \ll u^{kN}(b)$ and all other cases lead to the same contradiction in a similar way.

15.5 Notes and Comments

The results of Subsection 15.1.2, Section 15.2, and Section 15.4 are due to Holzman, Peleg, and Sudhölter (2007), who also show that the Mas-Colell bargaining set of a simple majority voting game with an odd number of players may be empty.

Section 15.3 is based on Peleg and Sudhölter (2005), who provide a different proof of a statement that is slightly weaker than the statement of Theorem 15.3.9.

Finally we remark that there are well-known voting rules other than simple majority voting. Any of these rules, together with a profile of linear orders on the alternatives, leads to some strategic game so that other NTU voting games arise via α-effectiveness. The two voting games that are associated with plurality voting and approval voting are introduced and discussed by Peleg and Sudhölter (2004). A structural difference between simple majority voting games and plurality (or approval) voting games is that the latter may depend on the utility representation of R^N in the sense that two plurality (or approval) voting games associated with R^N may not be derived from each other by a monotone transformation. Hence, nonemptiness of a bargaining set may depend on the utility profile of R^N for voting games associated with plurality or approval voting, whereas the nonemptiness of a bargaining set may only depend on R^N for simple majority voting games (see Remark 15.1.3).

16

Variants of the Davis-Maschler Bargaining Set for NTU Games

In this chapter a further generalization of the classical bargaining set to NTU games with coalition structures, the ordinal bargaining set, is discussed as well as the ordinal generalizations of the (semi-)reactive bargaining set and a generalization of the prekernel to NTU games.

Section 16.1 is devoted to an existence theorem for the ordinal bargaining set. A theorem of Billera is the basic tool to show nonemptiness.

In Section 16.2 the theorem of Billera is proved.

Finally, several NTU extensions of the reactive bargaining set, of the semi-reactive bargaining set, and of the prekernel are studied in Section 16.3. Adequate modifications of the existence proof of the ordinal prebargaining set also imply nonemptiness of the ordinal (semi-)reactive prebargaining set. Furthermore, we introduce the bilateral consistent prekernel, a subsolution of the cyclic prekernel, and discuss several illuminating examples.

16.1 The Ordinal Bargaining Set $\mathcal{M}^o$

This section is devoted to the ordinal bargaining set, a modification of $\mathcal{M}$, which coincides with $\mathcal{M}$ on TU games and which allows us to generalize Theorem 4.2.12. In order to introduce $\mathcal{M}^o$, let $(N, V, \mathcal{R})$ be an NTU game with coalition structure and let $x \in X = X(N, V, \mathcal{R})$. If $R \in \mathcal{R}$ and $k, \ell \in R$, $k \neq \ell$, such that k has a justified objection against ℓ at x, then we write $k \succ_x^{\mathcal{M}} \ell$ (compare with Definition 4.2.1) or, simply, $k \succ_x \ell$. We denote by ${}^t\!\succ_x^{\mathcal{M}} = {}^t\!\succ_x$ the *transitive closure* of $\succ_x$. That is, $k \,{}^t\!\succ_x \ell$ iff there exist players $i_0, \ldots, i_p \in R$ such that $i_0 = k$, $i_p = \ell$, and $i_{j-1} \succ_x i_j$ for all $j = 1, \ldots, p$.

Definition 16.1.1 (Asscher (1976)). *The* **ordinal prebargaining set** *of* $(N, V, \mathcal{R})$, $\mathcal{PM}^o(N, V, \mathcal{R})$, *is defined by*

$$\mathcal{PM}^o(N,V,\mathcal{R}) = \{x \in X \mid \forall k,\ell \in R \in \mathcal{R}, k \neq \ell : k \succ_x \ell \Rightarrow \ell \;{}^t\!\!\succ_x k\}$$

and the **ordinal bargaining set** *of* $(N,V,\mathcal{R})$, $\mathcal{M}^o(N,V,\mathcal{R})$, *is*

$$\mathcal{M}^o(N,V,\mathcal{R}) = \mathcal{PM}^o(N,V,\mathcal{R}) \cap I(N,V,\mathcal{R}).$$

Furthermore, the ordinal bargaining set and the ordinal prebargaining set of the NTU game (N,V), $\mathcal{M}^o(N,V)$ *and* $\mathcal{PM}^o(N,V)$ *are the ordinal bargaining set and the ordinal prebargaining set of the NTU game with coalition structure* $(N,V,\{N\})$.

Remark 16.1.2. (1) $\mathcal{PM}$ is a subsolution of $\mathcal{PM}^o$.

(2) On any set of NTU games with coalition structures $\mathcal{PM}^o$ and $\mathcal{M}^o$ satisfy COV.

Remark 16.1.3. Definition 16.1.1 can be given in a completely symmetric form:

$$\mathcal{PM}^o(N,V,\mathcal{R}) = \{x \in X \mid \forall k,\ell \in R \in \mathcal{R}, k \neq \ell : k \;{}^t\!\!\succ_x \ell \Rightarrow \ell \;{}^t\!\!\succ_x k\}.$$

In the case of TU games with coalition structures the ordinal bargaining set coincides with the bargaining set. Indeed, the following result holds.

Lemma 16.1.4 (Billera (1970a)). *Let* (N,v) *be a TU game,* (N,V_v) *be the associated NTU game, and* $\mathcal{R}$ *be a coalition structure for* N. *Then* $\mathcal{PM}(N,v,\mathcal{R}) = \mathcal{PM}^o(N,V_v,\mathcal{R})$.

Proof: By Remarks 15.1.1 and 16.1.2 (1),

$$\mathcal{PM}(N,v,\mathcal{R}) \subseteq \mathcal{PM}^o(N,V_v,\mathcal{R}).$$

If $x \in \mathcal{PM}(N,V_v,\mathcal{R})$, then, by Lemma 4.2.9, $\succ_x^{\mathcal{M}}$ is acyclic. Hence, $\succ_x^{\mathcal{M}} = \emptyset$ and, consequently, $x \in \mathcal{PM}^o(N,v,\mathcal{R})$. **q.e.d.**

Now we state the main result of this section.

Theorem 16.1.5 (Asscher (1976)). *If* $(N,V,\mathcal{R})$ *is an NTU game with coalition structure such that* $I(N,V,\mathcal{R}) \neq \emptyset$, *then* $\mathcal{M}^o(N,V,\mathcal{R}) \neq \emptyset$.

The foregoing theorem is a direct consequence of the following theorem. If $(N,V,\mathcal{R})$ is an NTU game with coalition structure and if $\alpha \in \mathbb{R}^N$, denote

$$X_\alpha(N,V,\mathcal{R}) = X(N,V,\mathcal{R}) \cap (\alpha + \mathbb{R}^N_+).$$

Theorem 16.1.6. *Let* $(N,V,\mathcal{R})$ *be an NTU game with coalition structure, let* $\alpha \in \mathbb{R}^N$, *and denote* $X_\alpha = X_\alpha(N,V,\mathcal{R})$. *If* $\alpha \leq (v^i)_{i\in N}$ *and* $X_\alpha \neq \emptyset$, *then* $\mathcal{PM}^o(N,V,\mathcal{R}) \cap X_\alpha \neq \emptyset$.

Indeed, Theorem 16.1.5 is a special case of Theorem 16.1.6, because $I(N, V, \mathcal{R}) = X_{(v^i)_{i \in N}}$.

We postpone the proof of Theorem 16.1.6 and we shall first present a corollary and three lemmata.

Corollary 16.1.7. *If* $(N, V, \mathcal{R})$ *is an NTU game with coalition structure, then* $\mathcal{PM}^o(N, V, \mathcal{R}) \neq \emptyset$.

Throughout this section we shall assume that $(N, V, \mathcal{R})$ is an NTU game with coalition structure and that $\alpha \in \mathbb{R}^N$ satisfies $\alpha^i \leq v^i$ for all $i \in N$ and $X_\alpha = X_\alpha(N, V, \mathcal{R}) \neq \emptyset$.

Lemma 16.1.8. *Let* $k, \ell \in R \in \mathcal{R}$, $k \neq \ell$. *Then* $G = \{x \in X_\alpha \mid k \succ_x^{\mathcal{M}} \ell\}$ *is open in the relative topology of* X_α.

Proof: Let $x \in G$. Then there exists a justified objection (P, y) of k against ℓ at x. Then there exists a neighborhood U of x in X_α such that if $\widehat{x} \in U$, then (P, y) is an objection of k against ℓ at $\widehat{x}$. We claim that there exists an open subset U' of U such that $x \in U'$ and (P, y) is a justified objection of k against ℓ at every element of U'. Assume, on the contrary, that there exist sequences $(x_j)_{j \in \mathbb{N}}$ and $(Q_j, z_j)_{j \in \mathbb{N}}$ such that $x_j \in U$, (Q_j, z_j) is a counterobjection to (P, y) at x_j for all $j \in \mathbb{N}$, and $\lim_{j \to \infty} x_j = x$. By the finiteness of $\mathcal{T}_{\ell k}(N)$ we may assume without loss of generality that $Q_j = Q$ for all $j \in \mathbb{N}$. Moreover, by (11.3.3), we may assume that $\lim_{j \to \infty} z_j = z$. These assumptions imply that (Q, z) is a counterobjection to (P, y) at x, which yields a contradiction. **q.e.d.**

Definition 16.1.9. *An* $|N| \times |N|$ *matrix of functions* $(d_{k\ell}(\cdot))_{k,\ell \in N}$ *is a* **demand matrix** *if, for all* $k, \ell \in N$, $d_{k\ell} : X_\alpha \to \mathbb{R}_+$ *is a continuous mapping and* $d_{k\ell}(x) > 0$ *iff* $k, \ell \in R \in \mathcal{R}$, $k \neq \ell$, *and* $k \succ_x^{\mathcal{M}} \ell$.

Remark 16.1.10. There exists a demand matrix. Indeed, let $R \in \mathcal{R}$, $k, \ell \in R$, $k \neq \ell$, and let $X_{k\ell} = \{x \in X_\alpha \mid k \not\succ_x^{\mathcal{M}} \ell\}$. Choose $x \in X_\alpha$ such that $x^\ell = \alpha^\ell$. Then $x \in X_{k\ell}$, because $(\{l\}, v^\ell)$ is a counterobjection to any objection of k against ℓ at x. Hence, $X_{k\ell} \neq \emptyset$. By Lemma 16.1.8, $X_{k\ell}$ is closed. Define $d_{k\ell}(x) = \rho(x, X_{k\ell})$, which is the distance between x and $X_{k\ell}$ (see (4.2.1)). Then $d_{k\ell}(\cdot)$ is continuous. Also, if $k \in R$ and $\ell \in N \setminus R$ or $\ell = k$, then define $d_{k\ell}(\cdot) = 0$. Then $(d_{k\ell})_{k,\ell \in N}$ is a demand matrix.

Let $(d_{k\ell}(\cdot))_{k,\ell \in N}$ be a demand matrix. For all $x \in X_\alpha(N, V, \mathcal{R})$ we denote

$$A_k(x) = \sum_{\ell \in N} (d_{\ell k}(x) - d_{k\ell}(x)) \text{ for all } k \in N. \tag{16.1.1}$$

Lemma 16.1.11. *Let* $R \in \mathcal{R}$ *and let* $x \in X_\alpha$. *If, for all* $k \in R$, $A_k(x) \geq 0$ $(A_k(x) \leq 0)$, *then* $A_k(x) = 0$ *for all* $k \in R$.

Proof: By definition $A_k(x) = \sum_{\ell \in R}(d_{\ell k}(x) - d_{k\ell}(x))$. Hence,

$$0 \leq (\geq) \sum_{k \in R} A_k(x) = \sum_{k \in R} \sum_{\ell \in R} (d_{\ell k}(x) - d_{k\ell}(x)) = 0.$$

Thus, $A_k(x) = 0$ for all $k \in R$. **q.e.d.**

The proof of the following lemma is based on a theorem due to Billera (1970b). The next section is devoted to the proof of this theorem (Theorem 16.1.13).

Lemma 16.1.12. *There exists $\xi \in X_\alpha$ such that $A_k(\xi) = 0$ for all $k \in N$.*

Proof: Let $E_k = \{x \in X_\alpha \mid A_k(x) \leq 0\}$. As $A_k(\cdot)$ is continuous, E_k is closed. Also, if $x \in X_\alpha$ such that $x^k \leq v^k$, then $d_{\ell k}(x) = 0$ for all $\ell \in N$. Hence, $E_k \supseteq \{x \in X_\alpha \mid x^k \leq v^k\} \supseteq \{x \in X_\alpha \mid x^k = \alpha^k\}$. Furthermore, by Lemma 16.1.11, for each $x \in X_\alpha$ and each $R \in \mathcal{R}$ there exists $k \in R$ such that $A_k(x) \leq 0$. Hence, for each $R \in \mathcal{R}$, $\bigcup_{k \in R} E_k = X_\alpha$. By Theorem 16.1.13, $\bigcap_{k \in N} E_k \neq \emptyset$. By Lemma 16.1.11, if $\xi \in \bigcap_{k \in N} E_k$, then $A_k(\xi) = 0$ for all $k \in N$. **q.e.d.**

Proof of Theorem 16.1.6: Let $\xi \in X_\alpha$ satisfy $A_k(\xi) = 0$ for all $k \in N$. We claim that $\xi \in \mathcal{PM}^o(N, V, \mathcal{R})$. Assume, on the contrary, that there exist $R \in R$ and $k, \ell \in N$, such that $k \succ_\xi^{\mathcal{M}} \ell$ and $\ell \ {}^t\!\not\succ_\xi^{\mathcal{M}} k$. Let now $R^+ = \{\ell\} \cup \{i \in R \mid \ell \ {}^t\!\succ_\xi^{\mathcal{M}} i\}$ and $R^- = R \setminus R^+$. Clearly, $\ell \in R^+$ and $k \in R^-$. Also, if $i \in R^+$ and $j \in R^-$, then $i \not\succ_\xi^{\mathcal{M}} j$ and, therefore, $d_{ij}(\xi) = 0$. We now compute

$$\begin{aligned} 0 = \textstyle\sum_{i \in R^+} A_i(\xi) &= \textstyle\sum_{i \in R^+} \sum_{j \in R} (d_{ji}(\xi) - d_{ij}(\xi)) \\ &= \textstyle\sum_{i \in R^+} \sum_{j \in R^-} (d_{ji}(\xi) - d_{ij}(\xi)) \\ &= \textstyle\sum_{i \in R^+} \sum_{j \in R^-} d_{ji}(\xi) > 0. \end{aligned}$$

The inequality is valid, because $d_{k\ell}(\xi) > 0$. Hence, the desired contradiction has been obtained. **q.e.d.**

Theorem 16.1.13 (Billera (1970b)). *Let $(N, V, \mathcal{R})$ be an NTU game with coalition structure and let $\alpha \in \mathbb{R}^N$ satisfy $\alpha^R \in V(R)$ for all $R \in \mathcal{R}$. If E_i, $i \in N$, are subsets of $X_\alpha(N, V, \mathcal{R})$ such that*

(1) E_i is closed for every $i \in N$,

(2) $E_i \supseteq \{x \in X_\alpha(N, V, \mathcal{R}) \mid x^i = \alpha^i\}$ for all $i \in N$, and

(3) $\bigcup_{i \in R} E_i = X_\alpha(N, V, \mathcal{R})$ for all $R \in \mathcal{R}$,

then $\bigcap_{i \in N} E_i \neq \emptyset$.

The proof of Theorem 16.1.13 is postponed to Section 16.2. We conclude this section by an example.

Example 15.1.2 (continued). We claim that

$$\mathcal{PM}^o(N,V,\mathcal{R}) = \left\{ x \in \mathbb{R}^N \,\middle|\, \begin{array}{l} x^1 + x^2 = 1, x^1 > 0, \\ x^2 > 0, x^3 = x^4 = 0 \end{array} \right\}. \tag{16.1.2}$$

The proof of (16.1.2) is left as Exercise 16.1.1.

Exercises

Exercise 16.1.1. Prove (16.1.2).

Exercise 16.1.2. Find an example of an NTU game whose ordinal prebargaining set does not contain any Pareto optimal element.

Exercise 16.1.3. Let $N = \{1, \ldots, 5\}$, let

$$S_1 = \{1,2,4,5\}, S_2 = \{3,4,5\}, S_3 = \{1,4,5\}, S_4 = \{1,2,3\}, S_5 = \{4,5\},$$

and let (N,V) be defined by $V(S_1) = \{x \in \mathbb{R}^{S_1} \mid 3x^4 + x(S_1) \leq 16\}$, $V(S_2) = \{x \in \mathbb{R}^{S_2} \mid 3x^5 + x(S_2) \leq 14\}$, $V(S_3) = \{x \in \mathbb{R}^{S_3} \mid x(S_3) \leq 6\}$, $V(N) = \{x \in \mathbb{R}^N \mid x(N) \leq 5\}$, $V(S_j) = \{x \in \mathbb{R}^{S_j} \mid x(S_j) \leq 15\}$, $j = 4,5$, and $V(S) = \{x \in \mathbb{R}^S \mid x(S) \leq 0\}$ for all other $\emptyset \neq S \subseteq N$. Show that $(1, \ldots, 1) \in \mathcal{M}^o(N,V)$.

Exercise 16.1.4. Let (N,V) be the NTU game defined in Exercise 16.1.3. Deduce that $\mathcal{PM}^o$ does not satisfy RGP on any set of non-levelled NTU games with coalition structures that contain $(N,V,\{N\})$.

16.2 A Proof of Billera's Theorem

In order to prove Theorem 16.1.13 let $(N,V,\mathcal{R})$ be an NTU game with coalition structure and $\alpha \in \mathbb{R}^N$ such that $\alpha^R \in V(R)$ for all $R \in \mathcal{R}$. Without loss of generality we shall assume throughout this section that $\alpha = 0 \in \mathbb{R}^N$ (otherwise replace $V(S)$ by $V(S) + \alpha^S$, $S \subseteq N$, and α by 0).

Note that under our assumption

$$X_\alpha = X_0(N,V,\mathcal{R}) = \prod_{R \in \mathcal{R}} \left(V(R)_e \cap \mathbb{R}^R_+ \right).$$

Let E^i, $i \in N$, satisfy the conditions of Theorem 16.1.13. It suffices to construct closed subsets $W(R)$ of $V(R)_e \cap \mathbb{R}^R_+$, $R \in \mathcal{R}$, such that

$$\bigcap_{i \in N} \left(E_i \cap \prod_{R \in \mathcal{R}} W(R) \right) \neq \emptyset. \tag{16.2.1}$$

To this end let $R \in \mathcal{R}$ and denote

$$P(R) = \{x^R \in V(R) \mid x^R \gg 0^R\}.$$

If $P(R) = \emptyset$, then define $W(R) = \{0^R\}$. If $P(R) \neq \emptyset$, then let $W(R) = (\overline{P(R)})_e$, where $\overline{P(R)}$ denotes the closure of $P(R)$.

By definition, $W(R)$ is closed for every $R \in \mathcal{R}$. We shall now prove that $W(R) \subseteq V(R)_e \cap \mathbb{R}^R_+$.

Lemma 16.2.1. *Let $R \in \mathcal{R}$. Then $W(R) \subseteq V(R)_e \cap \mathbb{R}^R_+$.*

Proof: If $P(R) = \emptyset$, then $0^R \in V(R)$, because $X_0 \neq \emptyset$. Hence, $W(R) = \{0^R\} \subseteq V(R)_e$. Thus, assume that $P(R) \neq \emptyset$. If $x^R \in V(R)$, $x^R \geq 0^R$, and $x^R \notin V(R)_e$, then there exists $y^R \in V(R)$ such that $y^R \gg x^R \geq 0$. Hence $y^R \in P(R)$ and $x^R \notin W(R)$. **q.e.d.**

The next three lemmata imply (16.2.1).

Lemma 16.2.2. *Let $R \in \mathcal{R}$ such that $P(R) \neq \emptyset$. Then $x^R \in W(R)$ iff $x^R \in \overline{P(R)}$ and there does not exist $y^R \in \overline{P(R)}$ such that $y^R > x^R$ and $x^i = 0$ for every $i \in \{j \in R \mid x^j = y^j\}$.*

Proof: If $x^R \in \overline{P(R)}$ satisfies the conditions of the lemma, then, clearly, $x^R \in W(R)$. Assume now that $x^R \in \overline{P(R)}$ and there exists $y^R \in \overline{P(R)}$ such that $y^R > x^R$ and $x^i = 0$ for all $i \in R$ such that $x^i = y^i$. Let

$$R^+ = \{i \in R \mid y^i > x^i\}.$$

As $y^R > x^R$, $R^+ \neq \emptyset$. Let $\varepsilon = \min_{i \in R^+}(y^i - x^i)$ and let

$$B(y^R, \varepsilon) = \{z^R \in \mathbb{R}^R \mid |z^i - y^i| \leq \varepsilon \; \forall i \in R\}.$$

As $y^R \in \overline{P(R)}$, there exists $z^R \in B(y^R, \varepsilon) \cap P(R)$. For each $i \in R \setminus R^+$, $z^i > 0 = x^i$. For each $i \in R^+$, $z^i - x^i = z^i - y^i + y^i - x^i > -\varepsilon + \varepsilon = 0$. Hence $z^R \gg x^R$, which proves that $x^R \notin W(R)$. **q.e.d.**

Let H_R, $R \in \mathcal{R}$, be defined by

$$H_R = \begin{cases} \{x \in \mathbb{R}^R_+ \mid x(S) = 1\} & \text{, if } P(R) \neq \emptyset, \\ \{0^R\} & \text{, if } P(R) = \emptyset. \end{cases}$$

Lemma 16.2.3. *Let $R \in \mathcal{R}$. Then there exists a continuous function $d_R : H_R \to \mathbb{R}_{++}$ such that $d_R(h)h \in W(R)$ for all $h \in H_R$ and such that $\varphi_R : H_R \to W(R)$, $h \mapsto d_R(h)h$, is a homeomorphism.*

Proof: If $W(R) = \{0\}$, let $d_R(0^R) = 1$. Assume now that $P(R) \neq \emptyset$. Then $0^R \notin W(R)$. Hence, for any $x \in W(R), \vartheta_R(x) = \frac{x}{x(R)} \in H_R$ so that $\vartheta_R : W(R) \to H_R$ is continuous.

We now show that ϑ_R is injective. Assume, on the contrary, that there exist $x, y \in W(R)$ such that $x \neq y$ and $x/x(R) = y/y(R)$. Without loss of generality we may assume that $y(R) > x(R)$. Therefore, $y > x$ and $x^i = 0$, whenever $x^i = y^i$. Hence $x \notin W(R)$ by Lemma 16.2.2, which is the desired contradiction.

We now show that ϑ_R is surjective. Let $h \in H_R$ and let

$$r(h) = \{th \mid t \in \mathbb{R}_{++}\}.$$

Then $r(h) \cap \overline{P(R)} \neq \emptyset$, because $\overline{P(R)} \neq \emptyset$. Let x be the vector with the maximal Euclidean norm in $r(h) \cap \overline{P(R)}$. We claim that $x \in W(R)$. Assume, on the contrary, that there exists $z \in \overline{P(R)}$ such that $z \gg x$. By comprehensiveness, $\{u \in \mathbb{R}^R_+ \mid u \leq z\} \subseteq \overline{P(R)}$. Let

$$t = \min\left\{ \frac{z^i}{x^i} \,\middle|\, i \in R, x^i > 0 \right\}.$$

Then $t > 1$ and $0^R \leq tx \leq z$. Hence, $tx \in \overline{P(R)}$. Also, $tx \in r(h)$, because $x \in r(h)$. In view of the fact that $\|tx\| = t\|x\| > \|x\|$, we have reached the desired contradiction.

As $W(R)$ is compact, ϑ_R is a homeomorphism. Define now $\varphi_R = \vartheta_R^{-1}$. Clearly, $\varphi_R(h) = d_R(h)h$ for some suitable function d_R which satisfies $d_R(h) > 0$ for all $h \in H_R$ (because $0 \notin W(R)$). Now, $d_R(h)$ can be expressed as

$$d_R(h) = d_R(h)\frac{\|h\|}{\|h\|} = \frac{\|d_R(h)h\|}{\|h\|} = \frac{\|\varphi_R(h)\|}{\|h\|};$$

hence it is continuous. **q.e.d.**

Let φ_R, $R \in \mathcal{R}$, be homeomorphisms that satisfy the conditions of Lemma 16.2.3. Define $\vartheta : \prod_{R\in\mathcal{R}} W(R) \to \prod_{R\in\mathcal{R}} H_R$ by $\vartheta = \prod_{R\in\mathcal{R}} \varphi_R^{-1}$. Let $\widetilde{E}_i = \vartheta(\widehat{E}_i)$ for all $i \in N$, where $\widehat{E}_i = E_i \cap \prod_{R\in\mathcal{R}} W(R)$. Then $\widetilde{E}_i$ is closed and $\widetilde{E}_i \supseteq \{x \in H(\mathcal{R}) \mid x^i = 0\}$ for all $i \in N$, where $H(\mathcal{R}) = \prod_{R\in\mathcal{R}} H_R$. Also, for each $R \in \mathcal{R}$, $\bigcup_{i\in R} \widetilde{E}_i = H(\mathcal{R})$. Therefore, in order to show (16.2.1), it suffices to show the following lemma.

Lemma 16.2.4 (Peleg (1967a)). *If E_i, $i \in N$, are subsets of $H(\mathcal{R})$ such that*

(1) E_i is closed for every $i \in N$,

(2) $E_i \supseteq \{x \in H(\mathcal{R}) \mid x^i = 0\}$ for all $i \in N$, and

(3) $\bigcup_{i\in R} E_i = H(\mathcal{R})$ for all $R \in \mathcal{R}$,

then $\bigcap_{i\in N} E_i \neq \emptyset$.

Proof: Let (N, v) be the TU game that is defined by $v(S) = 0$ for every $S \in 2^N \setminus \mathcal{R}$ and, for every $R \in \mathcal{R}$, by

$$v(R) = \begin{cases} 1 \text{ , if } P(R) \neq \emptyset, \\ 0 \text{ , if } P(R) = \emptyset. \end{cases}$$

Then $I(N, v, \mathcal{R}) = H(\mathcal{R})$. Hence the proof is completed by applying Lemma 4.2.11. **q.e.d.**

Exercises

Exercise 16.2.1. Show that $\mathcal{PM}^o$ is an ordinal solution on any set of NTU games with coalition structures.

Exercise 16.2.2. Show that the core is an ordinal solution on every set of NTU games with coalition structures.

16.3 Solutions Related to $\mathcal{M}^o$

For classes of TU games several bargaining sets are of interest (see Chapter 4). Moreover, the kernel and the prekernel may be regarded as close relatives of the (unrestricted) bargaining set. In the present section we shall discuss some possible extensions of the reactive bargaining set, the semi-reactive bargaining set, and the kernel for NTU games with coalition structures.

16.3.1 The Ordinal Reactive and the Ordinal Semi-Reactive Bargaining Sets

As we have already defined objections and counterobjections (in Section 15.1.1), we may now define the reactive and the semi-reactive prebargaining sets of an NTU game $(N, V, \mathcal{R})$ as in Definitions 4.4.2 and 4.4.4. Just the TU coalition function v has to be replaced by the NTU coalition function V wherever it occurs.

Clearly, for any NTU game $(N, V, \mathcal{R})$ with coalition structure,

$$\mathcal{PM}_r(N, V, \mathcal{R}) \subseteq \mathcal{PM}_{sr}(N, V, \mathcal{R}) \subseteq \mathcal{PM}(N, V, \mathcal{R}).$$

Hence, all these sets are empty for the game of Example 15.1.2. This fact motivates us to define the *ordinal* variants of the reactive bargaining set and of the semi-reactive bargaining set ($\mathcal{PM}_r^o(\cdot,\cdot,\cdot)$, $\mathcal{M}_r^o(\cdot,\cdot,\cdot)$ and $\mathcal{PM}_{sr}^o(\cdot,\cdot,\cdot)$, $\mathcal{M}_{sr}^o(\cdot,\cdot,\cdot)$), by literally copying Definition 16.1.1. Just the binary relation $\succ_x$ has to be defined by $\succ_x \; = \; \succ_x^{\mathcal{M}_r}$ or $\succ_x \; = \; \succ_x^{\mathcal{M}_{sr}}$, respectively. Note that ${}^t\!\succ_x$ remains the transitive closure of $\succ_x$.

Remark 16.3.1. Let $(N, v, \mathcal{R})$ be a TU game with coalition structure. Then the relations $\succ^{\mathcal{M}_r}$ and $\succ^{\mathcal{M}_{sr}}$ are subsets of $\succ_x^{\mathcal{K}}$. Hence Lemma 4.2.9 implies that

$$\mathcal{PM}_r^o(M, V_v, \mathcal{R}) = \mathcal{PM}_r(N, v, \mathcal{R}), \mathcal{PM}_{sr}^o(M, V_v, \mathcal{R}) = \mathcal{PM}_{sr}(N, v, \mathcal{R}).$$

We shall now prove the existence result for the ordinal reactive bargaining set and for the ordinal semi-reactive bargaining set.

Theorem 16.3.2. *Let $(N, V, \mathcal{R})$ be an NTU game with coalition structure, let $\alpha \in \mathbb{R}^N$, and denote $X_\alpha = X_\alpha(N, V, \mathcal{R})$. If $\alpha \leq (v^i)_{i \in N}$ and $X_\alpha \neq \emptyset$, then $\mathcal{PM}_r^o(N, V, \mathcal{R}) \cap X_\alpha \neq \emptyset$ and $\mathcal{PM}_{sr}^o(N, V, \mathcal{R}) \cap X_\alpha \neq \emptyset$.*

Proof: A careful inspection of the proof of theorem 16.1.6 shows that our theorem is shown as soon as the following variants of Lemma 16.1.8 and of Remark 16.1.10 are proved:

(1) If $k, \ell \in R \in \mathcal{R}, k \neq \ell$, then both

$$\{x \in X_\alpha \mid k \succ_x^{\mathcal{M}_r} \ell\} \text{ and } \{x \in X_\alpha \mid k \succ_x^{\mathcal{M}_{sr}} \ell\}$$

are open in the relative topology of X_α.

(2) There exist a demand matrix with respect to $\mathcal{M}_r$ and a demand matrix with respect to $\mathcal{M}_{sr}$. (Here a demand matrix *with respect to* $\mathcal{M}_r$ or $\mathcal{M}_{sr}$ is defined as in Definition 16.1.9; just $\succ_x^{\mathcal{M}}$ has to be replaced by $\succ_x^{\mathcal{M}_r}$ or $\succ_x^{\mathcal{M}_{sr}}$ respectively.)

The proof of (1) is similar to the proof of Lemma 16.1.8 and left as an exercise (Exercise 16.3.4). Moreover, analogously as in Remark 16.1.10, a demand matrix with respect to $\mathcal{M}_r$ or with respect to $\mathcal{M}_{sr}$ can be constructed. **q.e.d.**

Let $(N, V, \mathcal{R})$ be an NTU game with coalition structure. By Theorem 16.3.2, if $I(N, V, \mathcal{R}) \neq \emptyset$, then $\mathcal{M}_r^o(N, V, \mathcal{R}) \neq \emptyset$. This fact was already mentioned by Granot and Maschler (1997). The ordinal semi-reactive (pre)bargaining set was mentioned in Sudhölter and Potters (2001).

The inclusions of (4.4.5) are no longer valid in the NTU case. Indeed, by means of examples (see Exercise 16.3.5) it may be shown that neither the ordinal reactive (pre)bargaining set is a subsolution of the ordinal semi-reactive (pre)bargaining set or of the ordinal (pre)bargaining set, nor the ordinal semi-reactive (pre)bargaining set is a subsolution of the ordinal (pre)bargaining set.

16.3.2 Solutions Related to the Prekernel

In order to generalize the prekernel to NTU games with coalition structures, it is useful to define the individual excesses of a coalition at a payoff vector.

We shall first recall the definition of excess with respect to a TU game. Let (N, v) be a TU game, $x \in \mathbb{R}^N$, and $\emptyset \neq S \subseteq N$. Then, for all $k \in S$,

$$e(S, x, v) = v(S) - x(S) = \max\{t^k \in \mathbb{R} \mid (x^k + t^k, x^{S \setminus \{k\}}) \in V_v(S)\}.$$

Hence, if (N, V) is an NTU game and if $k \in S$, then the *individual excess* of k in S at x, $e^k(S, x, V)$, may be defined by

$$e^k(S, x, V) = \max\{t^k \in \mathbb{R} \mid (x^k + t^k, x^{S \setminus \{k\}}) \in V(S)\}$$

$(\max \emptyset = -\infty)$. By (11.3.1), (11.3.3), and (11.3.4), $e^k(S, x, V)$ is well defined and $e^k(S, x, V) < \infty$. Now, if $k, \ell \in N$, $k \neq \ell$, then the *maximum surplus* of k against ℓ at x, $s_{k\ell}(x, V)$, is given by

$$s_{k\ell}(x, V) = \max_{S \in \mathcal{T}_{k\ell}} e^k(S, x, V)$$

(see Definition 4.2.5).

Remark 16.3.3. Let $k \in S \subseteq N$ and assume that $V(S)$ is non-levelled. Then the function $g = e^k(S, \cdot, V) : \mathbb{R}^N \to \mathbb{R} \cup \{-\infty\}$ is continuous, that is, both g and $-g$ are lower semicontinuous (see Section 10.1 for the definition of lsc).

The straightforward proof of the foregoing remark is left to the reader (see Exercise 16.3.6).

Remark 16.3.4. For all $k, \ell \in N$, $k \neq \ell$,

$$s_{k\ell}(x, V) \geq e^k(\{k\}, x, V) = v^k - x^k;$$

thus $s_{k\ell}(x, V) \in \mathbb{R}$. Assume now that (N, V) is non-levelled. Then, by Remark 16.3.3, $s_{k\ell}(\cdot, V) : \mathbb{R}^N \to \mathbb{R}$ is a continuous function.

Throughout this section we restrict our attention to NTU games with coalition structures $(N, V, \mathcal{R})$) such that $V(R)$, $R \in \mathcal{R}$, is p-smooth (see Definition 14.2.1). For each $x \in \partial V(R)$ let $\lambda_x \in \Delta_+(R)$ be the normal vector. Note that $X = X(N, V, \mathcal{R}) = \prod_{R \in \mathcal{R}} \partial V(R)$. The *prekernel* of $(N, V, \mathcal{R})$, $\mathcal{PK}(N, V, \mathcal{R})$, is the set

$$\begin{aligned}&\mathcal{PK}(N, V, \mathcal{R}) \\ &= \{x \in X \mid \lambda_x^k s_{k\ell}(x, V) = \lambda_x^\ell s_{\ell k}(x, V)\ \forall \{k, \ell\} \in \mathcal{P}(R)\ \forall R \in \mathcal{R}\}.\end{aligned}$$

Hence, if $x \in X$ and the surpluses of partners k, ℓ (see Section 3.8) in $\mathcal{R}$ at x, weighted by their corresponding local transfer rates of utility, are "balanced" (equal), then x belongs to the prekernel of the NTU game with coalition structure.

Clearly, $\mathcal{PK}(N, V_v, \mathcal{R}) = \mathcal{PK}(N, v, \mathcal{R})$ for every TU game with coalition structure $(N, v, \mathcal{R})$. In contrast to the case of TU games, the prekernel may not be contained even in one of the ordinal bargaining sets (see the following example).

Example 16.3.5. Let $N = \{1,2,3\}$ and let (N, v) be the TU game that satisfies $v(\{1,j\}) = 5$, $j = 2,3$, $v(\{2,3\}) = 3$, and $v(S) = 0$ for all other coalitions. Let V be defined by $V(S) = V_v(S)$ for all $S \subsetneqq N$ and $V(N) = \{x \in \mathbb{R}^N \mid (1,3,3) \cdot x \leq 7\}$. Then $x = (1,1,1)$ is (the unique) element of $\mathcal{PK}(N, V)$. Also, $(\{1,2\}, (2,3))$ is a justified objection of 1 against 3 at x and every objection of 3 or of 2 against 1 can be countered via $(\{1,2\}, (2,3))$ or $(\{1,3\}, (2,3))$, respectively. Hence, x does not belong to any of the ordinal bargaining sets mentioned in Section 16.1 and Subsection 16.3.1.

Let $(N, V, \mathcal{R})$ be an NTU game with coalition structure, let $S \in 2^N \setminus \{\emptyset\}$, let $x \in X(N, V, \mathcal{R})$, and let $R \in \mathcal{R}$ such that $R' = R \cap S \neq \emptyset$. If $V(R)$ is p-smooth, then $V^{\mathcal{R}}_{S,x}(R')$ is p-smooth. Indeed, if $y \in \partial V^{\mathcal{R}}_{S,x}(R')$ and $z = (y, x^{R \setminus S})$, then $z \in \partial V(R)$. Thus, if λ_z is the normal to $V(R)$ at z, then $\lambda_z^{R'}/\lambda_z(R')$ is the normal to $V^{\mathcal{R}}_{S,x}(R')$ at y.

Remark 16.3.6. Let $\mathcal{U}$ be a set of players and let $\widehat{\Delta}$ be the set of games with coalition structures $(N, V, \mathcal{R})$ such that $N \subseteq \mathcal{U}$ and such that $V(R)$ is p-smooth for all $R \in \mathcal{R}$. Then $\mathcal{PK}$ satisfies RGP on $\widehat{\Delta}$.

Example 16.3.7. Let $N = \{1,2,3\}$ and denote by G the subgroup of permutations of N generated by the "cyclic" permutation defined by $\widehat{\pi}$, $\widehat{\pi}(1) = 2$, $\widehat{\pi}(2) = 3$, and $\widehat{\pi}(3) = 1$. Let $\widehat{S} = \{1,2\}$, let $\lambda_{\widehat{S}} = (1,2) \in \mathbb{R}^{\widehat{S}}$, and let (N, V) be the NTU game given by $V(\widehat{S}) = \{x \in \mathbb{R}^{\widehat{S}} \mid \lambda_{\widehat{S}} \cdot x \leq 2\}$, $V(\pi \widehat{S}) = \pi V(\widehat{S})$ for all $\pi \in G$, and $V(S) = \{x \in \mathbb{R}^S \mid x(S) \leq 0\}$ for all other coalitions. Then G is contained in (in fact coincides with) the symmetry group of (N, V). Hence, the NTU game is transitive (it has a transitive symmetry group).

Claim: $\mathcal{PK}(N, V) = \emptyset$.

In order to prove our claim, we assume, on the contrary, that there exists $x \in \mathcal{PK}(N, V)$. Then $x^3 = -x^1 - x^2$ by Pareto optimality (PO) of $\mathcal{PK}$. As $\mathcal{PK}$ satisfies anonymity (AN), we may assume that (a) $x^1 \leq 0$, $x^2 \leq 0$, and $x^3 \geq 0$, or (b) $x^1 \geq 0$, $x^2 \geq 0$, and $x^3 \leq 0$. Let $s_{k\ell} = s_{k\ell}(x, V)$ for all $k, \ell \in N$, $k \neq \ell$. In Case (a), s_{13} is attained by $\{1,2\}$ and s_{31} is attained by $\{3,2\}$; hence $s_{13} = 2 - 2x^2 - x^1 > \frac{2-x^2}{2} \geq \frac{2-x^2-2x^3}{2} = s_{31}$ implies the desired contradiction. In Case (b), s_{12} is attained by $\{1,3\}$ and s_{21} is attained by $\{2,3\}$; hence

$$s_{21} = 2 - x^2 - 2x^3 = 2 + x^2 + 2x^1 > \frac{2 - x^1 + x^2}{2} = \frac{2 - 2x^1 - x^3}{2} = s_{12}$$

implies the desired contradiction.

Example 16.3.7 motivates us to define the cyclic prekernel. Let $(N, V, \mathcal{R})$ be an NTU game with coalition structure such that $V(R)$ is p-smooth for all $R \in \mathcal{R}$. Furthermore, let λ_z denote the normal to $V(R)$ at $z \in \partial V(R)$. If $x \in X(N, V, \mathcal{R}) = X$, then we define $\succ^{\mathcal{K}}_x = \ \succ_x$ by $k \succ_x \ell$ if there exists $R \in \mathcal{R}$ such that $k, \ell \in R$, $k \neq \ell$, and, with $z = x^R$, $\lambda_z^k s_{k\ell}(x, V) > \lambda_z^\ell s_{\ell k}(x, V)$.

Also, let ${}^t\succ_x$ denote the transitive closure of $\succ_x$. Then the *cyclic prekernel* of $(N, V, \mathcal{R})$, $\mathcal{PK}^c(N, V, \mathcal{R})$, is given by

$$\mathcal{PK}^c(N, V, \mathcal{R}) = \{x \in X \mid \forall k, \ell \in R \in \mathcal{R}, k \neq \ell : k \succ_x \ell \Rightarrow \ell \;{}^t\!\succ_x k\}$$

(see Definition 16.1.1: Just the binary relation is suitably chosen for the prekernel).

The expression "cyclic" prekernel may be motivated as follows. Let $x \in X(N, V, \mathcal{R})$ and consider the directed graph whose vertices are the elements of N such that (k, ℓ), $k \neq \ell$, is an edge, if $k, \ell \in R \in \mathcal{R}$ and $k \succ_x^{\mathcal{K}} \ell$. Then $x \in \mathcal{PK}^c(N, V, \mathcal{R})$ iff every vertex either is an isolated point or belongs to a cycle of the graph.

Let $(N, v, \mathcal{R})$ be a TU game with coalition structure. By Lemma 4.2.7, $\succ_x^{\mathcal{K}}$ is acyclic for every $x \in X(N, v, \mathcal{R})$; hence

$$\mathcal{PK}(N, v, \mathcal{R}) = \mathcal{PK}^c(N, V_v, \mathcal{R}).$$

Example 16.3.7 (continued). Let $x = 0 \in \mathbb{R}^N$. Then $s_{12} = s_{23} = s_{31} = 1$ and $s_{21} = s_{32} = s_{13} = 2$; hence $1 \succ_x 3 \succ_x 2 \succ_x 1$ and $x \in \mathcal{PK}^c(N, V)$.

Let $d_{k\ell} : X \to \mathbb{R}$ be defined by $d_{k\ell}(x) = \lambda_z^k s_{k\ell}(x, V) - \lambda_z^\ell s_{\ell k}(x, V)$ if $k, \ell \in R \in \mathcal{R}$, $k \neq \ell$, where $z = x^R$, and $d_{k\ell}(x) = 0$ otherwise. Then $\left((d_{k\ell}(\cdot))_+\right)_{k,\ell \in N}$ is a demand matrix with respect to $\mathcal{PK}$. For $k \in N$, let $A_k^{\mathcal{K}} : X \to \mathbb{R}$ be defined by $A_k^{\mathcal{K}}(x) = \sum_{\ell \in N} d_{\ell k}(x)$. Clearly, if $A_k^{\mathcal{K}}(x) = 0$ for all $k \in N$, then $x \in \mathcal{PK}^c(N, V, \mathcal{R})$.

Definition 16.3.8. *Let $(N, V, \mathcal{R})$ be an NTU game with coalition structure such that $V(R)$ is p-smooth for all $R \in \mathcal{R}$. The* **bilateral consistent prekernel** *of $(N, V, \mathcal{R})$ is the set*

$$\mathcal{BCPK}(N, V, \mathcal{R}) = \{x \in X(N, V, \mathcal{R}) \mid A_k^{\mathcal{K}}(x) = 0 \; \forall \; k \in N\}.$$

Example 16.3.7 (continued). Note that $0 \in \mathcal{BCPK}(N, V)$.

The method by which to prove nonemptiness of the ordinal bargaining sets cannot be adopted to show that the cyclic prekernel is nonempty. In fact, it may not be possible to find $\alpha \in \mathbb{R}^N$ such that

$$X_\alpha = (\alpha + \mathbb{R}_+^N) \cap X \neq \emptyset$$

and $E_k = \{x \in X_\alpha \mid A_k^{\mathcal{K}}(x) \leq 0\} \supseteq \{x \in X_\alpha \mid x^k = \alpha^k\}$ for all $k \in N$. The following example shows that $\mathcal{PK}^c(N, V, \mathcal{R})$ may be empty.

Example 16.3.9. Let (N, V) be given by $N = \{1, 2, 3\}$ and

$$\begin{aligned} V(N) \quad &= \{x \in \mathbb{R}^N \mid x(N) \leq 1\}, \; V(\{k\}) = \{x \in \mathbb{R}^{\{k\}} \mid x^k \leq 0\}, \\ V(\{2,3\}) &= \{x \in \mathbb{R}^{\{2,3\}} \mid x^2 + x^3 \leq 1\}, \text{ and} \\ V(\{1,j\}) &= \{x \in \mathbb{R}^{\{1,j\}} \mid x^1 + 5x^j \leq 5\} \text{ for all } j = 2, 3. \end{aligned}$$

The straightforward proof that $\mathcal{PK}^c(N, V) = \emptyset$ is left as Exercise 16.3.7.

Clearly, the cyclic prekernel and the bilateral consistent prekernel satisfy COV and AN. Furthermore, let (N, V) be an NTU game with p-smooth $V(N)$. Then there exists $x = (t, \ldots, t) \in \partial V(N)$. If (N, V) is a transitive NTU game (an NTU game with a transitive symmetry group), then $x \in \mathcal{BCPK}(N, V)$.

Remark 16.3.10. Definition 16.3.8 and Example 16.3.9 are due to Orshan and Zarzuelo (2000), who consider only the case $\mathcal{R} = \{N\}$. Moreover, they show that the bilateral consistent prekernel exists for NTU games that satisfy certain additional technical properties.

Remark 16.3.11. Let $\mathcal{U}$ be a set of players and let $\widehat{\Delta}$ be the set of NTU games (N, V) such that $N \subseteq \mathcal{U}$ and $V(N)$ is a hyperplane (see (13.1.10) applied to N). Clearly, reduced games of games in $\widehat{\Delta}$ belong to $\widehat{\Delta}$. The bilateral consistent prekernel on $\widehat{\Delta}$ satisfies *bilateral consistency*: If $(N, V) \in \widehat{\Delta}$, $|N| \geq 2$, and $x \in \mathcal{BCPK}(N, V)$, then $\mathcal{BCPK}(T, V_{T,x})$ is a singleton $\{y_T\}$ for all $T \subseteq N$ such that $|T| = 2$ and for all $i \in N$,

$$\sum_{T \subseteq N : |T| = 2, T \ni i} y_T^i = (n-1)x^i.$$

The interpretation of bilateral consistency is similar to the interpretation of 2-consistency (see (14.1.3)). Just the σ-reduced games have to be replaced by the reduced games with respect to x.

Bilateral consistency may be used to axiomatize $\mathcal{BCPK}$. For a proof and discussion of bilateral consistency and for the axiomatization see Orshan and Zarzuelo (2000).

Exercises

Exercise 16.3.1. Prove that $\mathcal{PM}_r$, $\mathcal{PM}_r^o$, $\mathcal{PM}_{sr}$, and $\mathcal{PM}_{sr}^o$ are ordinal solutions.

Exercise 16.3.2. Let $N = \{1, \ldots, 6\}$, let

$$S_1 = \{1,3,4,5\}, S_2 = \{1,3,4,6\}, S_3 = \{2,3,6\}, S_4 = \{2,4,5\},$$

and let the TU game (N, v) be defined by $v(S_i) = 2$, $v(\{1,2\}) = 5$, $v(N) = 1$, and $v(S) = 0$ for all other subsets S of N. Let V be the NTU coalition function which differs from V_v only inasmuch as

$$V(S_1) = \{x \in \mathbb{R}^{S_1} \mid x^3 + x(S_1) \leq 2\}, V(S_2) = \{x \in \mathbb{R}^{S_2} \mid x^4 + x(S_2) \leq 2\}.$$

Prove that $(0,1,0,0,0,0) \in \mathcal{PM}_{sr}(N, V)$ and deduce that $\mathcal{PM}_{sr}$ does not satisfy RGP in general (consider the reduced game with players $1, \ldots, 4$).

Exercise 16.3.3. Let $\widehat{\Delta}$ be defined as in Exercise 15.1.4. Show that, on $\widehat{\Delta}$, $\mathcal{PM}_r$ satisfies RGP and that neither $\mathcal{PM}_r^o$ nor $\mathcal{PM}_{sr}^o$ satisfies RGP, provided that $|\mathcal{U}| \geq 5$ (see Exercise 16.1.4).

Exercise 16.3.4. Prove (1) of the proof of Theorem 16.3.2.

Exercise 16.3.5. Let $N = \{1, \ldots, 4\}$, let the TU game (N, v) be defined by

$$v(\{1,2\}) = 10, v(\{2,3\}) = v(\{2,4\}) = 5, v(\{1,3,4\}) = 8, v(N) = 2,$$

and $v(S) = 0$ for all other coalitions. Let V be the NTU coalition function that differs from V_v only inasmuch as

$$V(\{2,3,4\}) = \{x^{\{2,3,4\}} \mid x^2 \leq 2, x^3 \leq 4, x^4 \leq 4\}.$$

Furthermore, let $S_1 = \{1,3\}, S_2 = \{1,3,4\}, S_3 = \{2,3\}, S_4 = \{2,3,4\}$, let $x_1 = (2,0,4,0), x_2 = (2,0,2,2), x_3 = (0,1,3,0), x_4 = (0,2,5,1)$, and let (N, W) be given by $W(\{2,4\}) = -\mathbb{R}_+^{\{2,4\}}$, $W(S_i) = x^{S_i} - \mathbb{R}_+^{S_i}$ for $i = 1, \ldots, 4$, and $W(S) = V_v(S)$, otherwise. Show that

$$(1,1,0,0) \in \\ (\mathcal{M}_r^o(N,V) \cap \mathcal{M}_{sr}^o(N,V) \cap \mathcal{M}_r^o(N,W)) \setminus (\mathcal{M}^o(N,V) \cup \mathcal{M}_{sr}^o(N,W)).$$

Exercise 16.3.6. Prove Remark 16.3.3.

Exercise 16.3.7. Let (N, V) be given as in Example 16.3.9. Prove that $\mathcal{PK}^c(N, V) = \emptyset$.

Exercise 16.3.8. Let (N, V) be defined as in Exercise 13.3.2 (see also Exercise 14.1.3). Show that $(25, 45, 50) \in \mathcal{BCPK}(N, V)$.

Exercise 16.3.9. Let (N, V) be defined as in Exercise 13.3.3 and let U be the NTU coalition function that differs from V only inasmuch as $U(N) = \{x \in \mathbb{R}^N \mid x(N) \leq 1\}$. (1) Show that (N, V) arises from (N, U) by a monotone transformation for N. (2) Show that none of the following solutions is ordinal: the Shapley NTU value, the Harsanyi value, the consistent Shapley value (see Exercise 14.2.3), $\mathcal{PK}^c$, and $\mathcal{BCPK}$.

16.4 Notes and Comments

(1) Asscher (1977) introduced the *cardinal* prebargaining set of an NTU game with coalition structure and showed that it is a nonempty subset of the ordinal prebargaining set, provided that the game is convex-valued and non-levelled. The sizes of demands are taken into account in the definition of the cardinal bargaining set. In a completely analogous way one may define the cardinal reactive or semi-reactive (pre)bargaining set.

(2) Serrano and Shimomura (1998) axiomatized the prekernel on some sets of NTU games.

(3) In Section 16.3 we used individual excess functions to define "prekernel" concepts for NTU games. Kalai (1975) introduced more general excess functions for NTU games and discussed nucleoli resulting from fixed collections

of excess functions, one for each coalition. It is shown that the resulting pre-nucleoli, though nonempty, may depend on the choice of the excess functions and may not be single-valued.

References

ARROW, K. J. (1951): *Social Choice and Individual Values.* Wiley, New York.

ARTSTEIN, Z. (1971): "Values of games with denumerably many players", *International Journal of Game Theory*, 1, 27 – 37.

ASSCHER, N. (1976): "An ordinal bargaining set for games without side payments", *Mathematics of Operations Research*, 1, 381 – 389.

——— (1977): "A cardinal bargaining set for games without side payments", *International Journal of Game Theory*, 6, 87 – 114.

AUMANN, R. J. (1960): "Linearity of unrestrictedly transferable utilities", *Naval Research Logistics Quarterly*, 7, 281 – 284.

——— (1961): "The core of a cooperative game without side payments", *Transactions of the American Mathematical Society*, 98, 539 – 552.

——— (1967): "A survey of cooperative games without side payments", in Shubik (1967), pp. 3 – 27.

——— (1973): "Disadvantageous monopolies", *Journal of Economic Theory*, 6, 1 – 11.

——— (1985): "An axiomatization of the non-transferable utility value", *Econometrica*, 53, 599 – 612.

——— (1989): *Lectures on Game Theory*, Underground Classics in Economics. Westview Press, Boulder, CO.

AUMANN, R. J., AND J. H. DRÈZE (1974): "Cooperative games with coalition structures", *International Journal of Game Theory*, 3, 217 – 237.

AUMANN, R. J., AND M. MASCHLER (1964): "The bargaining set for cooperative games", in Dresher, Shapley, and Tucker (1964), pp. 443 – 476.

AUMANN, R. J., AND B. PELEG (1960): "Von Neumann-Morgenstern solutions to cooperative games without side payments", *Bulletin of the American Mathematical Society*, 66, 173 – 179.

AUMANN, R. J., AND L. S. SHAPLEY (1974): *Values of Non-Atomic Games.* Princeton University Press, Princeton, NJ.

BALINSKI, M., AND H. P. YOUNG (1982): *Fair Representation.* Yale University Press, New Haven, CT.

BANZHAF, J. R. (1965): "Weighted voting doesn't work: A mathematical analysis", *Rutgers Law Review*, 19, 317 – 343.

BILLERA, L. J. (1970a): "Existence of general bargaining sets for cooperative games without side payments", *Bulletin of the American Mathematical Society*, 76, 375 – 379.

——— (1970b): "Some theorems on the core of an n-person game without side-payments", *SIAM Journal on Applied Mathematics*, 18, 567 – 579.

——— (1972): "Global stability in n-person games", *Transactions of the American Mathematical Society*, 172, 45 – 56.

BILLERA, L. J., AND D. C. HEATH (1982): "Allocation of shared costs: A set of axioms yielding a unique procedure", *Mathematics of Operations Research*, 7, 32 – 39.

BILLERA, L. J., D. C. HEATH, AND J. RAANAN (1978): "Internal telephone billing rates - A novel application of non-atomic game theory", *Operations Research*, 26, 956 – 965.

BILLERA, L. J., AND L. S.-Y. WU (1977): "On a dynamic theory for the kernel of an n-person game", *International Journal of Game Theory*, 6, 65 – 86.

BIRD, C. G. (1976): "On cost allocation for a spanning tree: A game theoretic approach", *Networks*, 6, 335 – 350.

BONDAREVA, O. N. (1963): "Some applications of linear programming methods to the theory of cooperative games", *Problemi Kibernitiki*, 10, 119–139.

BORDER, K. C. (1985): *Fixed Point Theorems with Applications to Economics and Game Theory.* Cambridge University Press, Cambridge, New York.

CALVO, E., AND E. GUTIÉRREZ (1994): "Extension of the Perles-Maschler solution to n-person bargaining games", *International Journal of Game Theory*, 23, 325 – 346.

CHANG, C., AND Y.-A. HWANG (2003): "The Harsanyi-Shapley solution and independence of irrelevant expansions", *International Journal of Game Theory*, 32, 253 – 271.

CHANG, C., AND C. Y. KAN (1992): "The kernel, the bargaining set and the reduced game", *International Journal of Game Theory*, 21, 75 – 83.

CHARNES, A., AND K. O. KORTANEK (1967): "On balanced sets, cores, and linear programming", *Cahiers du Centre d'Etude de Recherche Operationelle (Institut de Statistique de l'Universite Libre de Bruxelles)*, 9, 32 – 43.

COLEMAN, J. S. (1971): "Control of collectivities and the power of a collectivity to act", in *Social Choice*, ed. by B. Lieberman, New York. Gordon and Breach.

DAVIS, M., AND M. MASCHLER (1965): "The kernel of a cooperative game", *Naval Research Logistics Quarterly*, 12, 223–259.

——— (1967): "Existence of stable payoff configurations for cooperative games", in Shubik (1967), pp. 39 – 52.

DE CLIPPEL, G., H. J. M. PETERS, AND H. ZANK (2004): "Axiomatizing the Harsanyi solution, the symmetric egalitarian solution and the consistent solution for NTU-games", *International Journal of Game Theory*, 33, 145 – 158.

DEBREU, G., AND H. E. SCARF (1963): "A limit theorem on the core of an economy", *International Economic Review*, 4, 235 – 246.

DERKS, J. J. M. (1992): "A short proof of the inclusion of the core in the Weber set", *International Journal of Game Theory*, 21, 149 – 150.

DRESHER, M., L. S. SHAPLEY, AND A. W. TUCKER (eds.) (1964): *Advances in Game Theory,* Vol. 52 of *Annals of Mathematics Studies*, Princeton, NJ. Princeton University Press.

DUBEY, P. (1975): "On the uniqueness of the Shapley value", *International Journal of Game Theory*, 4, 131 – 139.

EINY, E. (1988): "The Shapley value on some lattices of monotonic games", *Mathematical Social Sciences*, 15, 1 – 10.

EINY, E., R. HOLZMAN, AND D. MONDERER (1999): "On the least core and the Mas-Colell bargaining set", *Games and Economic Behavior*, 28, 181 – 188.

FRANKLIN, J. (1980): *Methods of Mathematical Economics.* Springer-Verlag, New York.

GEHRLEIN, W. V. (2002): "Condorcet's paradox in the likelihood of its occurrence: different perspectives on balanced preferences", *Theory and Decision*, 52, 171 – 199.

GÉRARD-VARET, L. A., AND S. ZAMIR (1987): "Remarks on the reasonable set of outcomes in a general coalition function form game", *International Journal of Game Theory*, 16, 123 – 143.

GILLIES, D. B. (1959): "Solutions to general non-zero-sum games", in Tucker and Luce (1959), pp. 47 – 85.

GRANOT, D. (1994): "On a new bargaining set for cooperative games", Working paper, Faculty of Commerce and Business Administration, University of British Columbia.

GRANOT, D., AND G. HUBERMAN (1981): "Minimum cost spanning tree games", *Mathematical Programming*, 21, 1 – 18.

——— (1984): "On the core and nucleolus of minimum cost spanning tree games", *Mathematical Programming*, 29, 323 – 347.

GRANOT, D., AND M. MASCHLER (1997): "The reactive bargaining set: Structure, dynamics and extension to NTU games", *International Journal of Game Theory*, 26, 75 – 95.

GREENBERG, J. (1985): "Cores of convex games without side payments", *Mathematics of Operations Research*, 10, 523 – 525.

GROTTE, J. H. (1976): "Dynamics of cooperative games", *International Journal of Game Theory*, 5, 27 – 64.

HARSANYI, J. C. (1963): "A simplified bargaining model for the n-person cooperative game", *International Economic Review*, 4, 194 – 220.

——— (1992): "Games and decision-theoretic models in ethics", in *Handbook of Game Theory*, ed. by R. J. Aumann, and S. Hart, Vol. 1, pp. 669 – 707, Amsterdam. Elsevier Science B. V.

HARSANYI, J. C., AND R. SELTEN (1988): *A General Theory of Equilibrium Selection in Games.* MIT Press, Cambridge, MA. With a foreword by Robert Aumann.

HART, S. (1985): "An axiomatization of Harsanyi's non-transferable utility solution", *Econometrica*, 53, 1445 – 1450.

——— (2005): "An axiomatization of the consistent non-transferable utility value", *International Journal of Game Theory*, 33, 355 – 366.

HART, S., AND M. KURZ (1983): "Endogenous formation of coalitions", *Econometrica*, 51, 1047 – 1067.

HART, S., AND A. MAS-COLELL (1988): "The potential of the Shapley value", in Roth (1988), pp. 127 – 137.

——— (1989): "Potential, value, and consistency", *Econometrica*, 57, 589 – 614.

HOLZMAN, R. (2000): "The comparability of the classical and the Mas-Colell bargaining set", *International Journal of Game Theory*, 29, 543 – 553.

HOLZMAN, R., B. PELEG, AND P. SUDHÖLTER (2007): "Bargaining sets of majority voting games", *Mathematics of Operations Research*, forthcoming, 20 pp.

HOUSMAN, D., AND L. CLARK (1998): "Core and monotonic allocation methods", *International Journal of Game Theory*, 27, 611 – 616.

HWANG, Y.-A., AND P. SUDHÖLTER (2000): "Axiomatizations of the core on the universal domain and other natural domains", *International Journal of Game Theory*, 29, 597 – 623.

ICHIISHI, T. (1981): "Super-modularity: Applications to convex games and to the greedy algorithm for LP", *Journal of Economic Theory*, 25, 283 – 286.

ISBELL, J. R. (1956): "A class of majority games", *The Quarterly Journal of Mathematics*, 7, 183 – 187.

——— (1969): "A counterexample in weighted majority games", *Proceedings of the American Mathematical Society*, 20, 590 – 592.

JUSTMAN, M. (1977): "Iterative processes with 'nucleolar' restrictions", *International Journal of Game Theory*, 6, 189 – 212.

KAHAN, J. P., AND A. RAPOPORT (1984): *Theories of Coalition Formation.* Lawrence Erlbaum, London.

KALAI, E. (1975): "Excess functions for cooperative games without sidepayments", *SIAM Journal on Applied Mathematics*, 29, 60 – 71.

KALAI, E., AND D. SAMET (1985): "Monotonic solutions to general cooperative games", *Econometrica*, 53, 307 – 327.

——— (1988): "Weighted Shapley values", in Roth (1988), pp. 83 – 99.

KALAI, E., AND M. SMORODINSKY (1975): "Other solutions to Nash's bargaining problem", *Econometrica*, 33, 513 – 518.

KALAI, E., AND E. ZEMEL (1982): "Generalized network problems yielding totally balanced games", *Operations Research*, 30, 998 – 1008.

KALAI, G., M. MASCHLER, AND G. OWEN (1975): "Asymptotic stability and other properties of trajectories and transfer sequences leading to the bargaining sets", *International Journal of Game Theory*, 4, 193 – 213.

KERN, R. (1985): "The Shapley transfer value without zero weights", *International Journal of Game Theory*, 14, 73 – 92.

KIKUTA, K. (1976): "On the contribution of a player to a game", *International Journal of Game Theory*, 5, 199 – 208.

——— (1997): "The kernel for reasonable outcomes in a cooperative game", *International Journal of Game Theory*, 26, 51 – 59.

KLEIN, E. (1973): *Mathematical Methods in Theoretical Economics.* Academic Press, New York.

KOHLBERG, E. (1971): "On the nucleolus of a characteristic function game", *SIAM Journal on Applied Mathematics*, 20, 62 – 66.

——— (1972): "The nucleolus as a solution of a minimization problem", *SIAM Journal on Applied Mathematics*, 23, 34 – 39.

KOPELOWITZ, A. (1967): "Computation of the kernels of simple games and the nucleolus of n-person games", Research memorandum 31, Department of Mathematics, The Hebrew University of Jerusalem.

KRUSKAL, JR, J. B. (1956): "On the shortest spanning subtree of a graph and the travelling salesman problem", *Proceedings of the American Mathematical Society*, 7, 48 – 50.

LEGROS, P. (1987): "Disadvantageous syndicates and stable cartels: The case of the nucleolus", *Journal of Economic Theory*, 42, 30 – 49.

LITTLECHILD, S. C. (1974): "A simple expression for the nucleolus in a special case", *International Journal of Game Theory*, 3, 21–29.

LITTLECHILD, S. C., AND G. OWEN (1973): "A simple expression for the Shapley value in a special case", *Management Science*, 20, 370 – 372.

LUCAS, W. F. (1981): "Applications of cooperative games to equitable allocation", in *Game Theory and its Applications*, ed. by W. F. Lucas, pp. 19 – 36, Providence, RI. American Mathematical Society.

LUCCHETTI, R., F. PATRONE, S. H. TIJS, AND A. TORRE (1987): "Continuity properties of solution concepts for cooperative games", *OR Spektrum*, 9, 101 – 107.

LUCE, R. D., AND H. RAIFFA (1957): *Games and Decisions.* John Wiley and Sons, New York.

MAS-COLELL, A. (1989): "An equivalence theorem for a bargaining set", *Journal of Mathematical Economics*, 18, 129 – 139.

MASCHLER, M. (1963): "n-person games with only 1, $n-1$, and n-person permissible coalitions", *Journal of Mathematical Analysis and Applications*, 6, 230 – 256.

——— (1964): "Stable payoff configurations for quota games", in Dresher, Shapley, and Tucker (1964), pp. 477 – 499.

——— (1966): "The inequalities that determine the bargaining set $M_1^{(i)}$", *Israel Journal of Mathematics*, 4, 127 – 134.

——— (1976): "An advantage of the bargaining set over the core", *Journal of Economic Theory*, 13, 184 – 192.

MASCHLER, M., AND G. OWEN (1989): "The consistent Shapley value for hyperplane games", *International Journal of Game Theory*, 18, 389 – 407.

——— (1992): "The consistent Shapley value for games without side payments", in *Rational Interaction: Essays in Honor of John Harsanyi*, ed. by R. Selten, pp. 5 – 12, New York. Springer-Verlag.

MASCHLER, M., AND B. PELEG (1966): "A characterization, existence proof and dimension bounds for the kernel of a game", *Pacific Journal of Mathematics*, 18, 289 – 328.

——— (1967): "The structure of the kernel of a cooperative game", *SIAM Journal on Applied Mathematics*, 15, 569 – 604.

——— (1976): "Stable sets and stable points of set-valued dynamic systems with applications to game theory", *SIAM Journal on Control and Optimization*, 14, 985 – 995.

MASCHLER, M., B. PELEG, AND L. S. SHAPLEY (1972): "The kernel and bargaining set for convex games", *International Journal of Game Theory*, 1, 73 – 93.

——— (1979): "Geometric properties of the kernel, nucleolus, and related solution concepts", *Mathematics of Operations Research*, 4, 303 – 338.

MASCHLER, M., J. A. M. POTTERS, AND S. H. TIJS (1992): "The general nucleolus and the reduced game property", *International Journal of Game Theory*, 21, 85 – 105.

MEGIDDO, N. (1974): "On the monotonicity of the bargaining set, the kernel, and the nucleolus of a game", *SIAM Journal on Applied Mathematics*, 27, 355 – 358.

——— (1978): "Cost allocation for Steiner trees", *Networks*, 8, 1 – 6.

MILNOR, J. W. (1952): "Reasonable outcomes for n-person games", Research memorandum 916, The Rand Corporation, Santa Monica, CA.

MIRMAN, L., AND Y. TAUMAN (1982): "Demand-compatible equitable cost-sharing prices", *Mathematics of Operations Research*, 7, 40 – 56.

MONJARDET, B. (1972): "Note sur les pouvoirs de vote au Conseil de Sécurité", *Mathématiques et Sciences Humaines*, 40, 25 – 37.

NASH, J. F. (1950): "The bargaining problem", *Econometrica*, 18, 155 – 162.

NEYMAN, A. (1989): "Uniqueness of the Shapley value", *Games and Economic Behavior*, 1, 116 – 118.

ORSHAN, G. (1993): "The prenucleolus and the reduced game property: Equal treatment replaces anonymity", *International Journal of Game Theory*, 22, 241 – 248.

——— (1994): "Non-symmetric prekernels", Discussion paper 60, Center for Rationality, The Hebrew University of Jerusalem.

ORSHAN, G., AND J. M. ZARZUELO (2000): "The bilateral consistent prekernel for NTU games", *Games and Economic Behavior*, 32, 67 – 84.

ORTMANN, K. M. (1998): "Conservation of energy in value theory", *Mathematical Methods of Operations Research*, 47, 423 – 449.

OSTMANN, A. (1987): "On the minimal representation of homogeneous games", *International Journal of Game Theory*, 16, 69 – 81.

OWEN, G. (1972): "Multilinear extensions of games", *Management Science*, 18, 64 – 79.

——— (1974): "A note on the nucleolus", *International Journal of Game Theory*, 3, 101 – 103.

——— (1975): "Evaluation of a presidential election game", *American Political Science Review*, 69, 947 – 953.

——— (1977): "Values of games with a priori unions", in *Essays in Mathematical Economics and Game Theory*, ed. by R. Henn, and O. Moeschlin, pp. 76 – 88, New York. Springer-Verlag.

PALLASCHKE, D., AND J. ROSENMÜLLER (1997): "The Shapley value for countably many players", *Optimization*, 40, 351 – 384.

PELEG, B. (1963): "Bargaining sets for cooperative games without side payments", *Israel Journal of Mathematics*, 1, 197 – 200.

——— (1964): "On the bargaining set M_0 of m-quota games", in Dresher, Shapley, and Tucker (1964), pp. 501 – 512.

——— (1965): "An inductive method for constructing minimal balanced collections of finite sets", *Naval Research Logistics Quarterly*, 12, 155 – 162.

——— (1966): "On the kernel of constant-sum simple games with homogeneous weights", *Illinois Journal of Mathematics*, 10, 39 – 48.

——— (1967a): "Equilibrium points for open acyclic relations", *Canadian Journal of Mathematics*, 19, 366 – 369.

——— (1967b): "Existence theorem for the bargaining set $M_1^{(i)}$", in Shubik (1967), pp. 53 – 56.

——— (1968): "On weights of constant-sum majority games", *SIAM Journal on Applied Mathematics*, 16, 527 – 532.

——— (1985): "An axiomatization of the core of cooperative games without side payments", *Journal of Mathematical Economics*, 14, 203 – 214.

——— (1986): "On the reduced game property and its converse", *International Journal of Game Theory*, 15, 187 – 200.

——— (1989): "An axiomatization of the core of market games", *Mathematics of Operations Research*, 14, 448 – 456.

——— (1993): "An axiomatization of the core of market games: A correction", *Mathematics of Operations Research*, 18, 765.

——— (1996): "A formal approach to Nash's program", Working paper 247, Institute of Mathematical Economics, University of Bielefeld.

——— (1997): "A difficulty with Nash's program: A proof of a special case", *Economics Letters*, 55, 305 – 308.

PELEG, B., AND P. SUDHÖLTER (2004): "Bargaining sets of voting games", Discussion paper, Center for Rationality and Interactive Decision Theory, The Hebrew University of Jerusalem.

——— (2005): "On the non-emptiness of the Mas-Colell bargaining set", *Journal of Mathematical Economics*, 41, 1060 – 1068.

Perles, M. A., and M. Maschler (1981): "The super-additive solution for the Nash bargaining game", *International Journal of Game Theory*, 10, 163 – 193.

Peters, H. J. M. (1992): *Axiomatic Bargaining Game Theory*, Vol. 9 of *Theory and Decision Library. Series C: Game Theory, Mathematical Programming and Operations Research.* Kluwer Academic Publishers Group, Dordrecht.

Postlewaite, A., and R. W. Rosenthal (1974): "Disadvantageous syndicates", *Journal of Economic Theory*, 9, 324 – 326.

Potters, J. A. M., J. H. Reijnierse, and M. Ansing (1996): "Computing the nucleolus by solving a prolonged simplex algorithm", *Mathematics of Operations Research*, 21, 757 – 768.

Potters, J. A. M., and P. Sudhölter (1999): "Airport problems and consistent allocation rules", *Mathematical Social Sciences*, 38, 83 – 102.

Rosenmüller, J. (1981): *The Theory of Games and Markets.* North Holland, Amsterdam.

——— (1987): "Homogeneous games: Recursive structure and computation", *Mathematics of Operations Research*, 12, 309 – 330.

——— (2000): *Game Theory. Stochastics, Information, Strategies and Cooperation*, Vol. 25 of *Theory and Decision Library. Series C: Game Theory, Mathematical Programming and Operations Research.* Kluwer Academic Publishers Group, Dordrecht.

Roth, A. E. (ed.) (1988): *The Shapley Value: Essays in Honor of Lloyd S. Shapley,* Cambridge. Cambridge University Press.

Scarf, H. E. (1967): "The core of an n-person game", *Econometrica*, 35, 50 – 69.

Schmeidler, D. (1969): "The nucleolus of a characteristic function game", *SIAM Journal on Applied Mathematics*, 17, 1163 – 1170.

Sen, A. K. (1970): *Collective Choice and Social Welfare.* Holden-Day, San Francisco.

Serrano, R., and K.-I. Shimomura (1998): "Beyond Nash bargaining theory: The Nash set", *Journal of Economic Theory*, 83, 286 – 307.

Serrano, R., and O. Volij (1998): "Axiomatizations of neoclassical concepts for economies", *Journal of Mathematical Economics*, 30, 87 – 108.

Shapley, L. S. (1953): "A value for n-person games", *Annals of Mathematics Studies*, 28, 307 – 318.

——— (1959): "The solutions of a symmetric market game", in Tucker and Luce (1959), pp. 145 – 162.

——— (1962a): "Simple games: An outline of the descriptive theory", *Behavioral Science*, 7, 59 – 66.

——— (1962b): "Values of games with infinitely many players", in *Recent Advances in Game Theory*, ed. by M. Maschler, pp. 113 – 118, Philadelphia. Ivy Curtis Press.

——— (1967): "On balanced sets and cores", *Naval Research Logistics Quarterly*, 14, 453 – 460.

——— (1969): "Utility comparisons and the theory of games", in *La Decision, aggregation et dynamique des ordres de preference*, pp. 251 – 263, Paris. Edition du Centre National de le Recherche Scientifique.

——— (1971): "Cores of convex games", *International Journal of Game Theory*, 1, 11 – 26.

Shapley, L. S., and M. Shubik (1963): "The core of an economy with nonconvex preferences", RM-3518, The Rand Corporation, Santa Monica, CA.

——— (1966): "Quasi-cores in a monetary economy with nonconvex preferences", *Econometrica*, 34, 805–827.

——— (1969a): "On market games", *Journal of Economic Theory*, 1, 9 – 25.

——— (1969b): "Pure competition, coalitional power, and fair division", *International Economic Review*, 10, 337 – 362.

——— (1972): "The assignment game I: The core", *International Journal of Game Theory*, 2, 111 – 130.

SHAPLEY, L. S., AND R. VOHRA (1991): "On Kakutani's fixed-point theorem, the K-K-M-S theorem and the core of a balanced game", *Economic Theory*, 1, 108 – 116.

SHARKEY, W. W. (1981): "Convex games without side payments", *International Journal of Game Theory*, 10, 101 – 106.

SHIMOMURA, K.-I. (1992): "Individual rationality and collective optimality", Ph.D. Thesis, University of Rochester.

——— (1997): "Quasi-cores in bargaining sets", *International Journal of Game Theory*, 26, 283 – 302.

SHUBIK, M. (ed.) (1967): *Essays in Mathematical Economics in Honor of Oskar Morgenstern*, Princeton, NJ. Princeton University Press.

SNIJDERS, C. (1995): "Axiomatization of the nucleolus", *Mathematics of Operations Research*, 20, 189 – 196.

SOBOLEV, A. I. (1975): "The characterization of optimality principles in cooperative games by functional equations", in *Mathematical Methods in the Social Sciences*, ed. by N. N. Vorobiev, Vol. 6, pp. 95 – 151, Vilnius. Academy of Sciences of the Lithuanian SSR, in Russian.

SOLYMOSI, T. (1999): "On the bargaining set, kernel and core of superadditive games", *International Journal of Game Theory*, 28, 229 – 240.

STEARNS, R. E. (1965): "The discontinuity of the bargaining set", Unpublished research memorandum.

——— (1968): "Convergent transfer schemes for N-person games", *Transactions of the American Mathematical Society*, 134, 449 – 459.

STRAFFIN, P. D., AND J. P. HEANEY (1981): "Game theory and the Tennessee Valley Authority", *International Journal of Game Theory*, 10, 35 – 43.

SUDHÖLTER, P. (1993): "Independence for characterizing axioms of the prenucleolus", Working paper 220, Institute of Mathematical Economics, University of Bielefeld.

——— (1996a): "The modified nucleolus as canonical representation of weighted majority games", *Mathematics of Operations Research*, 21, 734 – 756.

——— (1996b): "Star-shapedness of the kernel for homogeneous games", *Mathematical Social Sciences*, 32, 179 – 214.

——— (1997): "The modified nucleolus: Properties and axiomatizations", *International Journal of Game Theory*, 26, 147 – 182.

——— (2001): "Equal Treatment for both sides of assignment games in the modified least core", in *Power Indices and Coalition Formation*, ed. by M. Holler, and G. Owen, pp. 175 – 202, Dordrecht. Kluwer Academic Publishers.

SUDHÖLTER, P., AND B. PELEG (1998): "Nucleoli as maximizers of collective satisfaction functions", *Social Choice and Welfare*, 15, 383 – 411.

——— (2000): "The positive prekernel of a cooperative game", *International Game Theory Review*, 2, 287 – 305.

——— (2002): "A note on an axiomatization of the core of market games", *Mathematics of Operations Research*, 27, 441 – 444.

SUDHÖLTER, P., AND J. A. M. POTTERS (2001): "The semireactive bargaining set of a cooperative game", *International Journal of Game Theory*, 30, 117 – 139.

SUZUKI, M., AND M. NAKAYAMA (1976): "The cost assignment of the cooperative water resource development: A game-theoretical approach", *Management Science*, 22, 1081 – 1086.

TADENUMA, K. (1992): "Reduced games, consistency, and the core", *International Journal of Game Theory*, 20, 325 – 334.

TIJS, S. H., AND T. S. H. DRIESSEN (1986): "Game theory and cost allocation problems", *Management Science*, 32, 1015 – 1028.

TUCKER, A. W., AND R. D. LUCE (eds.) (1959): *Contribution to the Theory of Games IV,* Vol. 40, *Annals of Mathematics Studies*, Princeton, NJ. Princeton University Press.

VILKOV, V. B. (1977): "Convex games without side payments", *Vestnik Leningrad. Univ.*, 7, 21 – 24, in Russian.

VOHRA, R. (1991): "An existence theorem for a bargaining set", *Journal of Mathematical Economics*, 20, 19 – 34.

VON NEUMANN, J., AND O. MORGENSTERN (1953): *Game Theory and Economic Behavior*. Princeton University Press, Princeton, NJ, third edn.

VOORNEVELD, M., AND A. VAN DEN NOUVELAND (1998): "A new axiomatization of the core of games with transferable utility", *Economics Letters*, 60, 151 – 155.

WEBER, R. J. (1988): "Probabilistic values for games", in Roth (1988), pp. 101 – 119.

WESLEY, E. (1971): "An application of non-standard analysis to game theory", *Journal of Symbolic Logic*, 36, 385 – 394.

WINTER, E. (1988): "Cooperative games with a priori cooperation structure", Ph.D. Thesis, The Hebrew University of Jerusalem.

——— (2002): "The Shapley value", in *Handbook of Game Theory*, ed. by R. J. Aumann, and S. Hart, Vol. 3, pp. 2025 – 2053, Amsterdam. Elsevier Science B. V.

WU, L. S.-Y. (1977): "A dynamic theory for the class of games with nonempty core", *SIAM Journal on Applied Mathematics*, 32, 328 – 338.

YOUNG, H. P. (1985a): "Cost allocation", in *Fair Allocation*, ed. by H. P. Young, Vol. 33 of *Proceedings of Symposia in Applied Mathematics*, pp. 69 – 94, Providence, RI. American Mathematical Society.

——— (1985b): "Monotonic solutions of cooperative games", *International Journal of Game Theory*, 14, 65 – 72.

YOUNG, H. P., N. OKADA, AND T. HASHIMOTO (1982): "Cost allocation in water resources development", *Water Resources Research*, 18, 463 – 475.

ZHOU, L. (1994): "A new bargaining set of an n-person game and endogenous coalition formation", *Games and Economic Behavior*, 6, 512 – 526.

Author Index

Subject Index[1]

[1] If the page number is boldface, then the subject is defined or introduced on this page.

THEORY AND DECISION LIBRARY

SERIES C: GAME THEORY, MATHEMATICAL PROGRAMMING AND OPERATIONS RESEARCH

Editor: H. Peters, *Maastricht University, The Netherlands*

1. B. R. Munier and M. F. Shakun (eds.): *Compromise, Negotiation and Group Decision.* 1988 ISBN 978-90-277-2625-4
2. R. Selten: *Models of Strategic Rationality.* 1988 ISBN 978-90-277-2663-6
3. T. Driessen: *Cooperative Games, Solutions and Applications.* 1988 ISBN 978-90-277-2729-9
4. P. P. Wakker: *Additive Representations of Preferences.* A New Foundation of Decision Analysis. 1989 ISBN 978-0-7923-0050-2
5. A. Rapoport: *Experimental Studies of Interactive Decisions.* 1990 ISBN 978-0-7923-0685-6
6. K. G. Ramamurthy: *Coherent Structures and Simple Games.* 1990 ISBN 978-0-7923-0869-0
7. T. E. S. Raghavan, T. S. Ferguson, T. Parthasarathy and 0. J. Vrieze (eds.): *Stochastic Games and Related Topics. In Honor of Professor L. S. Shapley.* 1991 ISBN 978-0-7923-1016-7
8. J. Abdou and H. Keiding: *Effectivity Functions in Social Choice.* 1991 ISBN 978-0-7923-1147-8
9. H. J. M. Peters: *Axiomatic Bargaining Game Theory.* 1992 ISBN 978-0-7923-1873-6
10. D. Butnariu and E. P. Klement: *Triangular Norm-Based Measures and Games with Fuzzy Coalitions.* 1993 ISBN 978-0-7923-2369-3
11. R. P. Gilles and P. H. M. Ruys: *Imperfections and Behavior in Economic Organization.* 1994 ISBN 978-0-7923-9460-0
12. R. P. Gilles: *Economic Exchange and Social Organization.* The Edgeworthian Foundations of General Equilibrium Theory. 1996 ISBN 978-0-7923-4200-7
13. P .J.-J. Herings: *Static and Dynamic Aspects of General Disequilibrium Theory.* 1996 ISBN 978-0-7923-9813-4
14. E. van Dijk: *Social Ties and Economic Performance.* 1997 ISBN 978-0-7923-9836-3
15. W. Spanjers: *Hierarchically Structured Economies.* Models with Bilateral Exchange Institutions. 1997 ISBN 978-0-7923-4398-1
16. I. Curiel: *Cooperative Game Theory and Applications.* Cooperative Games Arising from Combinatorial Optimization Problems. 1997 ISBN 978-0-7923-4476-6
17. O.I. Larichev and H. M. Moshkovich: *Verbal Decision Analysis for Unstructured Problems.* 1997 ISBN 978-0-7923-4578-7
18. T. Parthasarathy, B. Dutta, J. A. M. Potters,T. E. S. Raghavan, D. Ray and A. Sen (eds.): *Game Theoretical Applications to Economics and Operations Research.* 1997 ISBN 978-0-7923-4712-5
19. A. M. A. Van Deemen: *Coalition Formation and Social Choice.* 1997 ISBN 978-0-7923-4750-7
20. M. 0. L. Bacharach, L.-A. Gérard-Varet, P. Mongin and H. S. Shin (eds.): *Epistemic Logic and the Theory of Games and Decisions.* 1997 ISBN 978-0-7923-4804-7
21. Z. Yang (eds.): *Computing Equilibria and Fixed Points.* 1999 ISBN 978-0-7923-8395-6

THEORY AND DECISION LIBRARY: SERIES C

22. G. Owen: *Discrete Mathematics and Game Theory.* 1999 ISBN 978-0-7923-8511-0
23. F. Patrone, I. Garcia-Jurado and S. Tijs (eds.): *Game Practice.* Contributions from Applied Game Theory. 1999 ISBN 978-0-7923-8661-2
24. J. Suijs: *Cooperative Decision-Making under Risk.* 1999 ISBN 978-0-7923-8660-5
25. J. Rosenmüller: *Game Theory: Stochastics, Information, Strategies and Cooperation.* 2000 ISBN 978-0-7923-8673-5
26. J. M. Bilbao: *Cooperative Games on Combinatorial Structures.* 2000 ISBN 978-0-7923-7782-5
27. M. Slikker and A. van den Nouweland: *Social and Economic Networks in Cooperative Game Theory.* 2000 ISBN 978-0-7923-7226-4
28. K. J. M. Huisman: *Technology Investment: A Game Theoretic Real Options Approach.* 2001 ISBN 978-0-7923-7487-9
29. A. Perea: *Rationality in Extensive Form Games.* 2001 ISBN 978-0-7923-7540-1
30. V. Buskens: *Social Networks and Trust.* 2002 ISBN 978-1-4020-7010-5
31. P. Borm and H. Peters (eds.): *Chapters in Game Theory.* In Honor of Stef Tijs. 2002 ISBN 978-1-4020-7063-1
32. H. Houba and W. Bolt: *Credible Threats in Negotiations.* A Game-theoretic Approach. 2002 ISBN 978-1-4020-7183-6
33. T. Hens and B. Pilgrim: *General Equilibrium Foundations of Finance: Structure of Incomplete Markets Models.* 2003 ISBN 978-1-4020-7337-3
34. B. Peleg and P. Sudhölter: *Introduction to the Theory of Cooperative Games.* 2007 Second Edition ISBN 978-3-540-72944-0
35. J. H. H. Thijssen: *Investment under Uncertainty, Coalition Spillovers and Market Evolution in a Game Theoretic Perspective.* 2004 ISBN 978-1-4020-7877-4
36. G. Gambarelli (ed.): *Essays on Cooperative Games.* In Honor of Guillermo Owen. 2004 ISBN 978-1-4020-2935-6
37. G. B. Asheim: *The Consistent Preferences Approach to Deductive Reasoning in Games.* 2006 ISBN 978-0-387-26235-2
38. U. Schmidt and S. Traub (eds.): *Advances in Public Economics: Utility, Choice and Welfare.* A Festschrift for Christian Seidl. 2005 ISBN 978-0-387-25705-1
39. T. S. H. Driessen, G. van der Laan, V. A. Vasil'ev and E. B. Yanovskaya (eds.): *Russian Contributions to Game Theory and Equilibrium Theory.* 2006 ISBN 978-3-540-31405-9
40. W. V. Gehrlein: *Condorcet's Paradox.* 2006 ISBN 978-3-540-33798-0
41. M. Abdellaoui, R. D. Luce, M. J. Machina, B. Munier (eds.): *Uncertainty and Risk.* 2007 ISBN 978-3-540-48934-4

Printing: Krips bv, Meppel
Binding: Stürtz, Würzburg